GW00697476

Recombinant Protein Protocols

METHODS IN MOLECULAR BIOLOGY™

John M. Walker, SERIES EDITOR

METHODS IN MOLECULAR BIOLOGY™

Recombinant Protein Protocols

Detection and Isolation

Edited by

Rocky S. Tuan

Thomas Jefferson University, Philadelphia, PA

Humana Press ✻ Totowa, New Jersey

Cover illustration: Fig. 3B from Chapter 27, "Hyperexpression of a Synthetic Protein-Based Polymer Gene," by Henry Daniell, Chittibabu Guda, David T. McPherson, Xiaorong Zhang, Jie Xu, and Dan W. Urry.

Cover design by Patricia F. Cleary.

For additional copies, pricing for bulk purchases, and/or information about other Humana titles, contact Humana at the above address or at any of the following numbers: Tel: 201-256-1699; Fax: 201-256-8341; E-mail: humana@interramp.com

Printed in the United States of America. 10 9 8 7 6 5 4 3 2 1

Library of Congress Cataloging in Publication Data

Main entry under title:

Methods in molecular biology™.

Recombinant protein protocols: detection and isolation/edited by Rocky S. Tuan
    p.   cm.—(Methods in molecular biology™; vol. 63)
  Includes bibliographical references and index.
  ISBN 0-89603-400-3 (combbound) (alk. paper); ISBN 0-89603-481-X (hardcover) (alk. Paper)
  1. Recombinant proteins—Laboratory manuals. I. Tuan, Rocky S. II. Series: Methods in molecular biology (Totowa, NJ); 63
TP248.65.P76R43   1997
660' .65—dc21
for Library of Congress   97-4023
                                                                 CIP

To Cecilia, Chuck, and my parents.

Preface

A major success story of modern molecular biology is the development of technologies to clone and express specific genes. Current applications of recombinant gene products cover a wide spectrum, including gene therapy, production of bioactive pharmaceuticals, synthesis of novel biopolymers, agriculture and animal husbandry, and so on. Inherent in bringing these applications to fruition is the need to design "expression constructs" that will permit the ready and specific detection and isolation of the defined recombinant gene products.

Recombinant Protein Protocols grows out of the need for a laboratory manual on the detection and isolation of recombinantly expressed genes that covers both the background information and the practical laboratory recipes for these analyses. In this book, detailed and contemporary protocols are collected to provide the reader with a wide-ranging number of methodologies to enhance the detection and isolation of their gene product(s) of interest. A large number of molecular tags and labels and their usage are described, including enzymes, ligand-binding moieties, immunodetectable molecules, as well as methods to detect interactive proteins, and gene expression-mediated alterations in cellular activity. Chapters on *in situ* detection of gene expression deal with technologies that are currently being applied to the study of gene function and activity. Highlights of applications for recombinant gene expression technologies are provided to give readers exciting perspectives on the future of such technologies.

Throughout *Recombinant Protein Protocols*, the authors have consistently striven for a balanced presentation of both background information and practical procedures for each of the methodologies treated. The reader is first guided through the necessary supporting background information and then presented with step-by-step specifics for each protocol, including reagents, instrumentation, and other requirements. It is anticipated that this highly practical format, a feature of the *Methods in Molecular Biology* series, will permit the reader to bring new concepts into personal practice in a most efficient manner.

The practice of molecular biology as a means to express recombinant genes continues to gain attention in basic biomedical research, as well as the biotechnology and pharmaceutical industries. For this reason, it is anticipated that the subjects covered here will continue to be developed, serving as the basis for more sophisticated and efficient methodologies in the future.

The preparation of *Recombinant Protein Protocols* would not have been possible without the outstanding work of the contributing authors, all of whom have been most tolerant of my persistent reminders. Dr. John M. Walker, the mastermind of the *Methods in Molecular Biology* series, was instrumental in initiating and guiding the project. The staff at the Humana Press showed great patience and provided excellent guidance and assistance. My wife, Cecilia, and my newborn son, Chuck, both tolerated my indulgence in the project, and always gave me the necessary emotional support throughout the preparation of the volume. Finally, the excellent secretarial assistance of Margaret Feoli, Susan Lowenstein, and in particular, Lynn Stierle, is gratefully acknowledged.

Rocky S. Tuan

Contents

Contents

xiii

Contributors

PATRICK A. BAEUERLE • *Institute of Biochemistry, Albert-Ludwigs-Universität Freiburg, Germany*

OMAR BAGASRA • *Section of Molecular Retrovirology, Division of Infectious Diseases, Department of Medicine, The Dorrance H. Hamilton Laboratories; Center of Human Retrovirology and Gene Therapy, Thomas Jefferson University, Philadelphia, PA*

PAUL BATES • *Department of Microbiology, University of Pennsylvania School of Medicine, Philadelphia, PA*

DANIEL R. BOGGS • *Gene Therapy Unit, Baxter Healthcare Corp., Round Lake, IL*

JAMES H. BRAUKER • *Gene Therapy Unit, Baxter Healthcare Corp., Round Lake, IL*

IRENA BRONSTEIN • *Tropix, Bedford, MA*

CHARLES R. CANTOR • *Center for Advanced Biotechnology and Departments of Biomedical Engineering and Pharmacology, Boston University, Boston, MA*

VICTORIA E. CARR-BRENDEL • *Gene Therapy Unit, Baxter Healthcare Corp., Round Lake, IL*

LUIGI CATANZARITI • *bioMérieux Vitek, Inc., Rockland, MA*

THOMAS M. S. CHANG • *Artificial Cells and Organs Research Centre, Faculty of Medicine, McGill University, Montreal, Quebec, Canada*

RONALD A. CONLON • *Department of Genetics, Case Western Reserve University, Cleveland, OH*

JOANNE CRUDELE • *Gene Therapy Unit, Baxter Healthcare Corp., Round Lake, IL*

HENRY DANIELL • *Molecular Genetics Program, Department of Botany and Microbiology, Auburn University, Auburn, AL*

JOHN J. FORTIN • *Tropix, Bedford, MA*

ROBIN E. GELLER • *Gene Therapy Unit, Baxter Healthcare Corp., Round Lake, IL*

URSULA A. GERMANN • *Vertex Pharmaceuticals Inc., Cambridge, MA*

MARYLOU G. GIBSON • *Gene Therapy Program, University of California, San Diego, CA*

ERICA A. GOLEMIS • *Fox Chase Cancer Center, Philadelphia, PA*

DEBYRA J. GROSKREUTZ • *Promega Corp., Oregon, WI*

CHITTIBABU GUDA • *Molecular Genetics Program, Department of Botany and Microbiology, Auburn University, Auburn, AL*

BRIAN A. HEMMINGS • *Friedrich Miescher-Institute, Basel, Switzerland*

KARMEN HODGES • *Immunex Corp., Seattle, WA*

DENNIS E. HRUBY • *Department of Microbiology, Oregon State University, Corvallis, OR*

CHRISTOPHER W. HSU • *Department of Orthopaedic Surgery, Thomas Jefferson University, Philadelphia, PA*

WILLIAM JAMES • *Sir William Dunn School of Pathology, University of Oxford, UK*

ROBERT C. JOHNSON • *Gene Therapy Unit, Baxter Healthcare Corp., Round Lake, IL*

STEVEN R. KAIN • *CLONTECH Laboratories Inc., Palo Alto, CA*

VLADIMIR KHAZAK • *Fox Chase Cancer Center, Philadelphia, PA*

JASPAL S. KHILLAN • *Department of Biochemistry and Molecular Pharmacology, Thomas Jefferson University, Philadelphia, PA*

PAUL KITTS • *CLONTECH Laboratories, Inc., Palo Alto, CA*

PEIYU LEE • *Department of Microbiology, Oregon State University, Corvallis, OR*

CECILIA W. LO • *Department of Biology, University of Pennsylvania, Philadelphia, PA*

LAURETTA LOWTHER • *Immunex Corp., Seattle, WA*

MICHAEL H. MALIM • *Departments of Microbiology and Medicine, Howard Hughes Medical Institute, University of Pennsylvania School of Medicine, Philadelphia, PA*

LAURA A. MARTINSON • *Gene Therapy Unit, Baxter Healthcare Corp., Round Lake, IL*

DAVID A. MARYANOV • *Gene Therapy Unit, Baxter Healthcare Corp., Round Lake, IL*

DAVID T. MCPHERSON • *Laboratory of Molecular Biophysics, School of Medicine, University of Alabama at Birmingham, AL*

JAMIL MOMAND • *Department of Cell and Tumor Biology, City of Hope National Medical Center, Duarte, CA*

MUHAMMAD MUKHTAR • *Section of Molecular Retrovirology, Division of Infectious Diseases, Department of Medicine, The Dorrance H. Hamilton Laboratories; Center of Human Retrovirology and Gene Therapy, Thomas Jefferson University, Philadelphia, PA*

PER-ÅKE NYGREN • *Department of Biochemistry and Biotechnology, Royal Institute of Technology, Stockholm, Sweden*

CORINNE E. M. OLESEN • *Tropix, Bedford, MA*

BERT W. O'MALLEY • *Department of Cell Biology, Baylor College of Medicine, Houston, TX*

GIORGIO PALÙ • *Universita'degli studi di Padova, Instituto di Microbiologia, Padova, Italy*

ZAHIDA PARVEEN • *University of Agriculture, Faisalabad, Pakistan*

ROGER J. POMERANTZ • *Section of Molecular Retrovirology, Division of Infectious Diseases, Department of Medicine, The Dorrance H. Hamilton Laboratories; Center of Human Retrovirology and Gene Therapy, Thomas Jefferson University, Philadelphia, PA*

SATYA PRAKASH • *Artificial Cells and Organs Research Centre, Faculty of Medicine, McGill University, Montreal, Quebec, Canada*

PAUL RIGGS • *New England BioLabs, Beverly, MA*

TAKESHI SANO • *Center for Advanced Biotechnology and Departments of Biomedical Engineering and Pharmacology, Boston University, Boston, MA*

M. LIENHARD SCHMITZ • *Institute of Biochemistry, Albert-Ludwigs-Universität Freiburg, Germany*

ELAINE T. SCHENBORN • *Promega Corp., Oregon, WI*

GEORG SCZAKIEL • *Deutsches Krebsforschungszentrum, Angewandte Tumorvirologie, Heidelberg, Germany*

BAHMAN SEPEHRNIA • *Department of Cell and Tumor Biology, City of Hope National Medical Center, Duarte, CA*

FARIDA SHAHEEN • *Section of Molecular Retrovirology, Division of Infectious Diseases, Department of Medicine, The Dorrance H. Hamilton Laboratories; Center of Human Retrovirology and Gene Therapy, Thomas Jefferson University, Philadelphia, PA*

KENNETH J. SHEPLEY • *Department of Orthopaedic Surgery, Thomas Jefferson University, Philadelphia, PA*

CASSANDRA L. SMITH • *Center for Advanced Biotechnology and Departments of Biomedical Engineering and Pharmacology, Boston University, Boston, MA*

STEFAN STÅHL • *Department of Biochemistry and Biotechnology, Royal Institute of Technology, Stockholm, Sweden*

RUTH SULLIVAN • *Department of Biology, University of Pennsylvania, Philadelphia, PA*

TRACY J. THOMAS • *Gene Therapy Unit, Baxter Healthcare Corp., Round Lake, IL*

SOPHIA Y. TSAI • *Department of Cell Biology, Baylor College of Medicine, Houston, TX*

ROCKY S. TUAN • *Department of Orthopaedic Surgery and Department of Biochemistry and Molecular Pharmacology, Thomas Jefferson University, Philadelphia, PA*

MATHIAS UHLÉN • *Department of Biochemistry and Biotechnology, Royal Institute of Technology, Stockholm, Sweden*

DAN W. URRY • *Laboratory of Molecular Biophysics, School of Medicine, University of Alabama at Birmingham, AL*

JOHN C. VOYTA • *Tropix, Bedford, MA*

YAOLIN WANG • *Department of Cell Biology, Baylor College of Medicine, Houston, TX*

JIE XU • *Laboratory of Molecular Biophysics, School of Medicine, University of Alabama at Birmingham, AL*

SUSAN K. YOUNG • *Gene Therapy Unit, Baxter Healthcare Corp., Round Lake, IL*

XIAORONG ZHANG • *Molecular Genetics Program, Department of Botany and Microbiology, Auburn University, Auburn, AL*

LI ZHU • *CLONTECH Laboratories Inc., Palo Alto, CA*

I

INTRODUCTION

1

Overview of Experimental Strategies on the Detection and Isolation of Recombinant Proteins and Their Applications

Rocky S. Tuan

1. Introduction

Recent advances in recombinant DNA technology have permitted the direct cloning of DNA fragments (either derived from naturally occurring or artificially designed gene sequences) into various cloning vectors including bacteriophages, plasmids, and viruses. Such recombinant constructs represent the basic reagents of molecular biology. A major application utilizing cloned DNA sequences is the expression of the cloned DNA into a protein product, i.e., the expression of recombinant genes. Because the cloned DNA sequences may be modified or altered, recombinant expression technology thus enables the investigator to "custom-design" the final protein product. Furthermore, most expression vectors are designed to allow the linking of various "tags" to the expressed recombinant protein to facilitate subsequent detection and isolation. This chapter provides a brief overview of the technologies currently employed in "tagging" expressed recombinant proteins and the corresponding detection and isolation methodologies, as well as some of the applications utilizing recombinant gene products.

2. Molecular Tags and Reporters

The basic strategy in "labeling" or "tagging" a cloned sequence is to place either upstream or downstream a translationally in-frame sequence corresponding to a polypeptide domain or protein that exhibits highly active or distinct properties not found in the host cell. In this manner, the recombinant hybrid protein, containing the tag and the desired expressed gene product, may be detected and/or isolated on the basis of the unique properties of the tag. In

From: *Methods in Molecular Biology, vol. 63: Recombinant Protein Protocols: Detection and Isolation* Edited by: R. Tuan Humana Press Inc., Totowa, NJ

some instances, an additional sequence corresponding to a specific protease cleavage site is inserted between the tag and the cloned sequence, such that treatment of the final recombinant hybrid protein with the appropriate protease produces the desired gene product from the tag. Chapter 2 (Groskreutz and Schenborn) in this book provides further background for the general rationale used in constructing an expression vector.

2.1. Enzymes

Owing to their ability to catalyze specific reactions yielding distinct, detectable products, enzymes are probably the most popular molecular tag for expression of recombinant genes. The most commonly used enzymes include: chloramphenicol acetyltransferase (CAT); firefly luciferase; β-galactosidase; alkaline phosphatase; and β-glucuronidase. Some of the key reasons for selecting these enzymes as functional labels include high signal-to-background ratios of the catalyzed reactions, high stability of enzyme activity, and the high sensitivity for detection. A number of methods are currently in use for the detection of enzyme activity, including standard colorimetric assays, more sensitive fluorescence- or luminescence-based procedures, chromogenic histochemistry, and immunohistochemistry or solution-phase immunoassays such as radioimmunoassay or enzyme-linked immunosorbent assay (ELISA). The specific characteristics of some of these enzymes and their respective detection protocols are presented in detail in a number of chapters in this book (Chapters 3, 4, 5, and 6).

2.2. Ligand-Binding Labels

Another type of molecular interaction that has been exploited to generate detectable activities in recombinant gene expression includes those involving specific, high-affinity ligand binding. In this manner, the recombinant product possesses the ability to interact with a specific ligand, which ideally is not a property of the host cell proteins. Using a labeled ligand, the corresponding recombinant product may be clearly identified. Alternatively, another reagent, either a protein or a chemical (which is itself labeled), may be used to detect the bound ligand, and thus the recombinant protein. In many instances, the ligand may be immobilized onto a solid support, such as chromatography resins and gels, to develop affinity fractionation methods for isolation and purification of the desired recombinant product. Examples of these protocols may be found in a number of chapters in this book, dealing with specific ligand-binding entities such as: maltose-binding protein, which allows purification of the chimeric protein on amylose columns; Protein A, which recognizes the Fc domain of immunoglobulin G; streptavidin, which binds with extremely high affinity and specificity to biotin; and hexahistidine peptide sequence, which has high metal affinity, i.e., applicable for affinity purification on nickel-nitrilotriacetate column. These topics are covered in detail in Chapters 9, 10, 11, and 12.

2.3. Expression-Coupled Gene Activation

Another means to detect recombinant gene expression, which has recently gained substantial popularity, is the coupling of recombinant gene expression to the transactivation of another unique gene. This approach, an example of which is the yeast two-hybrid system, is particularly useful for the detection of interacting proteins, and the assay is performed in vivo rather than in vitro, thus permitting the detection of such proteins in their native, biologically active state. The yeast two-hybrid system takes advantage of the fact that many eukaryotic transcription activators are made up of structurally separable and functionally independent domains. For example, the yeast transcriptional activator protein GAL4 contains a DNA-binding domain (DNA-BD), which recognizes a 17 base-pair DNA sequence, and an activation domain (AD). Upon DNA-BD binding to the specific upstream region of GAL4-responsive genes, the AD interacts with other components of the machinery to initiate transcription. Thus, both domains are needed in an interactive manner for specific gene activation to take place. In the popular yeast two-hybrid system, the two GAL4 domains are separately fused to genes encoding proteins that interact with each other, and these recombinant hybrid proteins are expressed in yeast. Interaction of the two-hybrid proteins brings the two GAL4 domains in close enough proximity to form a functional gene activator, resulting in the expression of specific reporter gene(s), thereby rendering the protein interaction, i.e., expression of the desired recombinant protein, phenotypically identifiable. In practice, the target protein gene is ligated to the DNA-BD in the form of an expression vector. The gene of interest, whose activity includes interaction with the target protein, is ligated into an AD vector. The two hybrid plasmids are then cotransformed into specialized yeast reporter strain. Expression of the desired gene thus activates a known GAL4 responsive gene(s) and confers specific phenotype to the host cell, which can be selectively identified. Protocols utilizing the two-hybrid system and its variants are described in several chapters in this book (Chapter 12, 15, and 16).

2.4. Immunospecific Detection

Another type of recombinant label or tag consists of components to which specific antibodies are available. In this manner, immunoassays and immunoaffinity chromatography may be used efficiently to detect and isolate, respectively, the recombinant protein. Momand and Sepehrnia (Chapter 14) illustrate how this principle may be exploited using recombinant p53 as an example, and

Olesen et al. (Chapter 7) present methodologies based on chemiluminescent immunoassays using enzyme-conjugated secondary antibodies. Recombinant protein expressing cells may also be cloned and detected by immunodetection methods as described by Gibson et al. (Chapter 8).

2.5. Expression-Coupled Alteration of Cellular Activity

The identification and cloning of multidrug resistance genes such as MDR1, which is responsible for the simultaneous resistance of cells to multiple structurally and functionally unrelated cytotoxic agents, offers the potential to use such genes in fusion gene constructs for the purpose of detection and isolation of the expressing cellular clone. Germann (Chapter 13) describes such an application pertaining to P-glycoprotein.

2.6. Labels with Unique Chemical Characteristics

The green fluorescent protein (GFP) of the jellyfish, *Aequora victoria*, fluoresces following the transfer of energy from the Ca^{2+}-activated photoprotein, aequorin. GFP has been cloned, and, when expressed in prokaryotic and eukaryotic cells, yields green fluorescence when excited by blue or ultraviolet light. Recent developments have focused on utilizing the GFP as a reporter gene to permit the detection of recombinant gene expression in vivo *(see below)*.

3. Detection of Gene Expression

The expression of specific genes in a recombinant form consisting of chimeric or hybrid labels has greatly facilitated their detection. A number of chapters in this book (Chapters 18, 19, 20, 21, 22, 23, and 24) focus on recent developments in the detection of gene expression *in situ*, utilizing DNA or RNA probes, as well as PCR amplification. Since *in situ* hybridization, *in situ* PCR, and label-specific histochemistry (e.g., β-galactosidase histochemistry), utilizing tissue section or whole-mount tissues and embryos, yield information on gene expression at the native, individual cell and tissue level, they are powerful techniques in gaining information on the functional aspects of gene expression. In particular, in transgenic experimentations, where a transgene is introduced into and expressed in an animal, the ability to correlate gene expression and altered cell/tissue phenotype allows the investigator to directly assess the function of the gene of interest. Examples of recombinant label-specific detection protocols include those based on β-galactosidase (Chapter 18) and the fluorescent jellyfish GFP (Chapter 24). *In situ* hybridization is based on hybridization to specific mRNA sequences by labeled DNA and RNA probes (Chapters 19, 20, 21, and 22), whereas sequence-specific amplification by *in situ* PCR provides both gene expression detection and cloning possibilities (Chapter 23). By coupling immunohistochemistry with *in situ* hybridization, it is possible to examine gene expression at both the protein and mRNA levels (Chapters 19 and 22).

4. Applications of Recombinant Gene Expression

The production of recombinant gene products is one of the major success stories of modern molecular biology. This book samples some of the applications to illustrate the present state and future potential of such a technology. Thus, in addition to answering fundamental questions related to regulation of gene expression, gene structure and function, and other basic issues of molecular biology, the technology of recombinant genes has revolutionized modern biotechnology and biomedicine. For example, gene therapy, which aims to compensate for gene defects and/or deliver therapeutic gene products for specific diseases resulting from defective genes, is critically and totally dependent on the design of expression and delivery vectors, which permit targeted and regulated expression of the cloned gene(s) (*see* Chapters 28, 29, 30, and 31). On the other hand, recombinant gene technology also makes it possible to custom-design gene sequences to "biosynthesize" novel biopolymers of unique physicochemical properties (*see* Chapter 27), which may be used for applications in biomaterial, pharmaceutical, agricultural, and other industries. Finally, it should be noted that the current state of recombinant technologies is capable of utilizing a wide spectrum of manufacturing units (the "bioreactors") for the custom-designed protein products, including *E. coli* (Chapter 26), yeast, cultured cells, and transgenic animals (Chapter 25).

II

Detection and Isolation of Expressed Proteins

2

Reporter Systems

Debyra Groskreutz and Elaine T. Schenborn

1. Introduction

Genetic reporter systems have contributed greatly to the study of eukaryotic gene expression and regulation. This chapter will describe what an ideal reporter system is and outline the many uses of genetic reporters. Furthermore, the currently available reporter genes and assays will be described in terms of their specific applications and limitations.

Reporter genes are most frequently used as indicators of transcriptional activity in cells (1). Typically, the reporter gene or cDNA is joined to a promoter sequence in an expression vector that is transferred into cells. Following transfer, the cells are assayed for presence of the reporter by directly measuring the amount of reporter mRNA, the reporter protein itself, or the enzymatic activity of the reporter protein. The ideal genetic reporter is not endogenously expressed in the cell type of interest, and the ideal reporter assay system has the characteristics of being sensitive, quantitative, rapid, easy, reproducible, and safe.

Reporter genes are used for both in vitro and in vivo applications (reviewed in ref. 2). Reporter systems are used to study promoter and enhancer sequences, trans-acting mediators such as transcription factors, mRNA processing, and translation. Reporters are also used to monitor transfection efficiencies, protein-protein interactions, and recombination events.

The E. coli enzyme, chloramphenicol acetyltransferase (CAT), was used in the first publication describing genetic reporter vector and assay systems designed for the analysis of transcriptional regulation in mammalian cells (3). Since that time, several reporter genes and assays have been developed and include β-galactosidase, luciferase, growth hormone (GH), β-glucuronidase (GUS), alkaline phosphatase (AP), and most recently, green fluorescent pro-

From: *Methods in Molecular Biology, vol. 63: Recombinant Protein Protocols: Detection and Isolation* Edited by: R. Tuan Humana Press Inc., Totowa, NJ

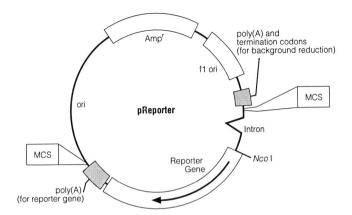

Fig. 1. Plasmid DNA map of a generic reporter expression vector, pReporter, highlighting regions of functional significance. Abbreviations: Amp^r, ampicillin resistance gene; f1 ori, origin of replication for filamentous phage; MCS, multiple cloning sites; poly (A), recognition sequence for polyadenylation; ori, plasmid origin of replication.

tein (GFP). These available genetic reporters and their potential applications are summarized in Section 4.

2. Ideal Characteristics of Reporter Vectors

The design of a typical reporter expression vector is described in Sections 2.1.–2.5. and summarized in Fig. 1.

2.1. Vector Backbone Considerations

In addition to the components required in all expression vectors for optimal protein production (*see* Chapter 25), there are additional considerations for designing reporter gene vectors. First, the ideal reporter vector contains no regulatory binding sites or sequences other than the ones knowingly inserted by the researcher. The presence of extraneous control elements can lead to artifactual results *(4,5)* such as increased or decreased background expression of the reporter. Although attempts can be made to remove known binding sites from the reporter vectors and to reduce read-through transcription originating upstream of the reporter gene, the practical likelihood is slim for removing all potential regulatory sequences from several kilobasepairs of DNA. Therefore, researchers using reporter vectors should be aware of this problem and use the proper vector controls in their experiments.

Reporter plasmid vectors are generally propagated in *E. coli*. Therefore, the plasmid backbone contains an origin for DNA replication (ori) and a gene for selection, typically the ampicillin-resistance gene. Presence of an origin of replication that allows a high copy number of plasmids in *E. coli* (e.g., the pUC ori) facilitates large-scale DNA preparations of plasmid DNA. Presence of an origin of replication for filamentous phage, such as the f1 ori, allows for the production of single-stranded DNA that is useful for mutagenesis and sequencing applications.

2.2. Reporter Gene and Flanking Regions

The coding and flanking regions of reporter genes can be altered for both improved performance and convenience. Altering naturally occurring targeting sequences of reporter proteins can be beneficial. Removal of the last 24 amino acids from the carboxyl end of alkaline phosphatase causes the reporter enzyme to be secreted directly from the cells, avoiding the need for cell lysis *(6)*. Additionally, elimination of the peroxisomal targeting sequence of luciferase allows luciferase to be transported to the cytosol rather than the peroxisomes *(7)*.

To maximize expression of the reporter gene, an optimal ribosomal binding sequence (GCCGCCA/GCCATG...) *(8)* may be placed at the 5' end of the reporter gene. This sequence has been shown to increase the efficiency of translation and it also produces a convenient *Nco*I site, which can be used to create N-terminal gene fusions with the reporter. Removing regulatory binding sites or altering cloning sites within reporter genes *(7,9)* can maximize performance and convenience, respectively.

2.3. Multiple Cloning Sites

Many reporter vectors contain two multiple cloning sites (MCS). One is located upstream of the reporter gene for cloning putative promoter or enhancer/promoter regions. The other MCS is located elsewhere in the plasmid for cloning regulatory elements, like enhancers, that act at a distance. Additional cloning sites can also be used to incorporate selectable gene markers for long-term expression of the reporter gene.

Many convenience features may be incorporated into an MCS. For example, the presence of a cleavage site for a restriction enzyme that generates blunt ends is useful because this allows any blunt-ended fragment to be inserted into the MCS. The MCS can be designed to also include a sequence cleaved by restriction enzymes that generate 3' overhanging ends at one or both ends of the MCS. A 3' overhanging end provides the opportunity to perform nested deletion analysis with exonuclease III *(10)*.

The nucleotide sequences of an MCS located 5' of the reporter gene should be designed to avoid potential hairpin loops and upstream ATG sequences that become part of the reporter mRNA. Either of these factors can decrease the efficiency of translation of the reporter message *(11–17)*.

2.4. Polyadenylation Signals

Polyadenylation has been shown to enhance mRNA stability and translation in mammalian cells *(18,19)*. Immediately following the reporter gene is a polyadenylation (poly A) sequence that signals the addition of 200–250 adenylate residues to the 3'-end of an RNA transcript *(20)*. Commonly used poly A sequences in reporter vectors are derived from the SV40 early and late genes, or the bovine growth hormone gene. For optimal expression, the SV40 late and the bovine growth hormone poly A sites have been shown to be five-, and threefold *(21,22)* more efficient than the SV40 early poly A at generating high levels of steady-state mRNA.

A poly A signal inserted 5' of the transcription unit can lower levels of gene expression originating from cryptic promoter sequences in the vector backbone. Elimination of this background expression effectively increases the sensitivity of the reporter system *(23)*. However, when incorporating two poly A signals within the same vector, nonhomologous regions containing poly A signals should be used to reduce the chances of recombination within the same vector. A further reduction in background reporter expression from spurious transcription within the vector backbone may be achieved at the translational level by incorporating stop codons in all three reading frames upstream of the reporter transcription unit.

2.5. Intron Effects on Reporter Gene Expression

The presence of an intron in the mRNA sequence has been shown to increase the level of expressed protein for particular cDNAs transfected into mammalian cells *(24–27)*. Many reporter genes are derived from bacterial genes containing no introns and, therefore, introns were added to vectors so that the reporter genes more closely resembled the pattern of exons and introns in mammalian genes. Early reporter vectors included the SV40 small-t antigen intron 3' of the reporter gene for the purpose of increasing message stability and protein expression. More recently, however, it has been demonstrated that the small-t intron in this position can lead to cryptic splicing of mRNA from sites within the reporter gene. Ironically, the presence of this intron reduces protein expression of CAT and β-galactosidase by 10-fold compared to the expression levels from the identical vectors lacking an intron *(26,28)*.

The effect of an intron may need to be determined empirically because it is dependent upon the specific gene sequence with which it is associated, and can be different for in vitro and in vivo applications. For transient transfection studies, a 5' intron increased luciferase expression three- to fivefold, although the same intron increased CAT expression over 20-fold *(29)*. Studies with transgenic animals have demonstrated that introns are generally required for protein expression in vivo, although the identical cDNA may not require an intron for protein expression when transfected into cells in vitro *(30–32)*.

3. Reporter Gene Applications

The following applications are a sampling of the types of studies in which reporter genes have played a significant role.

3.1. Transcriptional Control Element Testing

Analysis of *cis*-acting transcriptional elements is the most frequent application for reporter genes *(1)*. Reporter vectors allow functional identification and characterization of promoter and enhancer elements because expression of the reporter is correlated with transcriptional activity of the reporter gene. For these types of studies, promoter regions are cloned upstream of the reporter gene and enhancer elements are cloned upstream or downstream from the gene. The chimeric gene is introduced into cultured cells by standard transfection methods or into a germ cell to produce transgenic organisms. Using reporter gene technology, promoter and enhancer elements have been characterized that regulate cell-, tissue-, and developmentally-defined gene expression *(33,34)*.

Trans-acting factors can be assayed by cotransfer of the chimeric promoter element-reporter gene DNA together with another cloned DNA expressing a transacting protein or RNA of interest. The protein could be a regulatory transcription factor that binds to the promoter region of interest, cloned upstream of the reporter gene. For example, when tat protein is expressed from one vector in a transfected cell, the activity of the HIV-LTR linked to a reporter gene increases, and is reflected in the increase of reporter gene protein activity *(35)*.

Stable cell lines which integrate the chimeric reporter gene of interest into the chromosome can be selected and propagated when a selectable marker is included in a transfection vector. These types of engineered cell lines can be used for drug screening and to monitor the effect of exogenous agents and stimuli upon gene expression *(36)*. Reporter genes inserted into transgenic mice have also been developed as a system to monitor in vivo drug screening *(37)*.

3.2. Identification of Interacting Proteins

Interacting pairs of proteins can be identified in vivo using a clever system developed by Stanley Fields and coworkers *(38,39)*. Known as the two-hybrid system, the interacting proteins of interest are brought together as fusion partners—one is fused with a specific DNA binding domain and the other protein is fused with a transcriptional activation domain. The physical interaction of the two fusion partners is necessary for the functional activation of a reporter gene driven by a basal promoter and the DNA motif corresponding to the DNA binding protein. This system was originally developed with yeast, but has also been used in mammalian cells *(40)*.

3.3. Monitor Transfection Efficiency

The use of a control gene vector can be used to normalize for transfection efficiency or cell lysate recovery between treatments or transfection experiments *(41)*. Typically the control reporter gene is driven by a strong, constitutive promoter and is cotransfected with test vectors. The test regulatory sequences are linked to a different reporter gene so that the relative activities of the two reporter activities can be assayed individually. Control vectors can also be used to optimize transfection methods. Gene transfer efficacy is typically monitored by assaying reporter activity in cell lysates, or by staining the cells *in situ* to estimate the percentage of cells expressing the transferred gene *(42)*.

3.4. Viral Assays and Mechanisms of Action

Reporter genes have also been engineered into viral vectors. The reporter gene can be used to track the type of cells the virus infects, the timing and duration of expression for viral genes *(43)*, and viral latency *(44)*. By linking viral promoter and enhancer elements to reporter genes in viral or plasmid DNA vectors, an increased understanding of the control of viral genes has been possible *(45)*. This type of understanding has been applied to the development of genetically engineered reporter cell lines used for detection of virus, such as herpes simplex virus, in clinical samples *(46)*.

3.5. Other Cellular Processes

Vectors with reporter genes have been designed to monitor other processes in addition to transcriptional gene regulation. For example, recombination events *(47)*, gene targeting *(48)*, RNA processing *(49)*, and signal transduction pathways *(50,51)* in the cell have been studied using reporter genes.

4. Transfer and Detection of Reporter Genes

Reporter genes can be introduced into eukaryotic cells for either in vivo or in vitro applications. In vivo, the reporter genes can be inserted into host cells by viral infection, carrier-mediated transfection, or direct DNA uptake. In tissue culture, plasmid DNA can be directly injected into cells, but is generally transferred by calcium phosphate, DEAE-dextran, lipid-mediated, or electroporation methods. Following transfection of the DNA, detection of the reporter is required by measuring the reporter mRNA or protein. Detection of the mRNA is a more direct measure of reporter gene expression than protein detection since the effects of transcription are observed directly, avoiding possible artifacts that may be the result of downstream processing events such as translation or protein instability. Reporter mRNA can be detected by Northern blot analysis, ribonuclease protection assays, or reverse transcription PCR (RT-PCR) *(52)*. Though these assays are more direct than measuring protein expression, they are also very cumbersome. Therefore, many assays have been developed to measure the reporter protein rather than the mRNA.

Assays to detect the reporter proteins (Table 1) are very popular due to their tremendous ease of use and versatility. Reporter proteins can be assayed by detecting endogenous characteristics such as enzymatic activity or spectrophotometric characteristics, or indirectly with antibody-based assays. In general, enzymatic assays are quite sensitive due to the small amount of catalyst reporter enzyme required to generate the products of the reaction. One limitation of enzymatic assays can be a high background if there is endogenous enzymatic activity in the cell (e.g., β-galactosidase). Antibody-based assays are generally less sensitive, but will detect the reporter protein whether it is enzymatically active or not. Chemiluminescent technology can increase the sensitivity to the level of enzymatic assays. Antibody-based assays are also available to visualize reporter protein expression in cells via *in situ* cell staining and immunohistochemistry.

Sections 4.1.–4.8. provide brief descriptions of the most commonly used reporter genes and assays, together with their applications and limitations (Table 2).

4.1. Chloramphenicol Acetyltransferase (CAT)

4.1.1. Origin of Reporter Gene

Transposon 9 of *E. coli (53)*.

4.1.2. Protein Characteristics

CAT is a trimeric protein comprising three identical subunits of 25,000 Dalton *(54)*. The CAT protein is relatively stable in the context of mammalian cells, although the mRNA has a relatively short half-life, making the CAT reporter especially suited for transient assays designed to assess accumulation of protein expression *(55)*.

4.1.3. Enzymatic Reaction

CAT enzyme catalyzes the transfer of the acetyl group from acetyl-CoA to the substrate, chloramphenicol. The transfer occurs to the 3 position of chloramphenicol, and nonenzymatic rearrangement produces a 1-acetylated chloramphenicol species. Under high concentrations of the enzyme, a 1,3-diacetylated chloramphenicol product accumulates *(56)*.

4.1.4. Assay Formats

The enzyme reaction can be quantitated by incubating cell lysates with ^{14}C-chloramphenicol and following product formation by physical separation with

Table 1
Assay Formats Available with the Commonly Used Reporter Genes

Assay Format	CAT	Luciferase	β-Galactosidase	GUS	hGH	AP	SEAP	GFP
Isotopic	^{14}C- or ^{3}H-Cm or acetyl CoA	—	—	—	Iodinated (RIA)	—	—	—
Colorimetric	—	—	ONPG, CPRG	X-Gluc	—	PNPP FADP	PNPP FADP	—
Fluorescent	Bodipy-Cm	—	MUG	4-MUG FDG	—	+	+	+
Chemiluminescent	—	—	1,2-dioxetane-β-gal	1,2-dioxetane aryl glucuronide	—	+	+	—
Bioluminescent	—	Luciferin	—	—	—	+ (2 step assay)	+ (2 step assay)	—
ELISA	+	—	+	—	+	+	+	—
in situ staining	Antibody-based	Antibody-based	X-Gal (Histochemical)	X-Gluc	—	Limited; high endogenous levels	—	+ No substrate

Compounds listed in the table are substrates used in combination with the particular assay format and reporter gene protein.

Abbreviations: CAT, chloramphenicol acetyltransferase; GUS, β-glucuronidase; hGH, human growth hormone; AP, alkaline phosphatase; SEAP, secreted alkaline phosphatase; GFP, green fluorescent protein; Cm, chloramphenicol; RIA, radioimmunoassay; ONPG, o-nitrophenyl β-D-galactopyranoside; CPRG, chlorophenol red β-D-galactospyranoside; X-Gluc, 5-bromo-4-chloro-3-indolyl β-D-glucuronic acid; PNPP, p-nitrophenyl phosphate; FADP, flavin-adenine dinucleotide phosphate; 4-MUG, β-methyl umbelliferyl galactoside; FDG, fluorescein digalactoside; X-Gal, 5-bromo-4-chloro-3-indoyl β-D-galactoside. +, availability of assay; —, not available or not a commonly used assay.

**Table 2
Advantages and Limitations of the Commonly Used Reporter Genes**

Reporter gene	Advantages	Limitations
CAT	Widely accepted standard in literature Visual confirmation of enzyme activity	Relatively low sensitivity High costs for isotope and TLC systems
Luciferase	Fast and easy High sensitivity Large linear range	Requires luminometer for high sensitivity assays Relatively labile protein
β-Galactosidase	Easy to assay Variety of assay formats for use with cell extracts Widely used for *in situ* staining	Endogenous activity in some cell types Lower sensitivity in non-chemiluminescent assays
β-Glucuronidase (GUS)	Wide variety of formats Used for in situ staining Fusion proteins	Endogenous activity in mammalian cells
Human Growth Hormone (hGH)	Secreted Low background in most cells	RIA or EIA formats Low sensitivity
Alkaline Phosphatase (AP)	Wide variety of assay methods High sensitivity in some assays Large linear range in some assays	Endogenous activity in most cells
Secreted Alkaline Phosphatase (SEAP)	Secreted Low background activity	
Green Fluorescent Protein (GFP)	No substrates required Stable reporter protein *In situ* and in vivo applications	Low sensitivity and expression Expensive microscope required

Abbreviations: TLC, thin layer chromatography; RIA, radioimmunoassay; EIA, enzyme immunoassay.

thin layer chromatography (TLC). The TLC separates substrate and products, which are visualized by exposing the TLC plate to X-ray film or to phosphorimaging. This TLC assay, although rather tedious and not as sensitive as subsequently developed assays, allows a visual confirmation of the reaction, and has become a well-accepted standard for reporter gene assays. Quantitation of enzyme activity involves scraping the TLC spots corresponding to the substrate and products and counting the samples in a scintillation counter.

Alternative isotopic CAT assays have been developed which rely upon organic extraction of the more nonpolar products from the chloramphenicol substrate *(57)*. Acetyl CoA radiolabeled in the acetyl moiety allows the transfer of the radiolabel to the chloramphenicol substrate. The products can be preferentially extracted into an organic solvent, such as scintillation fluid *(58)*.

Fluorescent chloramphenicol is also available commercially, and serves as a nonisotopic alternative to the ^{14}C-chloramphenicol substrate *(59)*. Like the isotopic assay, the substrate and products are separated by TLC. Quantitation of enzyme activity can be achieved by use of a fluorescence scanner, or by scraping the "spots" and measuring with a fluorometer.

Antibodies to CAT also allow the expressed CAT protein to be quantitated by an ELISA format, to be identified in Western blots, and to be used in immunohistochemical applications.

4.2. Luciferase

4.2.1. Origin of Reporter Gene

The luciferase enzyme used most frequently for reporter gene technology is derived from the coding sequence of the *luc* gene cloned from the firefly *Photinus pyralis* (*60–62*; *see* ref. *62* for review).

4.2.2. Protein Characteristics

Monomer of 60,700 Dalton. Compared to CAT, the firefly luciferase protein has a shorter half-life in transfected mammalian cells *(55,63)*, making the luciferase reporter especially suited for transient assays designed to assess inducible and short-lived effects.

4.2.3. Enzymatic Reaction

The firefly luciferase enzyme catalyzes a reaction using D-luciferin and ATP in the presence of oxygen and Mg^{+2} resulting in light emission.

4.2.4. Assay Formats

The flash of light decays rapidly, in seconds, and is captured with a luminometer which measures integrated light output. The total amount of light measured during a given time interval is proportional to the amount of luciferase activity in the sample. The assay has been improved by including coenzyme A in the reaction which provides a longer, sustained light reaction *(64)*. The prolonged light output increases the sensitivity of the assay and allows more reproducible results to be obtained from assays using scintillation counters to measure the light output from a luciferase reaction.

The sensitivity of the luciferase assay is in the subattomole range, and approximately 30–1000X greater compared to the sensitivity of CAT assays *(63)*. An added advantage is that luciferase assay results can be obtained in minutes compared to hours, or even days, for the radioactive CAT assay. The linear range of the luciferase assay extends over an impressive seven orders of magnitude of firefly luciferase concentration. Luciferase has also been used for in vivo applications to study regulated reporter gene activity in whole organisms, such as plants as well as in single cells (65). Prior limitations for in vivo applications are being overcome by development of soluble forms of firefly luciferin which allow cell penetrance (66) and the instrumentation to detect single photons from microscopic samples.

4.3. β -Galactosidase

4.3.1. Origin of Reporter Gene

The *lacZ* gene which codes for the β-galactosidase enzyme from *E.coli* (67).

4.3.2. Protein Characteristics

Tetrameric enzyme with subunit size of 116,000 Dalton.

4.3.3. Enzymatic Reaction

β-Galactosidase catalyzes the hydrolysis of β-galactoside sugars such as lactose.

4.3.4. Assay Formats

The enzymatic activity in cell extracts can be assayed with various specialized substrates that allow enzyme activity quantitation with a spectrophotometer, fluorometer, or a luminometer. A major strength of this reporter gene is the ability to easily assay *in situ* expression with histochemical staining.

The substrate o-nitrophenyl- β-D-galactopyranoside (ONPG) is used in a standard colorimetric enzyme assay that is read with a spectrophotometer. This assay is easily adapted to the 96-well format for read-out with a standard microplate spectrophotometer, and can be extended to a dynamic range of six orders of magnitude for enzyme concentration (68). Chlorophenol red β-D-galactopyranoside (CPRG) is another colorimetric-based substrate, which provides approximately 10X higher sensitivity than ONPG (69). Fluorescent-based assays are available with substrates such as β-methyl umbelliferyl galactoside (MUG) and fluorescein digalactoside (FDG) that allow detection of enzyme activity in single cells and can be adapted for fluorescence activated cell sorting (70,71). Chemiluminescent assays display the highest sensitivity and largest dynamic range, and can be developed with 1,2-dioxetane substrates (72). The chemiluminescent assay format is similar in sensitivity to the bioluminescent luciferase assay.

In situ histochemical analyses use 5-bromo-4-chloro-3-indoyl β-D-galactoside (X-Gal) as a substrate. Enzymatic hydrolysis of this substrate, followed by oxidation produces a precipitate with a characteristic blue color. The histochemical staining can be used to monitor the percentage of cells effectively transfected in a particular experiment and expressing the reporter protein. The staining is also used to monitor localized and cell-specific expression of chimeric constructs in transgenic embryos and animals *(73,74)*.

The β-galactosidase reporter gene is frequently used as a control vector for normalizing transfection efficiency when cotransfected with chimeric DNAs linked to other reporter genes *(41)*. One potential limitation to this reporter gene is that certain mammalian cells have endogenous lysosomal β-galactosidase activity. Enzyme assays performed at a higher pH of 7.3–8.0, or with cell extracts heated at 50°C, preferentially favor the *E. coli* enzyme *(2,75)*. However, because of endogenous cellular β-galactosidase activity, it is important to include negative control extracts or cells which have not been transfected, as comparisons for the cell-free and *in situ* analyses.

4.4. β-Glucuronidase (GUS)

4.4.1. Origin of Reporter Gene

The *gus A* gene from *E. coli.*

4.4.2. Protein Characteristics

A tetrameric glycoprotein composed of four identical subunits of 68,000 Dalton is localized predominantly to the acidic environment of the lysosomes.

4.4.3. Enzymatic Reaction

GUS is an exoglycosidase which removes terminal β-glucorunic acid residues from the nonreducing end of glycosaminoglycans and other glyco-conjugates (*see* ref. *76*).

4.4.4. Assay Formats

β-Glucuronidase is used as a genetic reporter in both plant and animal cells. Like the β-galactosidase reporter, one of the principle advantages of GUS is the wide range of available assays for the enzyme. Several different colorimetric assays have been developed using a variety of β-glucuronides as substrates. The substrate X-gluc, for example, is a very popular colorimetric substrate that can also be used for histochemical staining of tissues and cells. In addition, fluorescent and chemiluminescent assays have been developed utilizing the substrates 4-MUG *(77)*, and 1,2-dioxetane aryl glucuronide substrate *(62)*, respectively. The sensitivity of the assays obtained with the above substrates vary greatly with the chemiluminescent assays being 100-fold more sensitive than the fluorescent assays, which can be 100–1000-fold more sensitive than the colorimetric assays.

In higher plants, GUS is most commonly used because most plants lack endogenous GUS activity, whereas virtually all mammalian tissues contain glucuronidases which aid in metabolism (*see* ref. *78*). The ubiquitous endogenous activity of GUS in mammalian cells has limited its use; however, the mammalian and bacterial GUS proteins can be distinguished from each other by their differing pH optima. In addition to the uses of GUS as a genetic reporter, GUS fusion proteins can be created to study protein transport and localization in both plant and animal cells *(76)*.

4.5. Human Growth Hormone (hGH)

4.5.1. Origin of Reporter Gene

Human growth hormone gene.

4.5.2. Protein Characteristics

The protein is 24–25,000 Dalton. Human growth hormone is normally expressed only in the somatotropic cells of the anterior pituitary, so this limited endogenous expression makes growth hormone an attractive choice as a genetic reporter for other mammalian cell types *(79)*.

4.5.3. Assay Formats

The hGH differs from the reporter proteins discussed previously in that it is secreted from the cells into the culture medium. The major advantage of secreted reporter proteins over intracellular reporters is that the cells do not have to be lysed for the reporter expression to be measured. Secreted reporters are particularly beneficial for kinetic analysis studies in which the various time points can be compared from the same dish of cells. The hGH is also commonly used as an internal control for normalizing the transfection efficiencies of different plates of cells within an experiment. The available assays for hGH are limited to ELISA and radioimmunoassay (RIA).

4.6. Alkaline Phosphatase (AP)

4.6.1. Origin of Reporter Gene

Mammalian; A generic term for a family of ubiquitously expressed orthophosphoric monoester phosphohydrolases which exhibit an alkaline pH optimum.

4.6.2. Protein Characteristics

Alkaline phosphatase AP is a relatively stable enzyme that has been utilized in many applications, including reporter analysis experiments *(80,81)*.

4.6.3. Enzymatic Reaction

The enzyme dephosphorylates a broad range of substrates at alkaline pH.

4.6.4. Assay Formats

One advantage of AP is that there are a number of developed assays, including 96-well formats, which serve a wide variety of user needs. A standard spectrophotometric assay is based on the hydrolysis of *p*-nitrophenyl phosphate (PNPP) by AP. This assay is inexpensive, rapid, and simple, but lacks the sensitivity obtained with other methods. Recently, a sensitive and flexible amplified colorimetric assay has been developed that uses flavin-adenine dinucleotide phosphate as a substrate for AP *(82)*. A two-step bioluminescent assay has also been developed *(83)* that provides sensitivity similar to that of the luciferase reporter. In this two-step approach, AP first hydrolyzes D-luciferin-O-phosphate to D-luciferin which then serves as a substrate for luciferase in the second step. There is also a single-step, chemiluminescent assay for AP *(84)*.

Because of its expression in virtually all mammalian cell types, the use of AP as a reporter can be limited owing to high background levels of endogenous AP. This background problem has been overcome with the development of secreted alkaline phosphatase *(6)*.

4.7. Secreted Alkaline Phosphatase (SEAP)

4.7.1. Origin of Reporter Gene

A form of the human placental alkaline phosphatase gene lacking 24 amino acids at the carboxy end of the protein.

4.7.2. Protein Characteristics

Removal of these amino acids prevents the enzyme from anchoring to the plasma membrane resulting in its secretion from the cells into the culture medium *(6)*. The SEAP is stable to heat and to the phosphatase inhibitor L-homoarginine, whereas endogenous AP is not.

4.7.3. Assay Formats

Treatment of cell lysates with heat or L-homoarginine inactivates background AP activity from within cells that may have entered the culture medium. Thus, the high background observed with the AP reporter system is essentially eliminated with the SEAP system. The combination of a secreted reporter protein, low endogenous reporter background, and a wide variety of easy and sensitive assays make SEAP a convenient and versatile reporter system. The assays for SEAP activity are identical to those described for AP.

4.8. Green Fluorescent Protein (GFP)

4.8.1. Origin of Reporter Gene

The green fluorescent protein from the jellyfish, *Aequorea victoria.*

4.8.2. Protein Characteristics

Green fluorescent protein is a 238 amino acid, 27,000 Dalton monomer that emits green light when excited with UV or blue light *(85)*. The active chromophore in GFP required for fluorescence is a hexa-peptide that contains a cyclized Ser-dehydroTyr-Gly trimer *(86)*.

4.8.3. Assay Formats

Light-stimulated GFP fluoresces in the absence of any other proteins, substrates, or cofactors. Therefore, unlike any of the other available reporters, GFP gene expression and localization can be monitored in living organisms and in live or fixed cells using only UV or blue-light illumination *(87,88)*. Additionally, GFP fusion proteins can be constructed to study protein transport and localization *(88)*. Other advantages of GFP are that it is very stable to heat, extreme pH, and chemical denaturants.

Current disadvantages of the GFP reporter system are low expression levels and the requirement for powerful fluorescent microscopes for detection. Being a new technology, however, advancements to GFP technology are inevitable.

References

1. Rosenthal, N. (1987) Identification of regulatory elements of cloned genes with functional assays. *Methods Enzymol.* **152,** 704–720.
2. Alam, J. and Cook. J. L. (1990) Reporter genes: application to the study of mammalian gene transcription. *Anal. Biochem.* **188,** 245–254.
3. Gorman, C. M., Moffat, L. F., and Howard, B. H. (1982) Recombinant genomes which express chloramphenicol acetyltransferase in mammalian cells. *Mol. Cell. Biol.* **2,** 1044–1051.
4. Boshart, M., Kluppel, M., Schmidt, A., Schutz, G., and Luckow, B. (1992) Reporter constructs with low background activity utilizing the *cat* gene. *Gene* **110,** 129–130.
5. Kushner, P. J., Baxter, J. D., Duncan, K. G., Lopez, G. N., Schaufele, F., Uht, R. M., Webb P., and West, B. L. (1994) Eukaryotic regulatory elements lurking in plasmid DNA: The activator protein-1 site in pUC. *Mol. Endo.* **8,** 405–407.
6. Berger, J., Hauber, J., Bauber, R., Geiger, R., and Cullen, B. (1988) Secreted placental alkaline phosphatase: a powerful new quantitative indicator of gene expression in eukaryotic cells. *Gene* **66,** 1–10.
7. Sherf, B. A. and Wood, K. V. (1995) Firefly luciferase engineered for improved genetic reporting. *Promega Notes* **49,** 14–21.
8. Kozak, M. (1989) The scanning model for translation: an update. *J. Cell. Biol.* **108,** 229–241.

9. Bonin, A. L., Gossen M., and Bujard, H. (1994) *Photinus pyralis* luciferase: vectors that contain a modified luc coding sequence allowing convenient transfer into other systems. *Gene* **141,** 75–77.
10. Henikoff, S. (1987) Unidirectional digestion with exonuclease III in DNA sequence analysis. *Methods Enzymol.* **155,** 156–165.
11. Fu, L. N., Ye, R. Q., Browder, L. W., and Johnston, R. N. (1991) Translational potentiation of messenger RNA with secondary structure in Xenopus. *Science* **251,** 807–810.
12. Kim, S. J., Park, K., Koeller, D., Kim K. Y., Wakefield, L. M., Sporn, M. B., and Roberts, A. B. (1992) Post-transcriptional regulation of the human transforming growth factor-beta 1 gene. *J. Biol. Chem.* **267,** 13,702–13,707.
13. Kozak, M. (1986) Influences of mRNA secondary structure on initiation by eukaryotic ribosomes. *Proc. Natl. Acad. Sci. USA* **83,** 2850–2854.
14. Kozak, M. (1989) Circumstances and mechanisms of inhibition of translation by secondary structure in eucaryotic mRNA's. *Mol. Cell. Biol.* **9,** 5134–5142.
15. Rao, C. D., Peck, M., Robbins, K. C., and Aaronson, S. A. (1988) The 5' untranslated sequence of the c-sis/platelet-derived growth factor 2 transcript is a potent translational inhibitor. *Mol. Cell. Biol.* **8,** 284– 292.
16. Kozak, M. (1984) Selection of initiation sites by eukaryotic ribosomes: effect of inserting AUG triplets upstream from the coding sequence of preproinsulin. *Nucleic Acids Res.* 12, 3873–3893.
17. Liu, C., Simonsen, C. C., and Levinson, A. D. (1984) Initiation of translation at internal AUG codons in mammalian cells. *Nature* **309,** 82–85.
18. Bernstein, P., and Ross, J. (1989) Poly(A), poly(A) binding protein and the regulation of mRNA stability. *Trends Biochem. Sci.* **14,** 373–377.
19. Jackson, R. J. and Standart, N. (1990) Do the poly(A) tail and 3' untranslated region control mRNA translation? *Cell* **62,** 15–24.
20. Proudfoot, N. J. (1991) Poly(A) signals. *Cell* **64,** 671–674.
21. Carswell, S. and Alwine, J. C. (1989) Efficiency of utilization of the simian virus 40 late polyadenylation site: effects of upstream sequences. *Mol. Cell. Biol.* **9,** 4248–4258.
22. Pfarr, D. S., Rieser, L. A., Woychik, R. P., Rottman, F. M., Rosenberg, M., and Reff, M. E. (1986) Differential effects of polyadenylation regions on gene expression in mammalian cells. *DNA* **5,** 115–122.
23. Araki, E., Shimada, F., Shichiri, M., Mori, M., and Ebina, Y. (1988) pSV00CAT: Low background CAT plasmid. *Nucleic Acids Res.* **16,** 1627–1630.
24. Gross, M. K., Kainz, M. S., and Merrill, G. F. (1987) Introns are inconsequential to efficient formation of cellular thymidine kinase mRNA in mouse L cells. *Mol. Cell. Biol.* **7,** 4576–4581.
25. Buchman, A. R. and Berg, P. (1988) Comparison of intron-dependent and intron-independent gene expression. *Mol. Cell. Biol.* **8,** 4395–4405.
26. Evans, M. J. and Scarpulla, R. C. (1989) Introns in the 3'-untranslated region can inhibit chimeric CAT and β-galactosidase gene expression. *Gene* **84,** 135–142.

27. Huang, M. T. F. and Gorman, C. M. (1990) Intervening sequences increase the efficiency of RNA 3' processing and accumulation of cytoplasmic RNA. *Nucleic Acids Res.* **18,** 937–947.

28. Huang, M. T. F. and Gorman, C. M. (1990) The simian virus 40 small-t intron, present in many common expression vectors, leads to aberrant splicing. *Mol. Cell. Biol.* **10,** 1805–1810.

29. Brondyk, B. (1995) pCI and pSI mammalian expression vectors. *Promega Notes* **49,** 7–11.

30. Brinster, R. L., Allen, J. M., Behringer, R. R., Gelinas, R. E., and Palmiter, R. D. (1988) Introns increase transcriptional efficiency in transgenic mice. *Proc. Natl. Acad. Sci. USA* **85,** 836–840.

31. Choi, T., Huang, M., Gorman, C., and Jaenisch, R. A. (1991) generic intron increases gene expression in transgenic mice. *Mol. Cell. Biol.* **11,** 3070–3074.

32. Palmiter, R. D., Sandgren, E. P., Avarbock, M. R., Allen, D. D., and Brinster, R. L. (1991) Heterologous introns can enhance expression of transgenes in mice. *Proc. Natl. Acad. Sci. USA* **88,** 478–482.

33. Dahler, A., Wade, R. P., Muscat, G. E. O., and Waters, M. J. (1994) Expression vectors encoding human growth hormone (hGH) controlled by human muscle-specific promoters: prospects for regulated production of hGH delivered by myoblast transfer or intravenous injection. *Gene* **145,** 305–310.

34. Wegner, R. H., Moreau, H., and Neilsen, P. J. (1994) A comparison of different promoter, enhancer, and cell type combinations in transient transfections. *Anal. Biochem.* **221,** 416–418.

35. Koken, S. E., van Wamel, J., and Berkhout, B. (1994) A sensitivie promoter assay based on the transcriptional activator Tat of the HIV-1 virus. *Gene* **144,** 243–247.

36. Himmler, A., Stratowa C., Czernilofsky, A. P. (1993) Functional testing of human dopamine D1 and D5 receptors expressed in stable cAMP-responsive luciferase reporter cell lines. *J. Recept. Res.* **13,** 79–94.

37. Mehtali, M., Munschy, M., Ali-Hadji, D., and Kieny, M. D. (1992) A novel transgenic mouse model for the in vivo evaluation of anti-human immunodeficiency virus type1 drugs. *AIDS-Res. Hum. Retroviruses* **8,** 1959–1965.

38. Chien, C. T., Bartel, P. L., Sternglanz, R., and Fields, S. (1991) The two-hybrid system: a method to identify and clone genes for proteins that interact with a protein of interest. *Proc. Natl. Acad. Sci. USA* **88,** 9578–9582.

39. Fields, S. and Song, O. (1989) A novel genetic systme to detect protein-protein interactions. *Nature* **340,** 245–246.

40. Fearon, E. R., Finkel, T., Gillison, M. L., Kennedy, S. P., Casella, J. F., Tomaselli, G. F., Morrow, J. S., and Van-Dang, C. (1992) Karyoplasmic interaction selection strategy: a general strategy to detect protein-protein interactions in mammalian cells. *Proc. Natl. Acad. Sci. USA* **89,** 7958–7962.

41. Hollon, T. and Yoshimura, F. K. (1989) Variation in enzymatic transient gene expression assays. *Anal. Biochem.* **182,** 411–418.

42. Lopata, M. A., Cleveland, D. W., and Sollner-Webb, B. (1984) High level tran-
    sient expression of a chloramphenicol gene by DEAE-dextran mediated DNA
    transfection coupled with a dimethyl sulfoxide or glycerol shock treatment.
    Nucleic Acids Res. **12,** 5707–5717.
43. Mittal, S. K., McDermott, M. R., Johnson, D. C., Prevec, L. and Graham, F. L.
    (1993) Monitoring foreign gene expression by a human adenovirus-based vector
    using the firefly luciferase gene as a reporter. *Virus Res.* **28,** 67–90.
44. Chen, B. K., Saksela, K., Andino, R., and Baltimore, D. (1994) Distinct modes of
    human immunodeficiency virus type 1 proviral latency revealed by superinfec-
    tion of nonproductively infected cell lines with recombinant lucirease-encoding
    viruses. *J. Virol.* **68,** 654–660.
45. Martin, M. E., Nicholas, J., Thompson, B. J., Newman, C., and Honess, R. W.
    (1991) Identification of a transactivating function mapping to the putative imme-
    diate-early locus of human herpesvirus 6. *J. Virol.* **65,** 5381–5390.
46. Stabell, E. C., Rourke, S. R., Storch, G. A., and Olivo, P. D. (1993) Evaluation of
    a genetically engineered cell line and a histochemical beta-galactosidase assay to
    detect simplex virus in clinical specifimens. *J. Clin. Microbiol.* **31,** 2796–2798.
47. Herzing, L. B. K. and Meyn, M. S. (1993) Novel lacZ-based recombination vec-
    tors for mammalian cells. *Gene* **137,** 163–169.
48. Vile, R. G. and Hart, I. R. (1993) In vitro and in vivo targeting of gene expression
    to melanoma cells. *Cancer Res.* **53,** 962–967.
49. Huang, M. T. F. and Gorman, C. M. (1990) Intervening sequences increase effi-
    ciency of RNA 3' processing and accumulation of cytoplasmic RNA. *Nucleic
    Acids Res.* **18,** 937–947.
50. Medema, R. H., de Laat, W. L., Martin, G. A., McCormick, F., and Bos, J. L.
    (1992) GTPase-activation protein SH2-SH3 domains induce gene expression in a
    Ras-dependent fashion. *Mol. Cell. Biol.* **12,** 3425–3430.
51. Sakoda, T., Kaibuchi, K., Kishi, K., Kishida, S., Doi, K., Hoshino, M., Hattori, S.,
    and Takai, Y. (1992) smg/rap/Krev-1 p21s inhibit the signal pathway to the c-fos
    promoter/enhancer from c-Ki ras p21 but not from c-far-1 kinase in NIH3T3 cells.
    Oncogene **7,** 1705–1711.
52. Morales, M. J. and Gottlieb, D. I. (1993) A polymerase chain reaction-based
    method for detection and quantitation of reporter gene expression in transient
    transfection assays. *Anal. Biochem.* **210,** 188–194.
53. Alton, N. K. and Vapnek, D. (1979) Nucleotide sequence analysis of the
    chloramphenicol resistance transposon Tn9. *Nature* **282,** 864–869.
54. Leslie, A. G. W., Moody, P. C. E., and Shaw, W. V. (1988) Structure of
    chloramphenicol acetyltransferase at 1. 75A resolution. *Proc. Nat. Acad. Sci. USA*
    85, 4133–4137.
55. Thompson, J. F., Hayes, L. S., and Lloyd, D. B. (1991) Modulation of fire-
    fly luciferase stability and impact on studies of gene regulation. *Gene* **103,**
    171–177.
56. Shaw, W. V. (1975) Chloramphenicol acetyltransferase from chloramphenicol-
    resistant bacteria. *Methods Enzymol.* **43,** 737–755.

57. Seed, B. and Sheen, J.-Y. (1988) A simple phase-extraction assay for chloramphenicol acyltransferase activity. *Gene* **67,** 271–277.
58. Neumann, J. R., Morency, C. A., and Russian, K. O. (1987) A novel rapid assay for chloramphenicol acetyltransferase gene expression. *BioTechniques* **5,** 444–447.
59. Hruby, D. E., Brinkley, J. M., Kang, H. C., Haugland, R. P., Young, S. L., and Melnor, M. H. (1990) Use of a fluorescent chloramphenicol derivative as a substrate for CAT assays. *BioTechniques* **8,** 170–171.
60. DeWet, J. R., Wood, K. V., Helinski, D. R., and DeLuca, M. (1985) Cloning of firefly luciferase cDNA and the expression of active luciferase in *Escherichia coli. Proc. Natl. Acad. Sci. USA* **82,** 7870–7873.
61. DeWet, J. R., Wood, K. V., DeLuca, M., Helinski, D. R., and Subramani, S. (1987) Firefly luciferase gene: Structure and expression in mammalian cells. *Mol. Cell. Biol.* **7,** 725–737.
62. Bronstein, I., Fortin, J., Stanley, P. E., Stewart, G. S. A. B., and Kricka, L. J. (1994) Chemiluminescent and bioluminescent reporter gene assays. *Anal. Biochem.* **219,** 169–181.
63. Pazzagli, M., Devine, J. H., Peterson, D. O., and Baldwin, T. O. (1992) Use of bacterial and firefly luciferases as reporter genes in DEAE-dextran-mediated transfection of mammalian cells. *Anal. Biochem.* **204,** 315–323.
64. Wood, K. V. (1991) in *Bioluminescence and Chemiluminescence: Current Status* (Stanley, P. E. and Kricka, L. J., eds.),Wiley, Chichester, pp. 11–14.
65. Langridge, W., Escher, A., Wang, G., Ayre, B., Fodor, I., and Szalay, A. (1994) Low-light image analysis of transgenic organisms using bacterial luciferase as a marker. *J. Biolumin. Chemilumin.* **9,** 185–200.
66. Craig, F. F., Simmonds, A. C., Watmore, D., McCapra, F., and White, M. R. H. (1992) Membrane-permeable luciferin esters for assay of firefly luciferase in live intact cells. *Biochem. J.* **276,** 637–641.
67. Hall, C. V., Jacob, P. E., Ringold, G. M., and Lee, F. (1983) Expression and regulation of *Escherichia coli lacZ* gene fusions in mammalian cells. *J. Molec. Applied Gen.* **2,** 101–109.
68. Marsh, J. (1994) Kinetic determination of cellular LacZ expression. *Genet. Anal. Tech. Appl.* **11,** 20–23.
69. Eustice, D. C., Feldman, P. A., Colberg-Poley, A. M., Buckery, R. M., and Neubauer, R. H. (1991) A sensitive method for the detection of β-galactosidase in transfected mammalian cells. *BioTechniques* **11,** 739–742.
70. Price, J., Turner, D., and Cepko, C. (1987) Lineage analysis in the vertebrate nervous system by retrovirus-mediated gene transfer. *Proc. Nat. Acad. Sci. USA* **84,** 156–160.
71. Krasnow, M. A., Cumberledge, S., Manning, G., Herzenberg, L. A., and Nolan, G. P. (1991) Whole animal cell sorting of Drosophila embryos. *Science* **251,** 81–85.
72. Jain, V. K. and Magrath, I. T. (1991) A chemiluminescent assay for quantitation of β-galactosidase in the femtogram range: application to quantitation of β-galactosidase in *lacZ*-transfected cells. *Anal. Biochem.* **199,** 119–124.

73. Sanes, J. R., Rubenstein, J. L. R., and Nicolas, J.-F. (1986) Use of a recombinant retrovirus to study post-implantation cell lineage in mouse embryos. *EMBO J.* **5**, 3133–3142.

74. Lim, K. and Chae, C. B. (1989) A simple assay for DNA transfection by incubation of the cells in culture dishes with substrates for beta-galactosidase. *BioTechniques* **7**, 576–579.

75. Young, D. C., Kingsley, S. D., Ryan, K. A., and Dutko, F. J. (1993) Selective inactivation of eukaryotic β-galactosidase in assays for inhibitors of HIV-1 TAT using bacterial β-galactosidase as a reporter enzyme. *Anal. Biochem.* **215**, 24–30.

76. Gallagher, S. R. (1992) *GUS Protocols: Using the GUS Gene as a Reporter of Gene Expression.* Academic Press, San Diego, CA.

77. Jefferson, R. A., Kavanagh, T. A., and Bevan, M. W. (1987) GUS fusions: β-glucuronidase as a sensitive and versatile gene fusion marker in higher plants. *EMBO J.* **6**, 3901–3907.

78. Paigen, K. (1989) Mammalian beta-glucuronidase: genetics, molecular biology, and cell biology. *Prog. Nucleic Acid Res. Mol. Biol.* **37**, 155–205.

79. Selden, R. F., Howie, K. B., Rowe, M. E., Goodman, H. M., and Moore, D. D. (1986) Human growth hormone as a reporter gene in regulation studies employing transient gene expression. *Mol. Cell. Biol.* 6, 3173–3179.

80. Henthorn, P., Zervos, P., Raducha, M., Harris, H., and Kadesch, T. (1988) Expression of a human placental alkaline phosphatase gene in transfected cells: use as a reporter for studies of gene expression. *Proc. Natl. Acad. Sci. USA* **85**, 6342–6346.

81. Yoon, K., Thiede, M. A., and Rodan, G. A. (1988) Alkaline phosphatase as a reporter enzyme. *Gene* **66**, 11–17.

82. Harbron, S., Eggelte, H. J., Fisher, M., and Rabin, B. R. (1992) Amplified assay of alkaline phosphatase using flavin-adenine dinucleotide phosphate as substrate. *Anal. Biochem.* **206**, 119–124.

83. Miska, W. and Geiger, R. (1987) Synthesis and characterization of luciferin derivatives for use in bioluminescence enhanced enzyme immunoassays. *J. Clin. Chem. Clin. Biochem.* **25**, 23–30.

84. Shaap, A. P., Akhavan, H., and Romano, L. J. (1989) Chemiluminescent substrates for alkaline phosphatase: application to ultrasensitive enzyme-linked immunoassays and DNA probes. *Clin. Chem.* **35**, 1863–1864.

85. Prasher, D. C., Eckenrode, V. K., Ward, W. W., Prendergast, F. G., and Mormier, M. J. (1992) Primary structure of the Aequorea victoria green fluorescent protein. *Gene* **111**, 229–233.

86. Cody, C. W., Prasher, D. C., Westler, W. M., Prendergast, F. G., and Ward, W. (1993) Chemical structure of the hexapeptide chromophore of the aequorea green-fluorescent protein. *Biochemistry* **32**, 1212–1218.

87. Chalfie, M., Tu, Y., Euskirchen, G., Ward, W. W., and Prasher, D. C. (1994) Green fluorescent protein as a marker for gene expression. *Science* **263**, 802–805.

88. Wang, S. and Hazelrigg, T. (1994) Implications for bcd mRNA localization from spatial distribution of exu protein in Drosophila oogenesis. *Nature* **369**, 400–403.

3

Detection of Recombinant Protein Based on Reporter Enzyme Activity: Chloramphenicol Acetyltransferase

Peiyu Lee and Dennis E. Hruby

1. Introduction

Genetic engineering technologies allow the construction of genetic chimeras between the promoter region of a gene of interest and a reporter gene as a means to study the regulation of eukaryotic gene expression at the transcriptional level. The genetic constructs in plasmid form are delivered back into cells to measure the expression of the reporter gene. A good reporter gene product possesses the following characteristics: the enzymatic activity is heat stable and protease resistant, and corresponds well to the strength of the promoter; background and/or interfering enzymatic activities are not present in mammalian cells; and, simple, sensitive, reproducible, and convenient enzymatic or immunological assays of the reporter gene product are available for the assessment of the promoter activity *(1)*. For many applications, the bacterial chloramphenicol acetyltransferase (CAT) satisfies these criteria *(2)*.

The CAT enzyme catalyzes the transfer of the acetyl group from acetyl coenzyme A (acetyl CoA) to one or both of the hydroxyl groups of chloramphenicol. Two basic types of CAT assays have been described, with various modifications. The standard and most frequently used method was established by Shaw *(3)* and subsequently adapted to eukaryotic systems by Gorman and her colleagues *(1)*. Starting with acetyl CoA and [14C]-chloramphenicol as the substrates, the acetylated [14C]-chloramphenicol products were subsequently separated from unreacted substrate by thin-layer chromatography (TLC) followed by autoradiography. The results are quantitated by densitometric scanning of the X-ray films or by scintillation counting of the compounds eluted from the TLC plates. One drawback of this procedure is that the quantification

From: *Methods in Molecular Biology, vol. 63: Recombinant Protein Protocols:*
Detection and Isolation Edited by: R. Tuan Humana Press Inc., Totowa, NJ

is laborious, particularly when large numbers of samples need to be analyzed, although it has an advantage in visual identification of the correct acetylated products. A fluorescent chloramphenicol derivative (Bodipy™ chloramphenicol) has been developed recently to avoid the use of radioactivity. The products can be detected by either UV illumination or dual-beam fluorimeter *(4,5)*.

The second method generally applies phase extraction methodology to separate the acetylated products from substrates on the basis of their partition preferences in the extraction solvent *(6)*. The reaction starts with radiolabeled acetyl CoA and unlabeled chloramphenicol. The labeled chloramphenicol derivatives are more hydrophobic than the labeled acetyl CoA, thereby they can be subsequently extracted into a specific organic solvent. The relative radioactivity of the product phase is quantitated by scintillation counting. This protocol in general is cheaper, easier, and less time consuming than the first method, but does not provide visual identification of the substrate and derived products.

2. Materials

1. Phosphate-buffered saline (PBS), pH 7.4; NaCl 8.0 g, KCl 0.2 g, KH_2PO_4 0.2 g, Na_2HPO_4 1.13 g/L. (Store at 4°C.)
2. 0.25M Tris-HCl, pH 7.8. (Store at 4°C.)
3. 1M Tris-HCl, pH 7.8. (Store at room temperature.)
4. Bacterial CAT (Sigma C-8413). (Stable at 4°C up to 2 yr.)
5. [^{14}C]Chloramphenicol (50 μCi/mmol; ~50 mCi/mL; New England Nuclear, NEC-408A). (Store at –70°C up to 9 mo.)
6. 1.0 mM Bodipy chloramphenicol in 0.1M Tris-HCl:methanol (9:1) (Molecular Probes, Eugene, OR). (Store in a light-proof bottle at 4°C. Stable for at least 6 mo.)
7. 4 mM Acetyl CoA (Sigma C-2056). (Made in water immediately before use. Stable in liquid for only 2 wk when stored at –20°C.)
8. Ethyl acetate (4°C).
9. Silica gel TLC plates (silica gel 1B, J. T. Baker #4-4462).
10. 90:10 (v/v) Chloroform/methanol.
11. TLC development chamber.
12. Luminous permanent acrylic paint (Palmer Paint Products).
13. X-ray films (Kodak XAR 5). (Store at 4°C.)
14. Methanol (spectrophotometric grade).
15. Scintillation cocktail (0.5% PPO, 0.03% POPOP in toluene).
16. [^{14}C] Acetyl CoA (4 mCi/mmol; 10 μCi/mL; New England Nuclear, Boston, MA NEC-313L). (Store at –70°C for up to 8 mo.)
17. Acetyl CoA dilution buffer: 0.5 mM acetyl CoA (Sigma C-2056), 250 mM Tris-HCl, pH 7.8. (Store at 4°C.)
18. 8 mM Chloramphenicol (Sigma 0378. Store at –20°C).

3. Methods

3.1. Preparation of Cell-Free Extracts for CAT Assays

Cytoplasmic cellular extracts are prepared from cells expressing the CAT gene. Interference of CAT activity by still-unidentified enzymatic activity in cell extracts has been reported *(7)*. However, heat treatment prior to the final centrifugation step has been reported to be sufficient for eliminating the inhibitory activity (*see* Note 1).

1. Scrape cells from 100-mm dishes at the appropriate time points and transfer to 15-mL conical tubes.
2. Pellet cells by centrifugation in a Beckman CS-6R centrifuge at 1400*g* for 5 min at 4°C.
3. Resuspend cell pellets in 5 mL of PBS. Centrifuge as above.
4. Resuspend cells in 100 µL of 0.25*M* Tris-HCl, pH 7.8. Transfer to 1.7-mL microfuge tubes.
5. Vortex and sonicate 6X (each time for 10 s) in a water bath sonicater. Place tubes on ice in between to prevent the cell extracts from being overheated.
6. Freeze and thaw 3X. (This is done by alternating between liquid nitrogen or a –70°C freezer and a 37°C water bath. Mix the lysates thoroughly by vortexing before each freezing step.)
7. Place cell extracts in a 65°C waterbath for 10 min. This is to inactivate possible inhibitory activities for CAT assay in the cell extracts.
8. Spin in a microfuge at 12,000*g* for 15 min at 4°C.
9. Collect the supernatants. Freeze at –20°C until ready to perform the assays (*see* Note 2).

3.2. Chloramphenicol Acetyl Transferase Assay

3.2.1. TLC-Based CAT Assay

The protocols using either [^{14}C]-chloramphenicol or Bodipy chloramphenicol are both described in this section. The CAT enzyme catalyzes the addition of an acetyl group from acetyl CoA to the 1- and 3-hydroxyl groups of chloramphenicol. At the end of the incubation period (*see* Notes 3 and 4), ethyl acetate is added to stop the reactions and extract both unacetylated and acetylated chloramphenicol. The volume of the ethyl acetate phase is brought down by drying. The substrate and products (1-acetate, 3-acetate, and 1,3-diacetate chloramphenicol) are subsequently separated on silica gel TLC plates. The results are visualized by autoradiography (radioactive substrate) or by UV illumination (fluorescent substrate) (Fig. 1). The substrate-to-product conversion is calculated after quantitation is determined by densitometric scanning of the X-ray film or by measuring the level of radioactivity (or fluorescence intensity) in the silica gel from the spots of interest on the TLC plate (*see* Note 5).

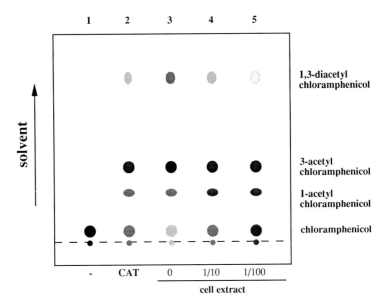

Fig. 1. Schematic illustration of typical results from TLC analysis of the products of CAT assays. Visualization of the results on the TLC plates after TLC analysis of the CAT assay reactions shows that three acetylated products are resolved from the substrate, with 1,3-diacetyl chloramphenicol migrating the longest distance, followed by 3- and 1- acetyl chloramphenicol. The unreacted chloramphenicol substrate also migrates slightly out of the origin. In the negative control lane (lane 1), where only reaction buffer is added to the reaction, no acetylated products are detected. All three products along with the substrate are detected when appropriate units of the commercially available bacterial CAT enzyme is added as the positive control (lane 2). When serial dilutions of the cell extract prepared from cells transfected with a chimeric CAT construct are subjected to CAT assay, different degrees of the chloramphenicol acetylation take place accordingly (lanes 3–5).

3.2.1.1. TLC-BASED CAT ASSAY USING ^{14}C-CHLORAMPHENICOL

1. Mix and preincubate the assay mixtures at 37°C for 5 min to reach equilibrium at this temperature:
   a. 20 µL cell extract (*see* Note 6).
   b. 100 µL 0.25*M* Tris-HCl, pH 7.8.
   c. 1 µCi of [^{14}C]chloramphenicol.
   The following controls should be included: 0.1 U of bacterial CAT (positive control) and reaction buffer (negative control).
2. Start the reaction by adding 20 µL of 4 m*M* acetyl CoA (*see* Note 7) to each reaction. Incubate at 37°C for 1 h (*see* Note 3).
3. Stop the reaction and extract the chloramphenicol with 1 mL of cold ethyl acetate (4°C) by vortexing for at least 30 s.

4. Transfer 900 μL of the top layer to fresh 1.7-mL microfuge tubes.
5. Dry samples in a Savat Speed Vac Concentrator.
6. Resuspend pellets in 30 μL of cold ethyl acetate.
7. With a #2 pencil, draw a line about 2 cm from one edge of the TLC plate to mark the starting line. Spot 5 mL of the samples at a time on the line. To minimize the size of the spot, dry each spot before spotting another 5 mL until the samples are used up. Allow at least 1.5-cm space between two samples. Dry completely.
8. Run samples on the TLC plates in a chamber pre-equilibrated with chloroform-methanol (90:10, ascending) about 1-cm deep. To keep the inside of the developing chamber saturated with eluent, put in several sheets of filter paper behind the TLC plate. Make sure the chamber is well sealed.
9. Remove the TLC plates from the chamber after the solvent front has traveled about 7/8 of the length (*see* Note 8).
10. Air dry the TLC plates in a fume hood.
11. Detection and quantitation (*see* Note 9): Autoradiography and densitometric scanning:
    a. Spot with a small amount of luminous paint for orientation. Wrap TLC plates in a piece of plastic wrap and place on X-ray films.
    b. The developed X-ray films can be analyzed using a densitometer to determine the relative intensity of each spot.
    Elution and Scintillation quantitation:
    a. For each sample, the spots corresponding to the acetylated products are identified by overlaying the X-ray films on the TLC plates and are marked with a #2 pencil. These spots are subsequently cut out from the TLC plates and combined together into a scintillation vial. Add an appropriate amount of scintillation cocktail and count.
    b. Cut out the unreacted substrate spot and count in the same manner.
12. Calculation of the substrate-product conversion: (*see* Notes 10 and 11).
    The results can be indicated as a percentage conversion of the input radioactive substrate.

$$\% \text{ conversion} = [C_{product}/(C_{substrate} + C_{product})] \times 100\%$$

C = scintillation counts (or densitometric intensity).

3.2.1.2. TLC-Based CAT Assay Using Bodipy Chloramphenicol (*see* Note 12)

1. Mix the following ingredients. Incubate at 37°C for 5 min.
   a. 60 μL cell extract.
   b. 10 μL of Bodipy chloramphenicol.
   The following controls should be included: 0.1 U of bacterial CAT (positive control) and reaction buffer (negative control).
2. Add 10 μL of 4 m*M* acetyl CoA (*see* Note 7). Continue the incubation for about 15 min to 5 h (*see* Note 3).
3. Follow steps 3–10 in Section 3.2.1.1. for the extraction of chloramphenicol and its derivatives as well as the TLC analysis of the reactions.

4. After the TLC development, direct visual inspection of the fluorescent substrate and products is possible because they are bright yellow in visible light. Ultraviolet illumination (wavelength at 366 or 254 nm) can be utilized to enhance the levels of fluorescence. Furthermore, photographs of the gels can be taken under UV illumination to be saved as permanent records for notebook purposes.

5. For quantitative analysis:
   a. Using a soft lead pencil, lightly circle the substrates and the acetylated products on the TLC plates under an UV illuminator.
   b. For each sample, cut out the product spots and combine them into a centrifuge tube. Cut out and place the substrate spot into another centrifuge tube.
   c. Add a constant vol (about 2.4 mL) of methanol to each tube and extract the compounds from the gel by vortexing for about 1 min.
   d. Centrifuge in Beckman CS-6R centrifuge at 1430g for 5 min to pellet the silica beads.
   e. Without disturbing the silica gel pellet, transfer 2 mL of the yellowish supernatant to measure the fluorescence in a 3-mL cuvet (*see* Note 13). Measure the fluorescence at 512 nm either by exciting the samples at 490 nm or by using fluorescein filters in a fluorimeter.
   f. If a fluorimeter is not available, the relative absorption of the substrate and product compounds at 505 nm could be measured in a spectrophotometer with less sensitivity.

6. Calculation of the substrate-product conversion:

$$\% \text{ conversion} = [C_{product}/(C_{substrate} + C_{product})] \times 100\%$$

   C = fluorescence (or absorbency) intensity.

3.2.2. Phase Extraction-Based CAT Assay Using [^{14}C] Acetyl CoA (see Notes 14 and 15)

In this assay system, following incubation of the radiolabeled acetyl CoA and cold chloramphenicol with cell extracts, separation of the radiolabeled products from the radiolabeled substrate relies on extraction of the reactions with ethyl acetate. Chloramphenicol and its acetylated derivatives are highly soluble in ethyl acetate whereas acetyl CoA is not. The final ethyl acetate phase is carefully removed and added into scintillation cocktail for counting (*see* Note 16).

1. The [^{14}C] acetyl CoA (*see* Note 17) was diluted 1:10 in acetyl CoA dilution buffer.

2. Mix the following ingredients together. Incubate at 37°C for 1 h (*see* Notes 3 and 4).
   a. 20 mL 8 mM chloramphenicol.
   b. 60 mL cell extract (*see* Note 18).
   c. 20 mL diluted [^{14}C] acetyl CoA.
   Include one reaction with 1 U of bacterial CAT as the positive control, and one reaction with water only as the negative control.

3. Extract the reactions with 100 μL of cold ethyl acetate by vortexing. Microfuge at 12,000g for 3 min at room temperature.

4. Transfer 75 μL of the ethyl acetate phase (top) to a scintillation vial. Be fairly exact with each sample to avoid transfer of the labeled acetyl CoA into the organic phase.
5. Repeat extraction with an additional 100 μL ethyl acetate.
6. Transfer another 75 μL of the ethyl acetate phase to the same scintillation vial.
7. Add scintillation flour and count in a liquid scintillation counter (*see* Note 9).

4. Notes

1. To test whether the cells in which the CAT gene is going to be expressed have any endogenous acetyl transferase or inhibitory activities, include the following reactions in a preliminary testing assay:
   a. Bacterial CAT and untransfected cell lysate (CAT inhibitory activity might be present in certain cells).
   b. Untransfected cell lysate (endogenous CAT activity).
   c. Bacterial CAT (positive control).
   d. Water (negative control).
   e. No chloramphenicol (for phase extraction-based CAT assay only; acetylation of other compounds which could result in [^{14}C]-acetyl-labeled compounds soluble in ethyl acetate would lead to over estimation of the CAT activity).
2. The preparation of 10 different cell extracts takes about 1–2 h to finish.
3. If little activity was detected when the maximal amount of cell extracts was added, try prolonging the incubation time to several h. (When longer periods of incubation time are required, use higher concentrations of acetyl CoA [up to 40 m*M*] as acetyl CoA is not stable under the assay conditions.) Alternatively, radiolabeled substrates with higher specific activities can be used to increase the sensitivity of the assays.
4. Keep the assay within linear range by dilution of cell extracts or varying the reaction time. The phase extraction method is linear only over a fivefold range of enzyme concentration. The incubation time could be tested by establishing an activity-vs-time plot to make sure the end-point activity falls within the linearity of the enzyme activity. For instance, set up a 560-μL reaction for the method described in Section 3.2.1.1. and remove 140 μL at four different evenly-distributed time points. Quantify the samples as described above and the activities should remain within the linear time range.
5. For 10 reactions, the time required for setting up the TLC-based CAT assay using either radioactive or fluorophore-conjugated chloramphenicol is about 30 min. After an appropriate reaction time, the ethyl acetate extraction, as well as setup and development of the TLC generally take approx 4 h. Detection of the results using fluorophore-conjugated chloramphenicol could be performed immediately. One or two days of exposure of the TLC, plates on films is usually enough for detection of the radioactive signals. The preparation for the scintillation counting or fluorimetric measurement takes about 1 h.
6. Different amount of cell extracts (up to 50 μL) can be assayed by adjusting the volume of 0.25*M* Tris-HCl to keep the total volume at 121 μL.

7. Fresh acetyl CoA is essential.
8. Take the TLC plates out of the chromatography chamber immediately before the solvent reaches the top to have the maximal separation of the products and to prevent diffusion of the spots.
9. If no CAT activity was detected, consider the following:
   a. One of the reagents is bad. (To be able to rule out this possibility, the bacterial CAT control should be used.)
   b. A positive plasmid control that is known to express sufficient amount of CAT protein should be included in the transient expression experiments as an indicator for the efficiency of transfection and for those constructs without any promoter activity.
10. Less than 20–30% of the conversion of chloramphenicol to acetyl chloramphenicol usually indicates that the assay was stopped within the linear range of the CAT enzyme activity. Samples with more than 30% conversion should be assayed again with different dilutions of the cell extracts.
11. The sensitivity of the assay is about 10^{-2} CAT units. Its linear response range is between 10^{-2} to 4×10^{-1} CAT units *(8)*.
12. Several assay systems have been developed to avoid the need for radiolabeled substrate. Besides Bodipy™ chloramphenicol derivative, one modification with similar sensitivity to the TLC-based assay is an assay based on HPLC separation *(2,8)*. An alternative method is the enzyme-linked immunosorbance assay using CAT-specific antibodies with comparable limit of detection to that of the radioactive tests (Boehringer Mannheim, Mannheim, Germany; *9*).
13. If 0.5-mL cuvets are available, use about 0.7 mL of methanol for extraction and transfer 0.5 mL of the supernatant after centrifugation for the measurement of fluorescence.
14. On the basis of the phase-extraction assay using radiolabeled acetyl CoA as the substrate, alternative protocols which employ direct extraction of the acetylated chloramphenicol derivatives into a nonpolar scintillation cocktail are also available *(10,11)*. These assays were reported to be able to detect CAT activity of 10^{-4}–10^{-5} CAT units. Likewise, simpler one-vial continuous extraction assays which can detect less than 2.5×10^{-3} CAT units were also developed. This method is based on the ability of the acetylated chloramphenicol products to diffuse into the water-immissible scintillation cocktail while the reactions are going *(12–15)*. Another modification for the phase-extraction method was the employment of [³H]-acetate instead of [³H]-acetyl CoA as the substrate. It has the advantage that the labile [³H]-acetyl CoA could be replenished by endogenous synthesis. Therefore, the reaction time could be prolonged for extracts with low CAT activity *(16)*.
15. One additional cheaper and easier phase extraction-based method which maintains similar sensitivity employs radiolabeled chloramphenicol instead of radiolabeled acetyl CoA *(17)*. Relying on the low specificity of CAT enzyme for the acyl donor, this method is based on separation of the hydrophobic butyrylated chloramphenicol products from unmodified chloramphenicol by their differential solubility in a mixture of tetramethylpentadecane and xylene which is subse-

quently added into scintillation cocktail for counting. A listed protocol of this method can be found in *Current Protocols* (Wiley, New York).

16. Setup of 10 reactions takes less than 30 min. Approximately 1 h is needed for the extraction steps.

17. Repeated freeze-thawing of [^{14}C] acetyl CoA should be avoided by aliquoting the stock into an amount suitable for single experiments.

18. If necessary, smaller amounts of cell extracts can be used to keep the reactions in the linear range of the enzyme activity. Bring the reaction volume up to 100 µL with 250 m*M* Tris-HCl, pH 7.8.

References

1. Gorman, C. M., Moffat, L. F., and Howard, B. H. (1982) Recombinant genomes which express chloramphenicol acetyltransferase in mammalian cells. *Mol. Cell. Biol.* **2,** 1044–1051.

2. Young, S. L., Jackson, A. E., Puett, D., and Melner, M. H. (1985) Detection of chloramphenicol acetyl transferase activity in transfected cells: a rapid and sensitive HPLC-based method. *DNA* **4,** 469–475.

3. Shaw, W. V. (1975) Chloramphenicol acetyltransferase from chloramphenicol-resistant bacteria. *Methods Enzymol.* **43,** 373–755.

4. Hruby, D. E., Brinkley, J. M., Kang, H. C., Haugland, R. P., Young, S. L., and Melner, M. H. (1990) Use of a fluorescent chloramphenicol derivative as a substrate for CAT assays. *BioTechniques* **8,** 170–171.

5. Young, S. L., Barbera, L., Kaynard, A. H., Haugland, R. P., Kang, H. C., Brinkley, M., and Melner, M. H. (1991) A nonradioactive assay for transfected chloramphenicol acetyltransferase activity using fluorescent substrates. *Anal. Biochem.* **197,** 401–407.

6. Sleigh, M. J. (1986) A nonchromatographic assay for expression of the chloramphenicol acetyltransferase gene in eucaryotic cells. *Anal. Biochem.* **156,** 251–256.

7. Crabb, D. W., Minth, C. D., and Dixon, J. E. (1989) Assaying the reporter gene chloramphenicol acetyltransferase. *Methods Enzymol.* **48,** 690–701.

8. Davis, A. S., M. R. Davey, R. C. Clothier, and E. C. Cocking. (1992) Quantification and comparison of chloramphenicol acetyltransferase activity in transformed plant protoplasts using high-performance liquid chromatography- and radioisotope-based assays. *Anal. Biochem.* **210,** 87–93.

9. Porsch, P., Merkelbach, S., Gehlen, J., and Fladung, M. (1993) The nonradioactive chloramphenicol acetyltransferase-enzyme-linked immunosorbent assay test is suited for promoter activity studies in plant protoplasts. *Anal. Biochem.* **211,** 113–116.

10. Nielson, D. A., Chang, T.-S., and Shapiro, D. J. (1989) A highly sensitive, mixed-phase assay for chloramphenicol acetyltransferase activity in transfected cells. *Anal. Biochem.* **179,** 19–23.

11. Sankaran, L. (1992) A simple quantitative assay for chloramphenicol acetyltransferase by direct extraction of the labeled product into scintillation cocktail. *Anal. Biochem.* **200,** 180–186.

12. Chauchereau, A., Astinotti, D., and Bouton, M.-M. (1990) Automation of a chloramphenicol acetyltransferase assay. *Anal. Biochem.* **188**, 310–316.
13. Eastman, A. (1987) An improvement to the novel rapid assay for chloramphenicol acetyltransferase gene expression. *BioTechniques* **5**, 730–732.
14. Neumann, J. R., Morency, C. A., and Russian, K. O. (1987) A novel rapid assay for chloramphenicol acetyltransferase gene expression. *BioTechniques* **5**, 444–447.
15. Martin, J. D. (1990) Application of the two-phase partition assay for chloramphenicol acetyl transferase (CAT) to transfection with simian virus 40-CAT plasmids. *Anal. Biochem.* **191**, 242–246.
16. Nordeen, S. K., Green, P. P., III, and Fowlkes, D. M. (1987) A rapid, sensitive, and inexpensive assay for chloramphenicol acetyltransferase. *DNA* **6**, 173–178.
17. Seed, B., and Sheen, J.-Y. (1988) A simple phase-extraction assay for chloramphenicol acyltransferase activity. *Gene* **67**, 271–277.

4

Human Placental Alkaline Phosphatase as a Marker for Gene Expression

Paul Bates and Michael H. Malim

1. Introduction

Much of current biomedical research requires that the expression of a gene (or DNA sequence) under investigation be detected by rapid and reliable means. To achieve this, it is frequently convenient to place a marker (or surrogate) gene under the genetic control of the regulatory sequences of interest. Expression of the marker gene is then monitored by analyzing accumulation of the encoded protein and this, in turn, serves as a measure of gene expression. Here, we describe two methodologies that utilize human placental alkaline phosphatase (hPLAP) as a marker gene. The first exploits a secreted version of alkaline phosphatase (SEAP) to measure gene expression in transfected cells *(1)*, whereas the second uses the naturally occurring membrane-bound form of the protein as a marker for retroviral infection in either tissue culture *(2)* or challenged animals *(3,4)*. In all cases, the advantages of using hPLAP include the rapidity of the detection procedure, its avoidance of radioisotopes, the relatively low cost of the reagents, the limited number of tissues and cell lines in which hPLAP is ordinarily expressed, and its high temperature stability (the other isozymes of alkaline phosphatase are relatively heat labile).

1.1. Secreted Alkaline Phosphatase (SEAP)

A typical application of this assay would be for the analysis of the *cis*-acting sequences and *trans*-acting factors that modulate the transcriptional activity of a promoter *(5)*. Secreted alkaline phosphatase is a convenient choice for this type of experiment because the preparation of cell lysates is not required and changes in expression level over time in a single sample can therefore be determined. A general purpose plasmid vector, pBC12/PL/SEAP *(6)*, that provides

From: *Methods in Molecular Biology, vol. 63: Recombinant Protein Protocols:*
Detection and Isolation Edited by: R. Tuan Humana Press Inc., Totowa, NJ

41

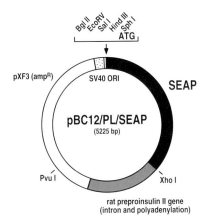

Fig. 1. Plasmid map of pBC12/PL/SEAP: The positions of the multiple cloning site, *SEAP* gene (solid box), rat preproinsulin II sequences (gray box), sequences for selection and propagation in bacteria (open box), and SV40 origin of replication (speckled box) are indicated.

a suitable backbone for such an analysis is depicted in Fig. 1, and a representative experiment that utilized it is shown in Fig. 2 (*see* Note 1 for further details). The vector pBC12/PL/SEAP contains a 489-amino acid version of hPLAP that was truncated by 24 residues at its carboxy-terminus *(1)* and is therefore efficiently secreted by expressing cells. Located 5' to the *SEAP* gene is a polylinker sequence that includes the ATG initiation codon and into which the promoter regions of genes can be readily inserted. Located 3' to *SEAP* are an intron and the polyadenylation signals of the rat preproinsulin II gene. In addition, the vector also contains sequences that confer replication and selection in bacteria as well as a minimal origin of DNA replication derived from simian virus 40 (SV40). This latter sequence is useful as it permits plasmid replication in cells expressing SV40 T antigen (for example, COS and 293T); this serves to amplify the overall extent of gene expression and thereby increase the sensitivity of this assay.

1.2. Membrane-Bound Alkaline Phosphatase (AP) as a Marker of Retroviral Infection

Moloney murine leukemia virus (MoMuLV) *(7)*, Rous sarcoma virus (RSV) *(3)*, and HIV-1 *(8)* vectors that harbor the native membrane bound form of hPLAP have all been developed. Typical applications of these vectors would include the analysis of viral infection (for example, for quantitating the efficiency of viral receptor-virion Env glycoprotein interactions; refer to Rong and Bates, 1995 *[2]*) and cell lineage mapping *(3,7)*. An experiment that uti-

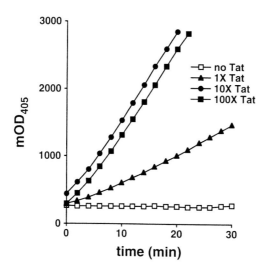

Fig. 2. Demonstration of *SEAP* as a reporter gene: 293T cells were transfected with ɔHIV-1/SEAP and a negative control vector (no Tat, open squares), 1X Tat (solid ⸬riangles), 10X Tat (solid circles), or 100X Tat (solid squares) and the levels of SEAP determined as described. The calculated levels of SEAP are: no Tat, 0.0 mOD_{405}/min; 1X Tat, 35.9 mOD_{405}/min; 10X Tat, 91.6 mOD_{405}/min; 100X Tat, 92.5 mOD_{405}/min.

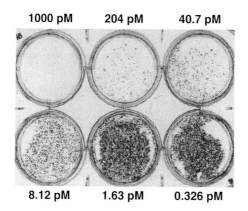

Fig. 3. Demonstration of *AP* as a histochemical marker for retroviral infection: ⸬urkey embryo fibroblasts were challenged with RCAS(A)-AP in the presence of the ɪndicated concentrations of inhibitor. The cultures were fixed and stained for AP ⸬xpression at 24 h as described. All six wells of the culture dish are shown.

ɪzed an RSV (subgroup A) vector in which hPLAP has replaced the *src* gene, ⸬rmed RCAS(A)-AP, is outlined in Note 4 and shown in Fig. 3; it is the procedure for such an experiment that is described here. Because RCAS(A)-AP still carries intact *gag, pol,* and *env* genes, as well as all the *cis*-acting sequences required for replication, stocks of infectious virus can be readily generated in avian cells. Importantly, even though this virus is unable to productively replicate in mammalian cells, it can infect such cells and efficiently express AP provided that the virus receptor, Tva, is present on the surface of the challenged cells.

2. Materials
2.1. Secreted Alkaline Phosphatase (SEAP)

1. Diethanolamine (cat no. D45-500; Fisher, Pittsburgh, PA), L-homoarginine (cat no. H-1007; Sigma, St. Louis, MO), *p*-nitrophenol phosphate (cat no. 104-0; Sigma.)
2. 2X SEAP buffer; to prepare 50 mL, mix the following with water and store at 4°C without autoclaving. L-homoarginine is included in the buffer as inhibits the activity of endogenous alkaline phosphatases but not hPLAP.

Stock	Amount	Concentration
Diethanolamine (100 % solution)	10.51 g	$2M$
$1M$ MgCl$_2$	50 μL	1 mM
L-homoarginine	226 mg	20 mM

3. 120 mM *p*-nitrophenol phosphate; make fresh at room temperature by dissolving 158 mg in 5 mL 1X SEAP buffer.
4. 65°C water bath.
5. Eppendorf microcentrifuge.
6. Vortex.
7. Flat-bottomed, 96-well microtiter plates.
8. Multichannel pipet.
9. For measuring light absorbance at 405 nm (OD$_{405}$), it is most convenient to use an enzyme-linked immunosorbent assay (ELISA) reader. An excellent machine for achieving this is the EL340 automated microplate reader from Bio-Tek Instruments (Winooski, VT). When linked to a computer, the Delta Soft II software is straightforward to use and can readily calculate the changes in OD$_{405}$ for multiple samples as they occur over time.

2.2. Membrane-Bound Alkaline Phosphatase (AP)

1. N,N-dimethylformamide (cat no. D-8654; Sigma), Fast Red TR salt (4-chloro-2-methylbenzenediazonium salt) (cat no. F-2768; Sigma), naphthol AS-BI phosphate (cat no. N-2250; Sigma), 4% paraformaldehyde (cat no. P-6148; Sigma; dissolved in phosphate buffered saline and stored at −20°C), phosphate-buffered saline (PBS), 50 mM Tris-HCl (pH 9.0).
2. AP stain; to 25 mL 50 mM Tris-HCL (pH 9.0), add 25 mg Fast Red TR, 12.5 mg naphthol AS-BI phosphate and 250 μL dimethyl formamide, mix and filter

through 3MM filter paper (Whatman Ltd., Maidstone, UK, *see* Note 5). The stain should be used within 10 min of preparation.

3. Preparation of RCAS(A)-AP virus stocks. Transfect chicken embryo fibroblasts (CEFs) with plasmid DNA containing the recombinant RCAS(A)-AP provirus using calcium phosphate. The CEFs are derived from 10-d line 0 embryos and are grown in Dulbecco's Modified Eagle's Medium supplemented with 5% fetal bovine serum, 1% chicken serum, 100 µg/mL streptomycin and 100 U/mL penicillin. Filter the virus containing medium through 0.22-µm filters and store at –80°C in 1-mL aliquots.

4. 37°C incubator
5. Low magnification inverted light microscope.

3. Methods
3.1. Secreted Alkaline Phosphatase

1. Collect 250 µL of culture supernatant from the samples of interest and transfer to Eppendorf tubes. For a representative experiment and an alternative use of this assay, refer to Notes 1 and 2.
2. Heat tubes at 65°C for 5 min to inactivate endogenous alkaline phosphatases that may be present in the samples.
3. Centrifuge at full speed in a microcentrifuge for 2 min at room temperature.
4. Transfer the supernatants to fresh Eppendorf tubes. These samples can be stored at –20°C indefinitely.
5. In an Eppendorf tube, add 100 µL of 2X SEAP buffer to 100 µL (or an empirically determined dilution) of sample. As a zero standard, make up one mix substituting the sample with water.
6. Mix by vortexing.
7. Transfer the contents of each tube (for example, 190 µL) to the well of a flat bottomed 96-well microtiter plate, taking care to avoid creating air bubbles.
8. Prewarm by incubating the plate at 37°C for 10 min.
9. During this incubation, make up the *p*-nitrophenol phosphate (the substrate) solution and prewarm it at 37°C.
10. Add 20 µL of the substrate solution to each well avoiding air bubbles; this is most easily accomplished by using a multichannel pipet.
11. As quickly as possible and using an ELISA reader, measure the OD_{405} at regular intervals (for example, every 2 min) over 30 min. The OD_{405} of the zero sample must be subtracted from all readings. During this time, continue to incubate the plate at 37°C; the reader recommended above has a 37°C incubator.
12. Calculate the level of SEAP activity for each sample at a point when the changes in OD_{405} are linear with respect to time. These values can be expressed as changes in OD_{405} per minute or as milliunits (mU) per mL. One mU is defined as the amount of SEAP that will hydrolyze 1.0 pmol of *p*-nitrophenol phosphate per minute; this equals an increase of 0.04 OD_{405} U/min. The specific activity of SEAP is 2000 mU/µg protein.

13. Refer to Note 3 for alternative SEAP detection methodologies that offer enhanced sensitivity.

3.2. Membrane-Bound Alkaline Phosphatase

1. Thaw the stock of RCAS(A)-AP at 37°C; once the virus has been thawed it should be used as soon as possible (*see* Note 6). Refreezing is not recommended as the infectious titer will decrease significantly.
2. Infect 100-mm cell monolayers with RCAS(A)-AP by adding the virus directly to the culture medium.
3. At 24–48 h postinfection, wash the cells twice with 1X PBS.
4. Fix the cells by adding enough 4% paraformaldehyde to cover the monolayer (*see* Note 7); for a 100-mm dish, this requires 2 mL.
5. Leave at room temperature for 2–4 min.
6. Wash the dish twice with 1X PBS and aspirate to remove excess liquid.
7. Add 2 mL AP stain per 100-mm dish.
8. Incubate the dish at 37°C for 30 min.
9. Wash the dish twice with 1X PBS.
10. Leave the dish at room temperature for 60 min or at 4°C overnight.
11. Determine the number of infected cells by counting the number or red staining cells using low magnification (for example, 40X) light microscopy. The level of background staining varies depending on the levels of endogenous AP activity on the cell surface. Importantly, this background can often be eliminated by incubating the fixed cells (or tissue sections) at 65°C for 30 min prior to AP staining.

4. Notes

1. The data from a typical experiment are shown in Fig. 2. The 35-mm monolayer cultures of the human embryonal kidney cell line 293T were transiently transfected with cesium chloride-purified plasmids using calcium phosphate. No Tat 3.5 μg phIV-1/SEAP + 3.5 μg pCMV/Il-2; 1 X Tat, 3.5 μg phIV-1/SEAP + 3.49 μg pCMV/Il-2 + 0.01 μg pcTAT; 10X Tat, 3.5 μg phIV-1/SEAP + 3.4 μg pCMV. Il-2 + 0.1 μg pcTAT; 100X Tat, 3.5 μg phIV-1/SEAP + 2.5 μg pCMV/Il-2 + 1.0 μg pcTAT. phIV-1/SEAP is a derivative of pBC12/PL/SEAP in which the human immunodeficiency virus type-1 (HIV-1) long terminal repeat (LTR) from −45 to +80 was inserted into the polylinker *(1)*. The pCMV/Il-2 *(9)* is a negative control vector used to maintain the level of transfected DNA as constant and pcTAT *(10)* expresses the HIV-1 transcriptional trans-activator protein Tat. A ~24 h posttransfection, culture supernatants were collected and 10 μL of each sample used for the analysis of SEAP.
2. In addition to using *SEAP* as the reporter gene, plasmids containing *SEAP* driven by a strong promoter are useful additions to transfection experiments as internal controls for transfection efficiency *(11)*.
3. Although SEAP does offer a number of advantages, the method described here can only reliably measure enzyme levels down to 50 pg/mL; this level of sensitivity is 10- to 50-fold lower than that for chloramphenicol acetyltransferase

(CAT). However, the use of bioluminescent and chemiluminescent substrates for monitoring SEAP (or AP) activity have been described and sensitivities down to 0.2 pg/mL can be attained *(1,12,13)*.

4. In the experiment shown in Fig. 3, 35-mm monolayers of turkey embryo fibroblasts were challenged with 5000 infectious units of RCAS(A)-AP in the presence of increasing (0.326–1000 pM) concentrations of an inhibitor of infection (in this case, a soluble form of Tva derived from a recombinant baculovirus). At ~24 h postinfection, the cells were fixed and stained for AP. The darkly staining AP-positive cells can be visualized by the naked eye and quantitated more accurately under 40X magnification.

5. The AP stain solution often contains particles, these must be removed by filtration prior to use or the results will be difficult to interpret.

6. Once thawed, the virus stock may be stored at 4°C for several days with a decrease in infectious titer of approx 10-fold. Prolonged storage at this temperature will result in more significant losses in infectious titer.

7. Extensive fixing of the cells before staining should be avoided since it can lead to a loss of AP activity.

References

1. Berger, J., Hauber, J., Hauber, R., Geiger, R., and Cullen, B. R. (1988) Secreted placental alkaline phosphatase, a powerful new quantitative indicator of gene expression in eukaryotic cells. *Gene* **66,** 1–10.

2. Rong, L. and Bates, P. (1995) Analysis of the subgroup A avian sarcoma and leukosis virus receptor, the 40-residue, cysteine-rich, low-density lipoprotein receptor repeat motif of Tva is sufficient to mediate viral entry. *J. Virol.* **69,** 4847–4853.

3. Fekete, D. M. and Cepko, C. L. (1993) Replication-competent retroviral vectors encoding alkaline phosphatase reveal spatial restriction of viral gene expression/transduction in the chick embryo. *Mol. Cell. Biol.* **13,** 2604–2613.

4. Federspiel, M. J., Bates, P., Young, J. A. T., Varmus, H. E., and Hughes, S. H. (1994) A system for tissue-specific gene targeting, transgenic mice susceptible to Subgroup A avian leukosis virus-based retroviral vectors. *Proc. Natl. Acad. Sci. USA* **91,** 11,242–11,245.

5. Fenrick, R., Malim, M. H., Hauber, J., Le, S.-Y., Maizel, J., and Cullen, B. R. (1989) Functional analysis of the Tat trans-activator of human immunodeficiency virus type 2. *J. Virol.* **63,** 5006–5012.

6. Cullen, B. R. and Malim, M. H. (1991) The HIV-1 Rev protein, prototype of a novel class of eukaryotic post-transcriptional regulators. *Trends Biochem. Sci.* **16,** 346–350.

7. Fields-Berry, S. C., Halliday, A. L., and Cepko, C. L. (1992) A recombinant retrovirus encoding alkaline phosphatase confirms clonal boundary assignment in lineage analysis of murine retina. *Proc. Natl. Acad. Sci. USA* **89,** 693–697.

8. He, J. and Landau, N. R. (1995) Use of a novel human immunodeficiency virus type 1 reporter virus expressing human placental alkaline phosphatase to detect an alternative viral receptor. *J. Virol.* **69,** 4587–4592.

9. Cullen, B. R. (1986) *Trans*-activation of human immunodeficiency virus occurs via a bimodal mechanism. *Cell* **46,** 973–982.
10. Malim, M. H., Hauber, J., Fenrick, R., and Cullen, B. R. (1988) Immunodeficiency virus *rev trans*-activator modulates the expression of the viral regulatory genes. *Nature* **335,** 181–183.
11. Malim, M. H., Böhnlein, S., Hauber, J., and Cullen, B. R. (1989) Functional dissection of the HIV-1 Rev *trans*-activator–derivation of a *trans*-dominant repressor of Rev function. *Cell* **58,** 205–214.
12. Bronstein, I., Fortin, J. J., Voyta, J. C., Jui, R.-R., Edwards, B., Olesen, C. E. M., Lijam, N., and Kricka, L. J. (1994) Chemiluminescent reporter gene assays, sensitive detection of the GUS and SEAP gene products. *BioTechniques* **17,** 172–177.
13. Kitamura, M., Maeda, M., and Tsuji, A. (1995) A new highly sensitive chemiluminescent assay of alkaline phosphatase using lucigenin and its application to enzyme immunoassay. *J. Bioluminescence Chemiluminescence* **10,** 1–7.

5

Use of Secreted Alkaline Phosphatase as a Reporter of Gene Expression in Mammalian Cells

Steven R. Kain

1. Introduction

The cDNA encoding secreted alkaline phosphatase (SEAP; *1*) is a powerful tool for investigating the function of known or presumed enhancer/promoter elements in transfected cells. As a reporter of gene expression, SEAP differs from intracellular reporters such as chloramphenicol acetyltransferase (CAT) and firefly luciferase in that SEAP is secreted from transfected cells. SEAP can thus be assayed using a small portion of the culture medium. The SEAP gene encodes a truncated form of human placental alkaline phosphatase (PLAP), an enzyme which normally resides on the cell surface. The SEAP gene lacks sequences encoding the membrane anchoring domain, thereby allowing the expressed protein to be efficiently secreted from cells. Levels of SEAP activity detected in the culture media are directly proportional to changes in intracellular concentrations of SEAP mRNA and protein *(2)*. SEAP has the unusual properties of being extremely heat-stable and resistant to the phosphatase inhibitor L-homoarginine. Therefore, endogenous alkaline phosphatase activity can be eliminated by treatment of samples at 65°C and incubation with this inhibitor. The secreted nature of SEAP provides several advantages for the use of this enzyme as a genetic reporter:

1. Preparation of cell lysates is not required.
2. The kinetics of gene expression can be easily studied using the same cultures by repeated collection of the medium.
3. Transfected cells are not disturbed for measurement of SEAP activity, and remain intact for further investigations such as RNA and protein analysis.

From: *Methods in Molecular Biology, vol. 63: Recombinant Protein Protocols: Detection and Isolation* Edited by: R. Tuan Humana Press Inc., Totowa, NJ

4. Background from endogenous alkaline phosphatase activity in the culture media is almost absent.
5. Sample collection and assay can be automated using 96-well microtiter plates.

The original assay for SEAP activity uses the chromogenic alkaline phosphatase substrate *p*-nitrophenyl phosphate (PNPP; *2*). This assay is fast, simple to perform, and inexpensive. However, the sensitivity of the assay is poor, and it has a narrow linear dynamic range. Sensitivity is improved using a two step bioluminescent assay for SEAP based on the hydrolysis of D-luciferin-O-phosphate *(1)*. The dephosphorylation reaction catalyzed by SEAP yields free luciferin, which in turn serves as the substrate for firefly luciferase. The sensitivity of this assay is roughly equivalent to that of the conventional bioluminescent assay for firefly luciferase. The most sensitive SEAP assays use chemiluminescent alkaline phosphatase substrates such as the 1,2-dioxetane CSPD *(3)*. Dephosphorylation of CSPD results in a sustained "glow" type luminescence which remains constant up to 60 min, and is easily detected using a luminometer or by exposure of X-ray film. These assays can detect as little as 10^{-15} g of SEAP in cell culture medium. In addition to enhanced sensitivity, the chemiluminescence assay for SEAP greatly increases the linear dynamic range of detection, which facilitates the analysis of a wide range of promoters. Lastly, chemiluminescent detection of SEAP is fast, easy to perform, and does not require the use of radioactive substrates or other hazardous compounds (Fig. 1).

The utility of the SEAP reporter assay is enhanced by the availability of expression vectors for this reporter protein. An integrated set of four SEAP reporter vectors has been designed to provide maximal flexibility in studying regulatory sequences from the researchers' gene of interest. Details concerning the format of these vectors is covered in the Section 2.

2. Materials

All materials required for the expression and detection of SEAP from transfected cells is available in the Great EscAPe SEAP Genetic Reporter System (Clontech Laboratories, Palo Alto, CA). Each of the SEAP reporter vectors are also available separately.

2.1. SEAP Reporter Vectors

pSEAP-Basic lacks eukaryotic promoter and enhancer sequences and has a multiple cloning site (MCS) that allows promoter DNA fragments to be inserted upstream of the SEAP gene. Enhancers can be cloned into either the MCS or unique downstream sites. The SEAP coding sequences are followed by an intron and polyadenylation signal from SV40 to ensure proper and efficient process-

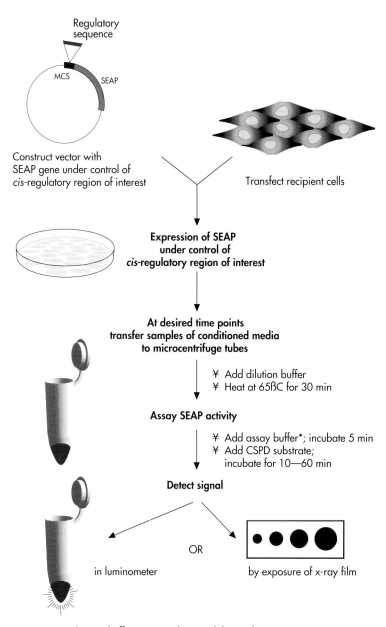

*Assay buffer contains the AP inhibitor L-homoarginine.

Fig. 1. Flow diagram for expression and detection of SEAP reporters in transfected cells.

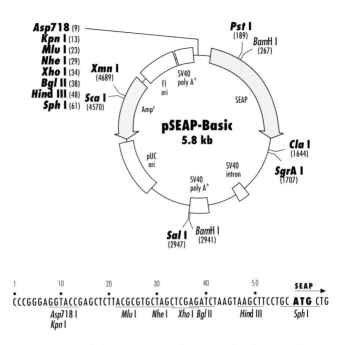

Fig. 2. Plasmid map of the pSEAP-Basic vector is shown. pSEAP-Basic lacks eukaryotic promoter and enhancer sequences. The MCS (shown below the plasmid map) allows promoter DNA fragments to be inserted upstream of the SEAP gene. Enhancers can be cloned into either the MCS or unique downstream sites. Unique restriction sites are in boldface on the plasmid map. All four Great EscAPe vectors contain an SV40 intron and the SV40 polyadenylation signal inserted downstream of the SEAP coding sequences to ensure proper and efficient processing of the transcript in eukaryotic cells. A second polyadenylation signal upstream of the MCS reduces background transcription *(7)*. The vector backbone contains an f1 origin for single-stranded DNA production and a pUC19 origin of replication and an ampicillin-resistance gene for propagation in *E. coli*. The MCS region is identical in all four vectors except for a 202-bp promoter fragment that has been inserted between the *Bgl*II and *Hind*III sites in pSEAP-Control and pSEAP-Promoter. The sequence of pSEAP-Basic has been deposited in GenBank (Accession # U09660).

ing of the SEAP transcript in eukaryotic cells. A second polyadenylation signal upstream of the MCS reduces background transcription *(6)*. The vector backbone also contains an f1 origin for single-stranded DNA production, a pUC19 origin of replication, and an ampicillin resistance gene for propagation in *Escherichia coli*. The map of the pSEAP-Basic vector is shown in Fig. 2.

pSEAP-Control is pSEAP-Basic with the SV40 early promoter inserted upstream of the SEAP gene and the SV40 enhancer (derived from the SV40

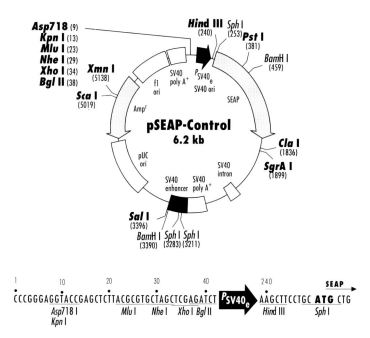

Fig. 3. Plasmid map of the pSEAP-Control Vector is shown. pSEAP-Control is pSEAP-Basic with the SV40 early promoter inserted upstream of the SEAP gene and the SV40 enhancer (derived from the SV40 early promoter) inserted downstream. pSEAP-Control should express SEAP in most cell types and provides an important positive control. Unique restriction sites are in boldface on the plasmid map. The MCS is shown below the plasmid map. The sequence of pSEAP-Control has been deposited in GenBank (Accession # U09661).

early promoter) downstream. pSEAP-Control should express SEAP in most cell types and provides an important positive control for transfection and expression of exogenous DNA. The map of the pSEAP-Control vector is shown in Fig. 3.

pSEAP-Enhancer contains the SV40 enhancer inserted downstream of the SEAP gene in pSEAP-Basic. pSEAP-Enhancer is intended for analyzing promoter sequences inserted into the MCS. The SV40 enhancer may increase transcription from these promoters—a useful feature when studying weak promoters.

pSEAP-Promoter contains the SV40 early promoter inserted upstream of the SEAP gene in pSEAP-Basic. pSEAP-Promoter is intended for analyzing enhancer sequences inserted either upstream or downstream of the SEAP gene.

2.2. Detection of SEAP Activity

2.2.1. Preparation of CSPD Chemiluminescent Substrate Solution

The CSPD Chemiluminescent substrate is provided in the Great EscApe kit as a 25 m*M* stock solution. This solution should be stored in the dark at 4°C, and is stable in this form for 6 mo. A final CSPD concentration of 1.25 m*M* is recommended for chemiluminescent detection of SEAP activity (20X dilution of CSPD stock solution). The stock CSPD is diluted with a chemiluminescent enhancer solution provided as a component of the Great EscApe kit. The dilution with enhancer should be performed just prior to use. Keep the diluted substrate solution on ice while performing the assay.

2.2.2. Preparation of Assay Buffer

The buffer consists of 2*M* diethanolamine, 1 m*M* MgCl$_2$, and 20 m*M* L-homoarginine (pH 9.8).

1. Add 50 mL ddH$_2$O to a clean 200-mL beaker.
2. Add 0.45 g L-homoarginine and 0.02 g MgCl$_2$ to the beaker while gently stirring on a magnetic stir plate. Keep stirring until powder is completely dissolved in water, then add 21 mL of diethanolamine to the solution.
3. Adjust the final volume to 100 mL with ddH$_2$O. Continue stirring.
4. Adjust the pH of the solution to pH 9.8 by adding concentrated HCl dropwise.
5. The assay buffer should be stored at 4°C, and is stable for up to 6 mo.

2.2.3. Preparation of 5X Dilution Buffer

The buffer consists of 0.75*M* NaCl, 0.2*M* Tris-HCl (pH 7.2).

1. Add 90 mL ddH$_2$O to a clean, 200 mL beaker.
2. Add 4.38 g NaCl and 2.42 g Tris Base to the beaker while gently stirring. Continue stirring, and adjust to pH 7.2 by adding concentrated HCl dropwise.
3. The 5X dilution buffer should be stored at 4°C, and is stable for up to 6 mo.
4. Prepare a 1X Dilution buffer as follows just prior to use, and equilibrate to room temperature (22–25°C): 5X dilution buffer (1 volume) and ddH$_2$O (4 volumes).

2.2.4. Control Placental Alkaline Phosphatase

This reagent is provided in the Great EscAPe kit at a concentration of 1.5 × 10^{-4} U/mL in 50% (v/v) glycerol, 200 m*M* Na$_2$HPO$_4$ (pH 7.2).

2.3. Equipment Required

Chemiluminescent detection of SEAP activity can be performed either with a tube luminometer, a microplate luminometer, or by exposure of X-ray film to reactions performed in white, opaque, 96-well, flat-bottomed microtiter plates. Detection by X-ray film exposure will also require film cassettes, and access to a darkroom for film development.

3. Methods
3.1. Transfection of Mammalian Cells with SEAP Expression Vectors
3.1.1. Methods of Transfection

The SEAP expression vectors may be transfected into eukaryotic cells by a variety of techniques, including those using calcium phosphate *(4)*, DEAE-dextran *(5)*, various liposome-based transfection reagents *(5)*, and electro-poration *(5)*. Each method is compatible with the chemiluminescent detection of SEAP activity, so the procedure of choice will depend primarily on the type of cell being transfected. Different cell lines may vary by several orders of magnitude in their ability to take up and express exogenously added DNA. Moreover, a method that works well for one type of cultured cell may be inferior for another. When working with a cell line for the first time, compare the efficiencies of several transfection protocols using the pSEAP-Basic and pSEAP-Control vectors as described in Section 3.4.

3.1.2. Transfection Considerations

Each different SEAP construct should be transfected (and subsequently assayed) in triplicate to minimize variability among treatment groups. The primary sources of such variability are differences in transfection efficiencies. When monitoring the effect of promoter and enhancer sequences on gene expression, it is critical to include an internal control that will distinguish differences in the level of transcription from variability in the efficiency of transfection. This is easily done by cotransfecting with a second plasmid that constitutively expresses an activity which can be clearly defined from SEAP. The level of the second enzymatic activity can then be used to normalize the levels of SEAP among different treatment groups. Reporter proteins frequently used for this purpose include *E. coli* β-galactosidase, which is expressed intracellularly, and human growth hormone (hGH), which is secreted *(5)*.

3.2. Preparation of Samples for Assay of SEAP

For transient transfection assays, maximal levels of SEAP activity are generally detected in the cell culture medium 48–72 h after cell transfection. This range is suggested only as a starting point, as optimal times for collecting samples will vary for different cell types, cell density, and the nature of the particular experimental conditions. The procedure below describes harvesting media from adherent cells. If working with suspended cell cultures, simply begin with 125 µL of the cell culture, pellet the cells by centrifugation, and proceed from step 1:

1. Remove 110 µL of cell culture medium and transfer to a microcentrifuge tube.
2. Centrifuge at 12,000g for 10 s to pellet any detached cells present in the culture medium.
3. Transfer 100 µL of supernatant to a fresh microcentrifuge tube.
4. Assay samples immediately, or store cell culture medium at –20°C until ready for assay.

3.3. Chemiluminescent Detection of SEAP Activity

3.3.1. General Concerns

It is important to stay within the linear range of the assay. High intensity light signals can saturate the photomultiplier tube in luminometers, resulting in false low readings. In addition, low intensity signals that are near to background levels may be outside the linear range of the assay. Therefore, the target amount of SEAP in the assay should be adjusted to bring the signal within the linear response capability of the assay. For signals that are too intense, this can be achieved by diluting the cell culture media with 1X dilution buffer. For low signals, the amount of SEAP may be increased by improving the transfection efficiency, starting with a greater number of cells, or increasing the volume of media assayed (*see* Section 4.2.). The linear range with the positive control placental alkaline phosphatase provided in the Great EscAPe kit is approximately 10^{-13}–10^{-9} g via detection with a Turner Designs Model 20e luminometer. A more extensive linear range may be obtained with other instruments.

3.3.2. Assay for SEAP Activity

As noted in Section 3.1.2., each construct should be transfected and subsequently assayed in triplicate.

1. Allow a sufficient amount of chemiluminescent enhancer and assay buffer required for the entire experiment to equilibrate to room temperature (22–25°C).
2. Prepare the required amount of 1X dilution buffer (*see* Section 2.2.3, step 4), and allow to equilibrate to room temperature.
3. Thaw samples of cell culture medium, and place 25 µL into a 0.5-mL transparent microcentrifuge tube.
4. Add 75 µL of 1X dilution buffer to each 25-µL sample and mix gently by hand (do not vortex).
5. Incubate the diluted samples for 30 min at 65°C using a heat block or water bath.
6. Cool samples to room temperature by placing on ice for 2–3 min, and then equilibrating to room temperature.
7. Add 100 µL of assay buffer to each sample and incubate for 5 min at room temperature.

8. Prepare 1.25 m*M* CSPD substrate by diluting 20X with chemiluminescent enhancer (*see* Section 2.2.1.).
9. Add 100 µL of the diluted substrate to each tube, and incubate for 10 min at room temperature (22–25°C).

Note: The chemiluminescent signal remains constant up to 60 min after the addition of the substrate solution, depending on the sample environment. Therefore, readings may be taken 10–60 min after the addition of CSPD. In order to optimize the assay sensitivity, it is recommended that measurements be performed at various incubation times between 10–60 min in order to determine the point of maximum light emission.

3.3.3. Chemiluminescence Detection Methods

3.3.3.1. DETECTION USING A TUBE LUMINOMETER

If the assay is performed in a tube suitable for luminometer readings, the sample may be placed directly in the instrument after step 9 of Section 3.3.2., and measurements taken following a minimum of 10 min incubation with the CSPD substrate. If the assay is not performed in a suitable tube, transfer the entire solution from this step to the appropriate luminometer tube and place in the instrument. Record light signals as 5–15 s integrals.

3.3.3.2. DETECTION USING A PLATE LUMINOMETER

The entire SEAP assay (Section 3.3.2.) may also be performed in 96-well flat-bottomed microtiter plates suitable for plate luminometers. Record light signals as 5–15 s integrals.

3.3.3.3. DETECTION BY EXPOSURE OF X-RAY FILM

If a luminometer is not available, light emission may be recorded by exposure of X-ray film from white opaque 96-well flat-bottomed microtiter plates. This detection procedure yields spots on the X-ray film, which can be quantitated by comparison to positive and negative control incubations (*see* Section 3.4.).

1. Perform the entire SEAP assay (*see* Section 3.3.2.) in white opaque 96-well flat-bottomed microtiter plates.
2. Following a minimum of 10 min incubation at step 9 of Section 3.3.2., overlay the microtiter plate with X-ray film, cover the film with plastic wrap, and place a heavy object such as a book on top to hold the film in place.
3. Expose the film 5–30 min at room temperature. It is critical for comparison between samples to remain within the linear response capability of the X-ray film. In order to avoid misleading results, it is recommended that several different film exposure times be utilized for each microtiter plate.

3.4. Proper Use of Controls

3.4.1. Negative Controls (see Section 4.3.)

A negative control is necessary to determine background signals associated with the cell culture media. This can be determined by assaying 25 μL of culture medium from cells transfected with the pSEAP-Basic vector, which contains the SEAP gene without a promoter or enhancer, or from untransfected cells. The values obtained from such controls should be subtracted from experimental results.

3.4.2. Positive Controls

3.4.2.1. POSITIVE CONTROL
FOR TRANSFECTION AND EXPRESSION OF EXOGENOUS DNA

A positive control is necessary to confirm transfection and expression of exogenous DNA and to verify the presence of active SEAP in the culture media. Expression and secretion of functional SEAP in transfected cells can be confirmed by assaying 25 μL of culture medium from cells transfected with the pSEAP-Control Vector, which contains the SEAP structural gene under transcriptional control of the SV40 promoter and enhancer. Cells transfected with this plasmid should yield high activity within 48–72 h after transfection (*see* Section 4.5.).

3.4.2.2. POSITIVE CONTROL FOR THE DETECTION METHOD

The positive control placental alkaline phosphatase can be used to confirm that the detection method is working. To do this, simply add 2 μL of the positive control phosphatase to 23 μL of culture medium from untransfected cells. This should give a strong positive signal. A dilution series of the positive control enzyme can also be used to determine the linear range of the assay (*see* Section 4.4.).

4. Notes
4.1. Determining Linear Range of the SEAP Assay

1. If in doubt about the linear range of the SEAP assay, prepare and assay a dilution series using the positive control placental alkaline phosphatase provided with the Great EscAPe kit. The linear range is approximately 10^{-13}–10^{-9} g of phosphatase using a Turner Designs Model 20e Luminometer. Approximately the same linear range is obtained using the exposure of X-ray film to detect the chemiluminescence signal.

4.2. Little or No Signal is Obtained from Transfected Cells

2. Ensure that the assay conditions are correct and that the detection method is working by assaying the positive control enzyme.

3. Ensure that the transfection efficiency has been optimized by using pSEAP-Control (or a suitable alternative) as an internal positive control for SEAP expression.
4. Increase the number of cells used in transfections by utilizing a larger diameter plate for adherent cells, or increase the volume of media for suspension cultures.
5. If background signals from negative controls (i.e., untransfected cells or cells transfected with pSEAP-Basic) are low, increase the volume of media assayed from experimental cultures. The volume used in step 3 of Section 3.3.2. may be increased to 50–75 µL, with 1X dilution buffer being added in step 4 of Section 3.3.2. to attain a final volume of 100 µL.
6. Increase the post-transfection time interval prior to collecting samples of media.
7. Ensure that the conditioned media does not contain an inhibitory activity by adding 2 µL of positive control placental alkaline phosphatase to 23 µL of culture medium at step 3 of Section 3.3.2.
8. For detection via exposure of X-ray film, try increasing the film exposure time.
9. For detection via a tube or plate luminometer, refer to the instrument instructions for means of increasing the sensitivity of light detection.

4.3. Background Signals are Excessive

10. Ensure that all intact cells and cellular debris are removed from the conditioned media by centrifugation as described in Section 3.2. This step is particularly important for suspension cultures.
11. Ensure that the diluted media samples are heated for the full 30 min at 65°C as specified in step 5 of Section 3.3.2.
12. The volume of media assayed from experimental cultures may be decreased if the signal is sufficiently high. The volume of media used in step 3 of Section 3.3.2. may be decreased, or diluted using 1X dilution buffer.
13. If possible, culture cells following the transfections in media containing minimal serum. Serum in excess of 10% (v/v) may cause excessive background.

4.4. Signal is Too High, Exceeding the Linear Range of the Assay

This problem is easily corrected by either assaying a lower volume of conditioned medium at step 3 of Section 3.3.2., or by dilution of the samples using 1X dilution buffer.

4.5. Effects of SV40 Large T Antigen (COS Cells)

The specific level of expression for the pSEAP vectors is likely to vary in different cell types. This may be particularly true for cell lines containing the SV40 large T antigen, such as COS cells. The large T antigen promotes replication of the SV40 origin, sequences of which are found in the promoter region of the pSEAP-Promoter and pSEAP-Control vectors. The combination of the large T antigen and SV40 origin leads to a higher copy number of these vectors in COS cells, which in turn may result in increased expression of the SEAP reporter gene relative to vectors lacking the SV40 origin.

References

1. Berger, J. Hauber, J., Hauber, R., Geiger, R., and Cullen, B. R. (1988) Secreted placental alkaline phosphatase: a powerful new quantitative indicator of gene expression in eukaryotic cells. *Gene* **66,** 1–10.
2. Cullen, B. R. and Malim, M. H. (1992) Secreted placental alkaline phosphatase as a eukaryotic reporter gene. *Mets. Enzymol.* **216,** 362–368.
3. Bronstein, I., Fortin, J., Stanley, P. E., Stewart, G. S. A. B., and Kricka, L. J. (1994) Chemiluminescent and Bioluminescent Reporter Gene Assays. *Anal. Biochem.* **219,** 169–181.
4. Chen, C. and Okayama, H. (1988) Calcium phosphate mediated gene transfer: A highly efficient transfection system for stably transforming cells with plasmid DNA. *BioTechniques* **6,** 632–638.
5. Sambrook, J. (1989) *Molecular Cloning: A Laboratory Manual*, Cold Spring Harbor Laboratory, Cold Spring Harbor, NY.
6. Araki, E., Shimada, F., Shichiri, M., Mori, M., and Ebina, Y. (1988) pSV00CAT: low Background CAT plasmid. *Nucleic Acids Res.* **16,** 1627.

6

Detection of β-Galactosidase and β-Glucuronidase Using Chemiluminescent Reporter Gene Assays

Corinne E. M. Olesen, John J. Fortin, John C. Voyta, and Irena Bronstein

1. Introduction

Reporter gene assays have become essential tools for the study of gene regulation. Several genes have been adapted as indicators of transcriptional activity with a variety of detection techniques including radioisotopic, colorimetric, luminescent, fluorescent, and immunoassay methods *(1–4)*. Chemiluminescent assays have been very effective in detecting enzyme labels in direct enzyme activity assays, immunoassays, and nucleic acid hybridization assays.

Chemiluminescent 1,2-dioxetane substrates for alkaline phosphatase, β-galactosidase (β-Gal), and β-glucuronidase (GUS) have been developed for detection of reporter enzyme activity. Enzymatic cleavage of corresponding chemiluminescent dioxetane substrates produces a destabilized dioxetane anion, which fragments with the production of light. These assays also incorporate Emerald chemiluminescence enhancing reagent, consisting of a water-soluble macromolecule and fluorescein, which is necessary for the rapid production of an intense light signal from the excited state generated in the enzyme-catalyzed reaction in solution. Energy transfer to the fluorescein emitter results in a shift of the light emission to 530 nm and enhanced signal intensity.

The bacterial β-galactosidase gene is a widely used reporter, and the gene product is commonly quantitated with a colorimetric substrate *(1)*. Two chemiluminescent assays for β-galactosidase incorporating 1,2-dioxetane substrates have been described *(5,6)*. Modifications of these assays based on Galacton and Galacton-Plus substrates enable extremely sensitive detection of β-galactosidase activity with the Galacto-Light and Galacto-Light Plus chemiluminescent reporter gene assays *(7)*. The Galacto-Light assay has been successfully

From: *Methods in Molecular Biology, vol. 63: Recombinant Protein Protocols: Detection and Isolation* Edited by: R. Tuan Humana Press Inc., Totowa, NJ

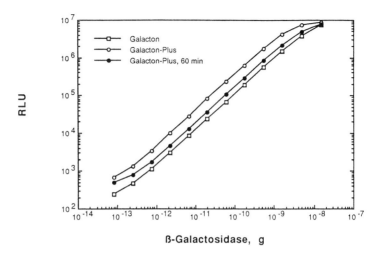

Fig. 1. Detection of purified β-galactosidase with Galacton and Galacton-Plus. 180 μL of Galacto-Light reaction buffer was added to 20 μL aliquots of purified β-galactosidase, diluted with 1% BSA in 0.1M sodium phosphate, pH 8.0. After a 60-min incubation at room temperature, 300 μL of light emission accelerator was added and the luminescence intensity measured immediately and at 60 min as a 5-s integral.

used to quantitate β-galactosidase activity in cultured cell extracts *(5,8–10)*, tissue extracts *(11)*, yeast extracts (D. Nathan and S. Linquist, personal communication), and bacteria *(12)*. Galacto-Light is suited for use with luminometers with automatic injectors and other instrumentation that measures light emission within a short period of time. The prolonged emission kinetics generated by Galacto-Light Plus make it ideal for use with either plate luminometers without automatic injection capabilities or with scintillation counters. These assays routinely enable the detection of extremely low levels (<10 fg) of purified enzyme *(7)*, and the dynamic range with each substrate is greater than five orders of magnitude, extending from 10^{-13}–10^{-8} g of enzyme (*see* Fig. 1).

The bacterial GUS gene has become widely used for the analysis of plant gene expression *(2)*, and preliminary results indicate that it is also a useful reporter gene for studies in mammalian cells *(13)*. Quantitative assays employ fluorogenic, chromogenic, and chemiluminescent substrates. The 1,2-dioxetane substrate, Glucuron, has been incorporated into the GUS-Light chemiluminescent reporter gene assay for GUS activity in plant or animal cell lysates *(4,7)*.

Measurements of purified GUS levels as low as 60 fg are performed reliably with the GUS-Light assay, and the linear range extends from 10^{-13}–10^{-7} g of enzyme. This assay enables the detection of 1.4 pg of purified enzyme with a signal-to-noise ratio (S/N) of 2. The sensitivity is at least 100-fold higher compared to the reported fluorescent assay detection limit, 0.02 ng *(14)*. GUS activity is detected over a wide range of total cellular protein, and the measurement of GUS activity with GUS-Light shows excellent correlation with GUS activity as determined by fluorescence assay with the MUG (4-methyl-umbelliferyl-β-D-glucuronide) substrate *(4)*.

A novel eukaryotic reporter gene for secreted placental alkaline phosphatase (SEAP) enables detection of the gene product in a sample of culture medium *(15)*. Quantitation of SEAP activity can be performed with the 1,2-dioxetane chemiluminescent substrate, CSPD *(16)*, which has been incorporated into the Phospha-Light assay (Tropix, Bedford, MA), *(3,4,7,17)*. A protocol for chemiluminescent detection of SEAP activity with CSPD is provided in Chapter 5 (Kain).

These chemiluminescent reporter gene assay systems are performed in microplate or tube luminometers, or in a scintillation counter, and offer rapid, highly sensitive, versatile alternatives to radioactive, colorimetric, and fluorescent detection systems. These assay systems provide additional flexibility in the choice of assays for transfection normalization, particularly with the bioluminescent luciferase reporter gene and expand the available repertoire of reporter gene assays for use with luminescence instrumentation.

2. Materials

2.1. Preparation of Extracts (β-Gal)

2.1.1. Cell Extract

Galacto-Light assay component *(1)* is available from Tropix. All solutions should be stored at 4°C.

1. Lysis solution: 100 m*M* potassium phosphate, pH 7.8, 0.2% Triton X-100. Dithiothreitol (DTT) should be added fresh prior to use to a final concentration of 1 m*M*.
2. 1X Phosphate-buffered saline (PBS): 58 m*M* Na_2HPO_4, 17 m*M* NaH_2PO_4, 68 m*M* NaCl, pH 7.3–7.4. Alternate PBS recipes may also be used.
3. Rubber policeman or scrapers.
4. Microcentrifuge tubes.

2.1.2. Tissue Extract

Components (1) and (4) from Section 2.1.1. and:

1. PMSF (Sigma, St. Louis, MO; P-7626) or AEBSF (Sigma, A-5938, *see* Note 1) and leupeptin protease inhibitors. PMSF is prepared as a 0.2*M* stock in isopropanol and stored at –20°C.
2. Microhomogenizer: Tissumizer (Tekmar, Cincinnati, OH) fitted with a microprobe.

2.1.3. Yeast Extract

Component (4) from Section 2.1.1. and:

1. Lysis solution: 100 m*M* potassium phosphate, pH 7.8, 20% glycerol. Just prior to use add: 1 m*M* DTT, 2 µg/mL aprotinin, 2 µg/mL leupeptin, and 2 m*M* AEBSF.

2. Glass tubes (10 × 75 mm).
3. Glass beads (Sigma, G-8772; 425–600 μm, acid-washed).

2.2. Chemiluminescent Detection (β-Gal)

Galacto-Light assay components (1–3) are available from Tropix. All solutions should be stored at 4°C.

1. Chemiluminescent substrate: Galacton or Galacton-Plus (100X concentrates).
2. Reaction buffer diluent: 100 mM sodium phosphate, pH 8.0, 1 mM MgCl$_2$.
3. Light emission accelerator: formulation containing Emerald enhancer.
4. Purified β-galactosidase (Sigma, G-5635).
5. Luminometer tubes (glass, 12 × 75 mm) or white microplates for chemiluminescence (Microlite 2; Dynatech Laboratories, Chantilly, VA).
6. Tube or microplate luminometer (or liquid scintillation counter).

2.3. Preparation of Extracts (GUS)

GUS-Light assay component (1) is available from Tropix. All solutions should be stored at 4°C.

1. Lysis solution (for plant cells or tissue): 50 mM sodium phosphate, pH 7.0, 10 mM EDTA, 0.1% sodium lauryl sarcosine, 0.1% Triton X-100. Fresh β-mercaptoethanol should be added prior to use to a final concentration of 10 mM. Lysis solution (for mammalian cells): 100 mM potassium phosphate, pH 7.8, 0.2% Triton X-100. Fresh dithiothreitol (DTT) should be added prior to use to a final concentration of 1 mM.
2. Rubber policeman or scrapers.
3. Microhomogenizer.
4. Microcentrifuge tubes.

2.4. Chemiluminescent Detection (GUS)

GUS-Light assay components (1–3) are available from Tropix. All solutions should be stored at 4°C.

1. Glucuron chemiluminescent substrate (100X concentrate).
2. Reaction buffer diluent: 0.1M sodium phosphate, pH 7.0, 10 mM EDTA.
3. Light emission accelerator: formulation containing Emerald enhancer.
4. Purified β-glucuronidase (Sigma, G-7896).
5. Luminometer tubes (glass, 12 × 75 mm) or white microplates for chemiluminescence detection (Microlite 2; Dynatech).
6. Tube or microplate luminometer (or liquid scintillation counter).

3. Methods

3.1. Preparation of Extracts (β-Gal)

3.1.1. Cell Extract

1. Aliquot the required amount of lysis solution, and add the appropriate amount of DTT (*see* Notes 2 and 3).

2. Wash cells twice with 1X PBS.
3. Add lysis solution to cover cells (250 µL of lysis solution/60-mm culture plate).
4. Detach cells from plate using a rubber policeman or equivalent. Nonadherent cells should be pelleted and sufficient lysis solution added to cover cells. Lyse cells by pipeting.
5. Transfer extract to a microcentrifuge tube and centrifuge at 12,500g for 2 min at 4°C to clarify, and transfer supernatant to a fresh microcentrifuge tube. Extracts may be used immediately or frozen at –70°C.
6. Optional: perform heat inactivation of endogenous β-galactosidase (*see* Section 3.1.4.).
7. Optional: perform protein assay with an aliquot of extract.

3.1.2. Tissue Extract

Preparation of tissue extract is according to Shaper *(11)*.

1. Aliquot the required amount of lysis solution, and add DTT to a final concentration of 1 mM, PMSF to 0.2 mM, and leupeptin to 5 µg/mL.
2. Remove tissues and homogenize for 20 s at 4°C in 1 mL of lysis solution.
3. Centrifuge in a microcentrifuge at 12,500g for 10 min at 4°C to clarify, and transfer supernatant to a fresh microcentrifuge tube.
4. Optional: perform heat inactivation of endogenous β-galactosidase activity (*see* Section 3.1.4.).
5. Recentrifuge (following heat inactivation) at 12,500g for 5 min at 4°C, and transfer supernatant to a fresh microcentrifuge tube. Extracts may be used immediately or stored at –70°C.
6. Optional: perform protein assay with an aliquot of extract.

3.1.3. Yeast Extract

Preparation of yeast extract is according to D. Nathan and S. Lindquist (personal communication). Lysates should be prepared on ice in a cold room.

1. Aliquot the required amount of lysis solution and add the appropriate amounts of DTT, aprotinin, leupeptin, and AEBSF.
2. Pellet 5×10^7 cells, wash once with water and recentrifuge to pellet. Cell pellets can be frozen at –70°C at this point.
3. Resuspend cell pellet in 150 µL lysis solution, transfer suspension to glass tube and add 150 µL glass beads (*see* Note 4).
4. Vortex for 3 min on high.
5. Centrifuge extract at 2500g for 5 min, and transfer supernatant to microcentrifuge tube. Extracts may be used immediately (amount can range from 10 µL of extract down to 10 µL of 1:100 dilution of extract) or stored at –70°C (*see* Note 5).
6. Optional: perform protein assay with an aliquot of extract (50 µL recommended).

3.1.4. Heat Inactivation of Endogenous β-Galactosidase Activity in Cell or Tissue Extract (Optional)

Some cell lines may exhibit relatively high levels of endogenous β-galactosidase activity. This may lead to high background that will decrease the overall assay sensitivity by lowering the signal to noise ratio (*see* Note 6). Selective heat inactivation of nonbacterial enzyme activity in cell extracts has been described *(18)*. This protocol has also been used for selective inactivation of nonbacterial enzyme activity in tissue extracts *(11)*.

1. Following extract preparation, incubate extract at 48°C for 50 min (or 60 min for tissue extracts).
2. Proceed with chemiluminescent detection (*see* Section 3.2.).

3.2. Chemiluminescent Detection (β-Gal)

All assays should be performed in triplicate.

1. Dilute Galacton (Galacton-Plus) substrate 1:100 with reaction buffer diluent to make reaction buffer. This mixture will remain stable for several months if stored uncontaminated at 4°C. It is recommended to only dilute the amount of substrate that will be used within a 2 mo period.
2. Warm the amount of reaction buffer required to room temperature.
3. Aliquot 2–20 μL of each cell extract (from transfected cells, mock-transfected, and mock-transfected containing control enzyme [add 1 μL of 10 U/mL purified β-galactosidase]) into luminometer tubes. Adjust the total volume with the appropriate lysis solution, such that the volume of extract (or extract plus lysis solution) in each tube is equivalent (*see* Notes 7 and 8).
4. Add 200 μL of reaction buffer to each tube and mix gently. Incubate for 10–60 min at room temperature (*see* Notes 9,10).
5. Place tube in luminometer. Inject 300 μL of accelerator (*see* Notes 3,11). After a 2–5-s delay following injection, measure light emission for 5 s (*see* Note 12).

3.3. Preparation of Extracts (GUS)

1. Aliquot the required amount of lysis solution and add the appropriate amount of β-mercaptoethanol or DTT (*see* Note 13).
2. Prepare sample (*see* Note 14): Rinse cells to remove culture media or prepare sample of plant tissue.
3. Add sufficient volume of lysis solution to cover cells (250 μL of lysis solution for a 60 mm culture plate should be adequate) or plant material (250 μL of lysis solution per 25 mg of plant tissue).
4. Detach cells from culture plate using a rubber policeman or equivalent device. For plant tissue, homogenize cells or tissue in a microhomogenizer.
5. Centrifuge sample in a microcentrifuge for 2 min to clarify.
6. Transfer supernatant to a fresh microcentrifuge tube.
7. Optional: perform protein assay with an aliquot of extract.

3.4. Chemiluminescent Detection (GUS)

All assays should be performed in triplicate.

1. Dilute Glucuron substrate 1:100 with reaction buffer diluent to make GUS reaction buffer. This mixture will remain stable for several weeks if stored uncontaminated at 4°C.
2. Warm the amount of GUS reaction buffer required for the entire experiment to room temperature.
3. Aliquot 2–20 μL of each cell extract (from transfected cells, mock-transfected, and mock-transfected containing control enzyme [add 1 μL {20 pg} of purified β-glucuronidase to extract]) into luminometer tubes. Adjust the total volume with the appropriate lysis solution, such that the volume of extract (or extract plus lysis solution) in each tube is equivalent (*see* Notes 7 and 14).
4. Add 180 μL of GUS reaction buffer to each tube and mix gently (*see* Note 8). Incubate for 60 min at room temperature (*see* Note 15).
5. Place tube in luminometer. Inject 300 μL of light emission accelerator (*see* Note 16). After a 2–5-s delay following injection, measure light emission for 5 s (*see* Note 12).

4. Notes

1. AEBSF, a water-soluble serine protease inhibitor, may be used instead of PMSF.
2. Alternative lysis buffers and lysis protocols may be used, particularly if assays for other cotransfected reporter genes require specific lysis buffers. The performance of alternative buffers should be compared with the Galacto-Light lysis solution to ensure optimum results.
3. Reducing agents such as β-mercaptoethanol or DTT will decrease the half-life of light emission of Galacton and particularly Galacton-Plus. If the extended half-life of light emission from Galacton-Plus is crucial to the assay, reducing agents should be omitted from the lysis solution. If they cannot be omitted, they should be minimized and care should be taken to ensure that the same concentration of reducing agent is present in each assay tube or well. This will ensure that the kinetics of light emission from each assay do not vary. If lysis buffer containing DTT has been used, the addition of hydrogen peroxide to the light emission accelerator to a final concentration of 10 mM (add 1 μL of 30% H_2O_2/1 mL of light emission accelerator) will prevent rapid decay of signal half-life.
4. The use of glass tubes for preparation of yeast extracts results in more efficient cell breakage than with microcentrifuge tubes. Glass beads are used directly from the bottle, without washing or reconstitution. To measure, mark desired volume on pipet tip, and scoop beads to level of mark.
5. Yeast cell extracts can be stored at –70°C; however, it is optimal to store cell pellets at –70°C instead and prepare extracts just before use.
6. Most cells and tissues have some level of endogenous β-galactosidase activity. Significant reductions of endogenous enzyme activity can be achieved by selective heat inactivation *(18)*. Endogenous enzyme activity is reduced at the pH of

the reaction buffer, whereas the transfected bacterial enzyme activity is only slightly affected *(5)*. It is important to determine the level of endogenous enzyme activity in extract from nontransfected cells to establish assay background.

7. The amount of cell extract required may vary depending on the level of expression and the instrumentation used. Use 5 μL of extract for positive controls and 10–20 μL of extract with cells containing a potentially low level of enzyme, however, it is essential to maintain a constant volume of extract and/or lysis solution in each reaction, since the DTT in the lysis solution can affect the chemiluminescent reaction. It is important to vary the amount of extract to keep the signal within the linear range of the assay.

8. Reaction component volumes can be scaled down to accommodate a microplate assay format or a luminometer with a smaller volume injector, although sensitivity may be affected slightly. It is recommended to keep the volume of cell extract 5–20 μL.

9. The incubation period with reaction buffer may be as short as 15 min, but the dynamic range of the assay may decrease. For yeast extracts, it is recommended that this incubation be only 15 min. After a 20-min incubation, the reaction may lose linearity, most likely, owing to proteolysis (D. Nathan and S. Lindquist, personal communication). For tissue extracts, a 10 min incubation has been used *(11)*.

10. Light intensities are time dependent. Reaction buffer should be added to samples within the same time frame as they are measured in the luminometer. For example, if it takes 10 s to complete a measurement, then reaction buffer should be added to tubes every 10 s.

11. If manual injection is used, then the light emission accelerator should be added in the same consistent time frame as the addition of reaction buffer. This is critical when using Galacton. Galacton substrate has a half-life of light emission of approximately 4.5 min after the addition of light emission accelerator. Galacton-Plus substrate emits light with a significantly longer half-life ($t_{1/2}$ = 180 min) after the addition of light emission accelerator.

12. A liquid scintillation counter may be used as an alternative to a luminometer *(19,20)*, however, the resulting assay sensitivity may be lower. When light emission reaches a maximum, the changes in signal intensity per unit time are at a minimum (steady-state or approaching steady-state). Therefore, the signal should be measured during this period. Some scintillation counters permit a measurement of chemiluminescence directly by turning off the coincidence circuit. If chemiluminescence is measured without turning off this circuit, a linear relationship between the light level and scintillation counts can be established by taking the square root of the counts per minute minus instrument background *(19)*.

$$\text{Actual} = (\text{measured-instrument background})^{1/2}$$

13. Alternate lysis buffers may be used, however, we recommend that their performance is compared with the GUS lysis buffers to ensure optimum results. A lysis buffer compatible with the luciferase assay, containing $0.1M$ potassium phos-

phate, 1 mM DTT, and 1 mg/mL BSA has equivalent performance to the GUS lysis buffer.

14. Bacterial contamination of plant material will cause high background. Best results will be obtained with sterile preparations. Chlorophyll in concentrated samples may interfere with signal intensity. Therefore, if high levels of chlorophyll are present, several dilutions of extract should be assayed.

15. The incubation with GUS reaction buffer may be as short as 15 min (especially if high levels of expression are expected), but the dynamic range of the assay may decrease. Light intensities are time dependent. Reaction buffer should be added to samples within the same time frame as they are measured in the luminometer. For example, if it takes 10 s to complete a measurement, then reaction buffer should be added to tubes every 10 s.

16. If manual injection is used, then the light emission accelerator should be added in the same consistent time frame as the addition of reaction buffer.

Acknowledgments

We thank Debbie Nathan and Susan Lindquist (University of Chicago, Chicago, IL) for the protocol for the preparation of yeast extracts and their use with the Galacto-Light assay.

References

1. Alam, J. and Cook, J. L. (1990) Reporter genes: application to the study of mammalian gene transcription. *Anal. Biochem.* **188,** 245–254.
2. Gallagher, S. R. (ed.), (1992) *GUS Protocols: Using the GUS Gene as a Reporter of Gene Expression.* Academic Press, San Diego.
3. Bronstein, I., Fortin, J., Stanley, P. E., Stewart, G. S. A. B., and Kricka, L. J. (1994a) Chemiluminescent and bioluminescent reporter gene assays. *Anal. Biochem.* **219,** 169–181.
4. Bronstein, I., Fortin, J. J., Voyta, J. C., Juo, R. R., Edwards, B., Olesen, C. E. M., Lijam, N., and Kricka, L. J. (1994b) Chemiluminescent reporter gene assays: Sensitive detection of the GUS and SEAP gene products. *BioTechniques* **17,** 172–177.
5. Jain, V. K. and Magrath, I. T. (1991) A chemiluminescent assay for quantitation of β-galactosidase in the femtogram range: Application to quantitation of β-galactosidase in *lacZ*-transfected cells. *Anal. Biochem.* **199,** 119–124.
6. Beale, E. G., Deeb, E. A., Handley, R. S., Akhaven-Tafti, H., and Schaap, A. P. (1992) A rapid and simple chemiluminescent assay for *Escherichia coli* β-galactosidase. *BioTechniques* **12,** 320–322.
7. Bronstein, I., Fortin, J., Voyta, J. C., Olesen, C. E. M., and Kricka, L. J. (1994c) Chemiluminescent reporter gene assays for β-galactosidase, β-glucuronidase and secreted alkaline phosphatase, in *Bioluminescence and Chemiluminescence: Fundamentals and Applied Aspects* (Campbell, A. K., Kricka, L. J., and Stanley, P. E., eds.), Wiley, Chichester, UK, pp. 20–23.

8. Lee, S. W., Trapnell, B. C., Rade, J. J., Virmani, R., and Dichek, D. A. (1993) In vivo adenoviral vector-mediated gene transfer into balloon-injured rat carotid arteries. *Circulation Research* **73**, 797–807.

9. O'Connor, K. L. and Culp, L. A. (1994) Quantitation of two histochemical markers in the same extract using chemiluminescent substrates. *BioTechniques* **17**, 502–509.

10. McMillan, J. P. and Singer, M. F. (1993) Translation of the human LINE-1 element, L1Hs. *Proc. Natl. Acad. Sci. USA* **90**, 11,533–11,537.

11. Shaper, N., Harduin-Lepers, A., and Shaper, J. H. (1994) Male germ cell expression of murine β4-galactosyltransferase. A 796-base pair genomic region, containing two cAMP-responsive element (CRE)-like elements, mediates male germ cell-specific expression in transgenic mice. *J. Biol. Chem.* **269**, 25,165–25,171.

12. Nelis, H. J. and Van Poucke, S. O. (1993) Comparison of chemiluminogenic, fluorogenic, and chromogenic substrates for the detection of total coliforms in drinking water. *Proc. the Water Quality Technology Conference (AWWA)*, 1663–1673.

13. Gallie, D. R., Feder, J. N., and Walbot, V. (1992) GUS as a useful reporter gene in animal cells, in *GUS Protocols: Using the GUS Gene as a Reporter of Gene Expression* (Gallagher, S. R., ed.), Academic Press, San Diego, CA, pp. 181–188.

14. Rao, A. G. and Flynn, P. (1992) Microtiter plate-based assay for β-D-glucuronidase: a quantitative approach, in *GUS Protocols: Using the GUS Gene as a Reporter of Gene Expression* (Gallagher, S. R., ed.), Academic Press, San Diego, CA, pp. 89–99.

15. Berger, J., Hauber, J., Hauber, R., Geiger, R., and Cullen, B. R. (1988) Secreted placental alkaline phosphatase: a powerful new quantitative indicator of gene expression in eukaryotic cells. *Gene* **66**, 1–10.

16. Bronstein, I., Juo, R. R., Voyta, J. C., and Edwards, B. (1991) Novel chemiluminescent adamantyl 1,2-dioxetane enzyme substrates, in *Bioluminescence and Chemiluminescence: Current Status* (Stanley, P. and Kricka, L. J., eds.), Wiley, Chichester, UK, pp. 73–82.

17. Jones, R. E., Defeo-Jones, D., McAvoy, E. M., Vuocolo, G. A., Wegrzyn, R. J., Haskell, K. M., and Oliff, A. (1991) Mammalian cell lines engineered to identify inhibitors of specific signal transduction pathways. *Oncogene* **6**, 745–751.

18. Young, D. C., Kingsley, S. D., Ryan, K. A., and Dutko, F. J. (1993) Selective inactivation of eukaryotic β-galactosidase in assays for inhibitors of HIV-1 TAT using bacterial β-galactosidase as a reporter enzyme. *Anal. Biochem.* **215**, 24–30.

19. Nguyen, V. T., Morange, M., and Bensaude, O. (1988) Firefly luciferase luminescence assays using scintillation counters for quantitation in transfected mammalian cells. *Anal. Biochem.* **171**, 404–408.

20. Fulton, R. and Van Ness, B. (1993) Luminescent reporter gene assays for luciferase and β-galactosidase using a liquid scintillation counter. *BioTechniques* **14**, 762–763.

7

Chemiluminescent Immunoassay for the Detection of Chloramphenicol Acetyltransferase and Human Growth Hormone Reporter Proteins

Corinne E. M. Olesen, John C. Voyta, and Irena Bronstein

1. Introduction

Chemiluminescent substrates have been successfully used for highly sensitive quantitation of reporter enzyme activity (*see* Chapters 5 and 6) and in enzyme-linked immunoassays for highly sensitive detection of many analytes (*1-6*). CSPD chemiluminescent 1,2-dioxetane substrate (*7*) for alkaline phosphatase has been used with Sapphire-II enhancing reagent for enzyme-linked immunoassay detection of chloramphenicol acetyltransferase and human growth hormone reporter gene products. These chemiluminescent reporter gene assay systems provide nonradioactive, simple, sensitive detection methods, performed in microplate or tube luminometers.

One of the most widely used reporter genes encodes the bacterial enzyme chloramphenicol acetyltransferase (CAT) (*8*), and numerous CAT vector constructs are available. Traditional assays for CAT activity involve radioisotopes and laborious thin layer chromatography or differential extraction techniques, followed by isotopic detection (*9*). More recently, fluorescent CAT substrates and assays (*10*) and immunoassays (*11,12*) have been developed. The chemiluminescent alkaline phosphatase (AP) substrate, CSPD, has been used with sandwich enzyme-linked immunoassay detection for CAT protein.

This chemiluminescent immunoassay enables detection of 5 pg/mL of purified CAT protein. At a signal-to-noise ratio (S/N) of 2, sensitivities of approx 17–45 pg/mL are achieved. In the linear assay range (10–1000 pg/mL), the simultaneous incubation of CAT protein and antibody-AP conjugate protocol

From: *Methods in Molecular Biology, vol. 63: Recombinant Protein Protocols: Detection and Isolation* Edited by: R. Tuan Humana Press Inc., Totowa, NJ

results in the highest sensitivity. The dynamic range extends beyond the linear assay range (greater than three orders of magnitude). Although the signal is highest with a two-step detection, incorporating a biotinylated detector antibody and streptavidin-AP, the sensitivity achieved with a detector antibody-AP conjugate is superior. The sensitivity of colorimetric assays ranges from 50–200 pg/mL. Thus, this assay combines a short protocol and higher sensitivity than colorimetric ELISAs and rivals the sensitivity of fluorescent and radioisotopic CAT enzyme activity assays.

The human growth hormone (hGH) reporter gene product offers the advantage of being a secreted protein, which permits detection in a sample of cell culture medium and allows cells to remain viable for further experimentation. It is a convenient control for normalizing expression of a second reporter gene. Quantitative assays for hGH have nearly exclusively been performed by radioimmunoassay methods *(9)*, and several ELISA-based assays have recently been introduced commercially.

CSPD has been used in a sandwich enzyme-linked immunoassay detection for hGH, which enables the detection of as little as 3.5 pg/mL of purified hGH. This concentration is routinely detectable above background, and at a S/N = 2, a sensitivity of 50 pg/mL is achieved. The assay exhibits a linear dynamic range of over four orders of magnitude. This chemiluminescent assay format provides higher sensitivity detection and a greater dynamic range than obtained with both colorimetric ELISA assays (5–1000 pg/mL reported sensitivities for commercial assays, two orders of magnitude) and radioimmunoassays (100 pg/mL sensitivity, 2.5 orders of magnitude).

These chemiluminescent reporter gene immunoassays offer rapid, sensitive alternatives to radioactive, colorimetric, and fluorescent detection systems for both CAT enzyme activity and immunoassays and hGH immunoassays. For both, immunoassay detection with a chemiluminescent substrate provides comparable or superior sensitivity and dynamic range than other detection methods.

2. Materials
2.1. Preparation of Cell Extracts by Freeze/Thaw Lysis (CAT)

1. 10X Phosphate buffered saline (PBS): $0.58M$ Na_2HPO_4, $0.17M$ NaH_2PO_4, $0.68M$ NaCl. Alternate PBS recipes may also be used.
2. Rubber policeman or scraper.
3. TEN buffer: 40 mM Tris-HCl, 1 mM EDTA, 150 mM NaCl, pH 7.8.
4. 250 mM Tris-HCl, pH 7.8.
5. Optional: phenylmethyl sulfonyl fluoride (PMSF; prepared as a $0.2M$ stock in isopropanol), dithiothreitol (DTT), and aprotinin.

2.2. Immunoassay Detection (CAT or hGH)

2.2.1. One-Step Immunoassay Detection

Reagents (1–4) are available from Tropix (Bedford, MA). All solutions should be prepared with Milli-Q H_2O.

1. CSPD chemiluminescent substrate: 25 mM (60X concentrate). Store at 4°C.
2. Sapphire-II chemiluminescence enhancer (10X concentrate). Store at 4°C.
3. I-Block blocking reagent: purified casein. Store dry at room temperature.
4. Diethanolamine (DEA): 99%. Store at room temperature. If material solidifies, warm to 37°C to melt.
5. 10X PBS (*see* Section 2.1.).
6. Blocking buffer: 0.2% I-Block, 0.05% Tween-20, 1X PBS. Add I-Block to 1X PBS and microwave for approximately 70 s. Alternatively, heat to 70°C for 5 min on a magnetic stir plate. The solution will remain slightly opaque. Cool to room temperature and add Tween-20.
7. Wash buffer: 0.05% Tween-20, 1X PBS.
8. Assay buffer: 0.1M diethanolamine, 1 mM MgCl$_2$, pH 9.5. Dissolve DEA in H_2O (4.79 mL or 5.25 g per 500 mL) and adjust pH to 9.5 with concentrated HCl. Add MgCl$_2$. Addition of MgCl$_2$ prior to pH adjustment will cause precipitation.
9. White multiwell strips (Microlite 2; Dynatech, Chantilly, VA). Alternatively, the assay can be formatted with coated polystyrene tubes or beads.
10. Anti-CAT capture antibody (polyclonal anti-CAT; 5 Prime - 3 Prime, Boulder, CO), or anti-hGH capture antibody (polyclonal, Medix Biotech, Foster City, CA).
11. Anti-CAT (polyclonal, 5 Prime–3 Prime) alkaline phosphatase conjugate or anti-hGH (monoclonal, Medix) alkaline phosphatase conjugate. Conjugates were prepared with alkaline phosphatase (Biozyme, San Diego, CA) using the heterobifunctional cross-linker SPDP (Pierce, Rockford, IL) and purified by FPLC. These reagents are under development (Tropix) and are not currently commercially available.
12. Purified CAT enzyme (5 Prime–3 Prime; Boehringer Mannheim, Indianapolis, IN), or purified hGH protein (Scripps Laboratories, San Diego, CA).
13. Microplate or tube luminometer.

2.2.2. Two-Step CAT Immunoassay Detection

Materials as for Section 2.2.1., except for reagent (11), and in addition:

1. Biotinylated anti-CAT detector antibody (Fc-biotinylated polyclonal anti-CAT; 5 Prime–3 Prime).
2. Avid*x*-AP (streptavidin-alkaline phosphatase conjugate, Tropix).

3. Methods

3.1. Preparation of Cell Extracts by Freeze/Thaw Lysis (CAT)

Protocol is according to the Boehringer Mannheim CAT ELISA Protocol *(13)*.

1. Precool required volumes of all buffers to 4°C.

2. For suspension cultures: Centrifuge the cell suspension for 10 min at 250*g*, and discard culture medium. Resuspend cell pellet in 5 mL of 1X PBS. Repeat the centrifugation and wash steps two more times. Resuspend the cell pellet in 1 mL of TEN buffer and transfer to a microcentrifuge tube. For adherent cultures (60 mm culture plate): Aspirate medium and wash cells three times with 5 mL of 1X PBS, carefully removing all PBS after last wash. Add 750 µL of TEN buffer to cells and incubate on ice for 10 min. Remove cells from culture plate with rubber policeman or scraper and transfer to a microcentrifuge tube. Rinse culture dish with 400 µL of TEN buffer and add to tube.

3. Centrifuge cells for 30 s and carefully remove supernatant. Resuspend cell pellet in 150 µL 250 m*M* Tris-HCl, pH 7.8.

4. Freeze cell suspension in dry ice/ethanol for 5 min.

5. Thaw cell suspension in 37°C water bath for 5 min. Repeat freeze/thaw cycle four times.

6. Centrifuge suspension for 10 min in a microcentrifuge (4°C) to clarify. Transfer supernatant to fresh tube (*see* Note 1).

3.2. Immunoassay Detection (CAT or hGH)

3.2.1. One-Step Immunoassay Detection

All assays should be performed in duplicate or triplicate.

1. Coat multiwell strips overnight at room temperature with 50 µL/well of 10 µg/mL anti-CAT or anti-hGH capture antibody, diluted in 1X PBS (*see* Note 2).

2. Discard coating solution, and block wells for 2 h at 37°C with 250 µL/well of blocking buffer.

3. Wash wells 3X with wash buffer (*see* Note 3).

All following incubations are performed at room temperature with shaking at 150 rpm.

4. Incubate wells for 1 h with 100 µL of purified CAT enzyme (5-10,000 pg/mL; *see* Note 4) or purified hGH (5–50,000 pg/mL; *see* Note 5), serially diluted in blocking buffer, or 100 µL of cell extract (CAT) or culture medium (hGH) diluted in blocking buffer.

5. Wash wells 3X with wash buffer.

6. Incubate wells for 1 h with 100 µL of anti-CAT-AP, diluted 1:1000 in blocking buffer, or 100 µL of anti-hGH-AP, diluted 1:500 in blocking buffer (*see* Note 6).

7. Wash wells 3X with wash buffer.

8. Prepare required volume of CSPD/Sapphire-II solution (10% Sapphire-II, 17 µL CSPD per 1 mL, diluted in assay buffer).

9. Wash wells 1X with assay buffer.

10. Add 100 µL of CSPD/Sapphire-II solution to each well and incubate at room temperature for 10 min.

11. Measure light emission at 10 min intervals (*see* Note 7).

3.2.2. Two-Step CAT Immunoassay Detection

Perform Section 3.2.1. through step 5.

6a. Incubate wells for 1 h with 100 μL biotinylated anti-CAT, diluted 1:3000 in blocking buffer.
6b. Wash wells 3X with wash buffer.
6c. Incubate wells for 45 min with 100 μL streptavidin-AP conjugate, diluted 1:10,000 in blocking buffer.

Proceed with step 7 of Section 3.2.1.

4. Notes

1. Extracts should be used immediately or stored at −70°C. Extracts should be frozen in dry ice/ethanol before transfer to −70°C to avoid degradation of CAT protein. Prolonged storage at 4°C is not recommended. Protease inhibitors may be added to stabilize extracts: 0.2 mM PMSF, 5 mM DTT, and 5 μg/mL aprotinin.
2. Capture antibodies can be monoclonal or polyclonal and should be used for coating at a concentration of 0.2–10 μg/mL *(14)*. The performance and optimal coating concentration for different antibody preparations should be determined.
3. To wash wells, solution is squirted vigorously into each well with a plastic wash bottle. Wash solution is then flicked out of wells. Following each final wash, the multiwell strips are blotted onto a clean piece of absorbent material to remove remaining wash solution. Alternatively, an automated plate washer could be used.
4. The one-step detection (Section 3.2.1.) can be shortened by simultaneous incubation of CAT enzyme and anti-CAT alkaline phosphatase conjugate in wells. In this case, 50 μL of CAT enzyme dilution and 50 μL of conjugate (diluted 1:500) are both added to the well together and incubated for 1 h. The CAT enzyme dilutions and the conjugate dilution are prepared as 2X such that the final concentrations of CAT and conjugate are identical for both the sequential and simultaneous incubations.
5. The detection can be shortened by simultaneous incubation of hGH and anti-hGH alkaline phosphatase conjugate in wells. In this case, 50 μL of hGH dilution and 50 μL of conjugate (diluted 1:250) are both added to the well together and incubated for 1 h. The hGH dilutions and the conjugate dilution are prepared as 2X such that the final concentrations of hGH and conjugate are identical for both the sequential and simultaneous incubations. Simultaneous hGH capture and detector conjugate incubations result in higher sensitivity compared to sequential incubations when the assay is performed with purified protein.
6. The dilution of a particular antibody-alkaline phosphatase conjugate should be optimized for maximal signal/noise.
7. Maximum light emission is reached 10–20 min following addition of CSPD. Measure light emission at 10 min intervals until maximum light output is achieved.

References

1. Bronstein, I., Voyta, J. C., Thorpe, G. H. G., Kricka, L. J., and Armstrong, G. (1989) Chemiluminescent assay of alkaline phosphatase applied in an ultrasensitive enzyme immunoassay of thyrotropin. *Clin. Chem.* **35**, 1441–1446.
2. Thorpe, G. H. G., Bronstein, I., Kricka, L. J., Edwards, B., and Voyta, J. C. (1989) Chemiluminescent enzyme immunoassay of α-fetoprotein based on an adamantyl dioxetane phenyl phosphate substrate. *Clin. Chem.* **35**, 2319–2321.
3. Bronstein, I., Voyta, J. C., Vant Erve, Y., and Kricka, L. J. (1991) Advances in ultrasensitive detection of proteins and nucleic acids with chemiluminescence: novel derivatized 1,2-dioxetane enzyme substrates. *Clin. Chem.* **37**, 1526–1527.
4. Albrecht, S., Ehle, H., Schollberg, K., Bublitz, R., and Horn, A. (1991) Chemiluminescent enzyme immunoassay of human growth hormone based on adamantyl dioxetane phenyl phosphate substrate, in *Bioluminescence and Chemiluminescence: Current Status* (Stanley, P. and Kricka, L. J., eds.), John Wiley, Chichester, England, pp. 115–118.
5. Nishizono, I., Iida, S., Suzuki, N., Kawada, H., Murakami, H., Ashihara, Y., and Okada, M. (1991) Rapid and sensitive chemiluminescent enzyme immunoassay for measuring tumor markers. *Clin. Chem.* **37**, 1639–1644.
6. Kricka, L. J. (1993) Ultrasensitive immunoassay techniques. *Clin. Biochem.* **26**, 325–331.
7. Bronstein, I., Juo, R. R., Voyta, J. C., and Edwards, B. (1991) Novel chemiluminescent adamantyl 1,2-dioxetane enzyme substrates, in *Bioluminescence and Chemiluminescence: Current Status* (Stanley, P. and Kricka, L. J., eds.), John Wiley, Chichester, England, pp. 73–82.
8. Gorman, C. M., Moffat, L. F., and Howard, B. H. (1982) Recombinant genomes which express chloramphenicol acetyltransferase in mammalian cells. *Mol. Cell. Biol.* **2**, 1044–1051.
9. Alam, J. and Cook, J. L. (1990) Reporter genes: Application to the study of mammalian gene transcription. *Anal. Biochem.* **188**, 245–254.
10. Young, S. L., Barbera, L., Kaynard, A. H., Haugland, R. P., Kang, H. C., Brinkley, M., and Melner, M. H. (1991) A nonradioactive assay for transfected chloramphenicol acetyltransferase activity using fluorescent substrates. *Anal. Biochem.* **197**, 401–407.
11. Burns, D. K. and Crowl, R. M. (1987) An immunological assay for chloramphenicol acetyltransferase. *Anal. Biochem.* **162**, 399–404.
12. Koch, W. and Bürgelt, E. (1992) The CAT ELISA: An alternative nonradioactive approach for measuring gene expression. *BMBiochemica* **9**, 10a,b.
13. Boehringer Mannheim CAT ELISA Protocol, Cat. No. 1363 727.
14. Hornbeck, P. (1994) Assays for antibody production, in *Current Protocols in Immunology* (Coligan, J. E., Kruisbeek, A. M., Margulies, D. H., Shevach, E. M., and Strober, W., eds.), Wiley, New York, p. 2.1.10.

8

Detection and Selection of Cultured Cells Secreting Recombinant Product by Soft Agar Cloning and Antibody Overlay

Marylou G. Gibson, Karmen Hodges, and Lauretta Lowther

1. Introduction

Detection of expression of recombinant genes in transfected cells can be combined with cell cloning to achieve pure cell populations secreting high levels of heterologous protein. The combination of cloning and selection speeds up the generation of permanent cell lines, increases the number of potential insertion events that can be screened, and reduces the tedium of sequentially cloning then analyzing the production levels of each clone.

Several direct cloning methodologies utilizing immunodetection have been reported. They involve capturing the secreted antigen on a nitrocellulose membrane replica followed by identification of positive clones by enzyme-linked immunosorbent assay or radioimmunodiffusion *(1,2)*. These techniques can efficiently detect low-frequency transformants.

The soft agar technique *(3)*, which this chapter describes, takes advantage of the fact that transformed cell clones grow as unattached balls in soft agar, and the product that the clones secrete slowly diffuses away from the colony. When a specific concentration of polyvalent antibody to the product is added to these cultures, an antigen/antibody complex forms that will aggregate and precipitate. These visible precipitates can be visually enhanced with microscopy. The degree of precipitation is relative to the amount of product secreted by the clone. Clones surrounded by these precipitates can be easily picked and transferred to another culture vessel.

This technique was adapted from the method of Coffino and Scharff *(4)* that described the identification and cloning of myeloma cell colonies secreting immunoglobulin heavy and light chains in soft agar. Myeloma and Chinese

From: *Methods in Molecular Biology, vol. 63: Recombinant Protein Protocols: Detection and Isolation* Edited by: R. Tuan Humana Press Inc., Totowa, NJ

hamster ovary-based production lines are routinely used as the recipient cell to generate permanent recombinant cell lines producing proteins of biological significance at economical scale. These lines producing immunoglobulin related proteins, cytokines, or growth factors are easily cloned using this soft agar technique, and resulting cell lines frequently have two- threefold improved specific production rates. The technique is also adaptable to the selection of secretor yeast clones.

2. Materials

1. SeaPlaque Agarose, (FMC BioProducts, Rockland, ME).
2. Tissue culture grade water.
3. 0.2 μm filter unit (e.g., Corning product # 25932-200, Corning, NY) for volumes 20 mL or greater; syringe tip Sterile Acrodisc, 0.2 μm (Gelman Sciences, Ann Arbor, MI) for smaller volumes.
4. Growth media: DMEM/F12 or IMDM or suitable equivalent for user's cells with fetal bovine serum and selective drugs if needed.
5. Two water baths adjustable to 37 and 45°C.
6. 60 × 15 mm plastic tissue culture dishes.
7. CO_2 incubator (5–10% CO_2) maintained at 37°C.
8. Stock cultures of cell lines of interest in logarithmic growth phase.
9. Polyclonal antiserum to the recombinant antigen.
10. Inverted microscope with 4X phase objective and phase ring. (e.g., Nikon Diaphot with an E Plan 4/0.1 objective and PH 2 or 3 phase ring; Olympus IM with a 4X, 0.1 objective and a 20X phase ring).
11. Gilson p200 Pipetman and sterile yellow pipet tips.
12. Various sterile pipets for tissue culture and sterile tissue culture grade 200–250-mL bottles.

3. Methods

3.1. Preparation of Agarose Solutions

1. Prepare a solution of 5% SeaPlaque agarose in tissue culture grade water in a 200-mL bottle. Add 5 g agarose to 100 mL of water and autoclave for 20 min on liquid cycle to solubilize and sterilize. Can be prepared ahead and stored at room temperature.
2. Within several hours or less of plate preparation and cell cloning, warm 100 mL of complete growth media to 45°C. Simultaneously microwave the 5% agarose solution until completely liquefied. Do not allow agarose solution to boil over; 30–60 s is usually sufficient.
3. Warm up the filter unit in a 37°C incubator.
4. Prepare 0.5% SeaPlaque agarose in complete tissue culture growth medium by adding 10 mL of molten agarose cooled to 45°C to the complete prewarmed tissue culture media. Mix and immediately filter. Place filtered agarose solution at 37°C while setting up for Sections 3.2. and 3.3.

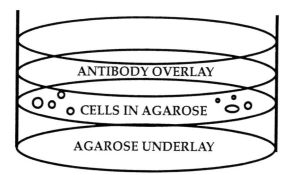

Fig. 1. Schematic diagram of the soft agar selective cloning technique. As described in ref. *4*, cells secreting recombinant product are plated as a single cell suspension in soft agar in complete selective media over an agarose in media underlay. A layer of antigen-specific antibody in medium with agarose is added several days after plating.

3.2. Preparation of Base Layer

1. A base layer of agarose in each culture dish is required so that cell colonies grow as unattached balls. Add 3 mL of filtered agarose solution to each 60×15 mm tissue culture dish and swirl to cover the entire bottom of the dish (agarose underlay, Fig. 1).
2. Solidify for 10 min at 4°C. Twenty dishes are usually sufficient to screen a potential cell line.

3.3. Cloning of Cells in Agarose (see Note 1)

1. Cells in logarithmic growth phase are gently trypsinized if growing as attached culture and counted. Cells routinely grown in suspension should be pipetted or gently trypsinized to achieve a single-cell suspension.
2. Cells are diluted in growth medium to a density of 1×10^5. Two milliliters of cells at this dilution is sufficient. Add 0.3 mL of diluted cells to 30 mL of filtered agarose solution (cells in agarose, Fig. 1).
3. Gently add 1 mL of cell/agarose solution to 10 of the dishes and 2 mL of agarose solution to the other 10 dishes. Tip gently to spread solution over the entire bottom of the dish.
4. Gel agarose layer at 4°C for 10 min. Place cultures in CO_2 37°C incubator.
5. Check for cell growth after 2–3 d.

3.4. Overlay with Antibody in Agarose (see Notes 2–5)

1. After cells begin to form 4–8 cell colonies, plates can be overlaid with antibody solution.
2. Prepare an appropriate amount of 0.5% agarose in growth medium as described above. Prepare 1 mL per each 60-mm dish to be overlaid.
3. To prepare overlay for 20 dishes add 1.5–2.0 mL of sterile filtered specific antiserum to 20 mL of agarose solution. Filter using a 0.2-μ filter unit.
4. Set aside several dishes that will not receive antibody overlay as negative controls.

Fig. 2. Photomicrography of CHO clones secreting various levels of recombinant product detected by the soft agar selective cloning technique. Photos were taken at day 10 using pseudo darkfield microscopy. Panel A shows a positive clone with a dense precipitin halo and a smaller negative clone. The distance between these clones is sufficient for the larger to be picked independently. Panel B shows the type of halo seen with clones secreting less than 1.0 μg product/mL into spent media. Panel C shows two different colony morphologies arising from a single CHO transformed cell pool. No precipitate is observed.

5. Gently add the antibody/agarose solution on top of the existing layers of solidified agarose (antibody overlay, Fig. 1). After the overlay has hardened at 4°C, return the cultures to the 37°C CO_2 incubator.

3.5. Colony Selection (see Notes 6 and 7)

1. Precipitate rings can be visualized from 3–10 d after antibody overlay.
2. An inverted microscope described above should relay an image that makes the cell clones appear silvery and the background black. Precipitate around positive clones will appear as a silver halo (Fig. 2A). Screen the entire dish for clones with significant halos and circle with a marker those well isolated colonies that will be selected.

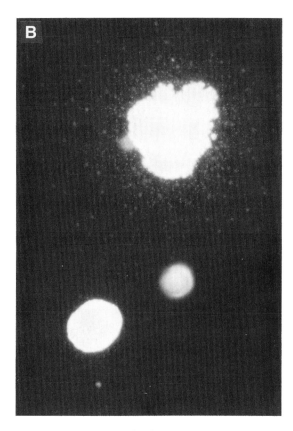

Fig. 2B

3. Fill 24-well dishes with 1 mL selective growth medium per well.
4. Move the microscope into a tissue culture hood and turn off the hood light. View the clones to be picked and draw up colonies with a Pipetman set at 50–75 μL. Clone picking and transfer are monitored microscopically to ensure that single clones are transferred. Transfer colonies into individual wells of a 24-well tissue culture dish.
5. Change media after 5 d and assay production levels of individual clones 2–4 d after monolayers become confluent.

4. Notes

1. Depending on the plating efficiency of your cell line, you may have to adjust the number of cells in the agarose added to each plate. Do not exceed 2 mL of the cell/agarose solution, because the cell clones will grow in different planes to make screening with the microscope difficult. The seeding density described presumes

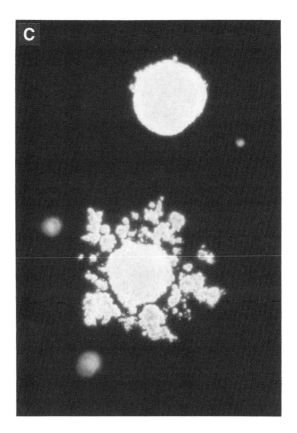

Fig. 2C

about a 30% plating efficiency in agarose. Perform agarose mixing and plating swiftly before the agarose solution hardens.

2. Antibody can be incorporated into the cell/agarose solution and added at d 0. The beginner should confirm cell growth and plating efficiency before overlaying cultures with potentially valuable antibody. Antibody is easily added in another layer of agarose anytime between d 3 and 7.

3. A high titer polyclonal antibody or a mixture of antibodies can be used to overlay cultures and detect product. Rabbit, sheep, and goat antisera have generated similar precipitates. If a polyclonal antibody is made against a peptide included in the larger secreted product, it may not be sufficiently polyvalent to cause aggregation and precipitation. Antibodies to several peptide epitopes can be mixed to achieve precipitates. The mixing ratio for each antiserum must be empirically determined. Cocktails of monoclonal antibodies could work if the mixing ratio to achieve aggregation can be determined.

4. Some commercially available antibodies contain sodium azide at 0.1%. If such antibody is diluted at least 1/50, the azide should have little or no effect on cell growth and secretion in culture.
5. The lower limit dilution of antiserum described frequently works, but higher dilutions of other antisera (e.g., 0.4 mL antiserum per 20 mL of agarose solution) have shown results. The best working dilution of antiserum must be empirically determined.
6. Colonies can be selected based on a variety of features. Large colonies with large halos can yield significant product in confluent 24-well dishes (1–10 µg/mL) (Fig. 2A). Smaller colonies with dense halos may yield cell lines with higher production rates per cell. Colonies with sparse halos (Fig. 2B) can yield between 0.1–0.8 µg product/mL spent medium in 24-well dishes. Depending on the expression stability of the line being cloned, it is usual to observe anywhere from 10–95% halo positive clones.
7. Colonies emerging from the same cell pool can have different morphologies in soft agar. Figure 2C shows a tight ball colony and a colony with migrating cells. No precipitate is visible.

References

1. Gherardi, E., Pannell, R., and Milstein, C. (1990) A single-step procedure for cloning and selection of antibody-secreting hybridomas. *J. Immunol. Meth.* **126,** 61–68.
2. Walls, J. D. and Grinnell, B. W. (1990) A rapid and versatile method for the detection and isolation of mammalian cell lines secreting recombinant proteins. *BioTechniques* **8,** 138–142.
3. Gibson, M. G., Ashey, L., Newell, A. H., and Bradley, C. (1993) Soft agar cloning with antibody overlay to identify chinese hamster ovary clones secreting recombinant products. *BioTechniques* **15,** 594–597.
4. Coffino, P. and Scharff, M. D. (1971) Rate of somatic mutation in immunoglobulin production by mouse myeloma cells. *Proc. Natl. Acad. Sci. USA* **68,** 219–223.

9

Detection and Isolation of Recombinant Protein Based on Binding Affinity Reporter

Maltose Binding Protein

Paul Riggs

1. Introduction

The pMAL-2 vectors (Fig. 1) provide a method for expressing and purifying a protein produced from a cloned gene or open reading frame. The cloned gene is inserted downstream from the *malE* gene of *E. coli*, which encodes maltose-binding protein (MBP), resulting in the expression of an MBP fusion protein *(1,2)*. The method uses the strong "tac" promoter and the *malE* translation initiation signals to give high-level expression of the cloned sequences *(3,4)*, and a one-step purification of the fusion protein using MBP's affinity for maltose *(5)*. The vectors express the *malE* gene (with or without its signal sequence) fused to the *lacZ*a gene. Restriction sites between *malE* and *lacZ*α are available for inserting the coding sequence of interest. Insertion inactivates the β-galactosidase α-fragment activity of the *malE-lacZ*a fusion, which results in a blue to white color change on X-gal plates when the construction is transformed into an α-complementing host such as TB1 *(6)* or JM107 *(7)*. The vectors carry the *lacI*^{q} gene, which codes for the Lac repressor. This keeps expression from P$_{tac}$ low in the absence of IPTG (isopropyl-β-D-thiogalactoside) induction. The pMAL-2 vectors also contain the sequence coding for the recognition site of the specific protease factor Xa *(9,10)*, located just 5' to the polylinker insertion sites. This allows MBP to be cleaved from the protein of interest after purification. Factor Xa cleaves after its four amino acid recognition sequence, so that few or no vector-derived residues are attached to the protein of interest, depending on the site used for cloning. A purification example is shown in Fig. 2.

From: *Methods in Molecular Biology, vol. 63: Recombinant Protein Protocols: Detection and Isolation* Edited by: R. Tuan Humana Press Inc., Totowa, NJ

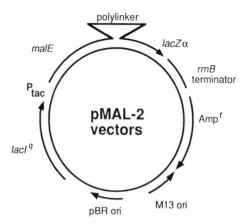

pMAL-c2, -p2 polylinker:

```
                ┌─Sac I ─┐
malE... TCG AGC TCG AAC AAC AAC AAC AAT AAC AAT AAC AAC AAC CTC GGG
```

```
       ┌────Xmnl────┐  ┌EcoRI ┐┌BamH I ┐┌Xba I ┐ ┌Sal I ┐ ┌Pst I ┐    ┌Hind III┐
ATC GAG GGA AGG ATT TCA GAA TTC GGA TCC TCT AGA GTC GAC CTG CAG GCA AGC TTG...lac₄
Ile Glu Gly Arg
                    ↘︎____ Factor Xa cleavage site
```

Fig. 1. pMAL-2 vectors are shown. pMAL-c2 (6646 bp) has an exact deletion of the *malE* signal sequence. pMAL-p2 (6721 bp) includes the *malE* signal sequence. Arrows indicate the direction of transcription. Unique restriction sites are indicated.

The pMAL vectors come in two versions: pMAL-c2, which lacks the N-terminal signal sequence normally present on MBP, and pMAL-p2 which includes the N-terminal signal sequence. Fusion proteins expressed from pMAL-c2 are expressed in the cytoplasm of *Escherichia coli*. These constructions generally give the highest levels of expression. Fusion proteins expressed from pMAL-p2 include a signal peptide on pre-MBP which directs fusion proteins through the cytoplasmic membrane into the periplasm. For fusion proteins that can be successfully exported, this allows folding and disulfide bond formation to take place in the periplasm of *E. coli*, as well as allowing purification of the protein from the periplasm *(8a)*.

To produce a fusion protein in the pMAL-2 vectors, the gene or open reading frame of interest must be inserted into the pMAL-2 vectors so that it is in the same translational reading frame as the vector's *malE* gene. The vectors have a polylinker containing an *Xmn*I site for cloning fragments directly down-

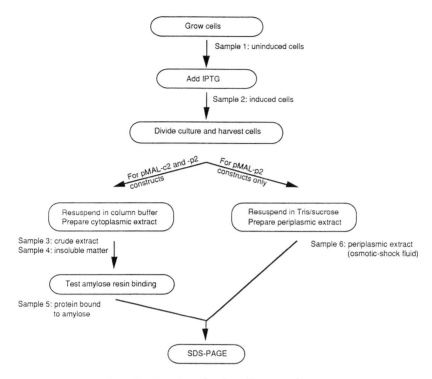

Fig. 2. Flowchart for the pilot experiment.

stream of the factor Xa site, as well as an *Eco*RI site in the same reading frame as λgt11. A number of other restriction sites are also available for cloning fragments downstream of the *Xmn*I site or for directional cloning of a blunt/sticky-ended fragment. Inserts cloned into the *Xmn*I site produce a protein of interest that, after factor Xa cleavage, contains no vector-derived amino acids *(9,10)*. Factor Xa will not cleave fusion proteins that have a proline or arginine immediately following the arginine of the factor Xa site, so the first three bases of the insert should not code for arg or pro when cloning into the *Xmn*I site. If the sequence of interest was identified as a *lacZ* fusion in λgt11, it can be subcloned into the *Eco*RI site of the pMAL-2 polylinker directly. If the sequence is from another source, several strategies may be employed to create an appropriate fragment to subclone. It is assumed that the sequence of interest includes a translational stop codon at its 3' end; if not, one should be engineered into the cloning strategy. Alternatively, a linker containing a stop codon can be inserted into one of the downstream polylinker sites.

Section 3. contains a section on cloning, a small-scale pilot experiment to diagnose the behavior of a particular fusion protein, two methods for the affinity purification of the fusion protein, a short method for regeneration of the amylose resin column, and methods for cleavage of the fusion protein with the specific protease factor Xa and separation of the target domain from MBP.

2. Materials

In addition to the materials listed below, the standard reagents for molecular biology such as restriction enzymes and buffers, $0.5M$ EDTA, phenol, chloroform, T4 DNA ligase, solutions for determining protein concentration (e.g., by the Bradford or Lowry method), and the materials and buffers for SDS-PAGE are required.

2.1. Construction of the Fusion Plasmid

1. DNA fragment to be subcloned, preferably with a blunt end at the 5' end of the gene and a sticky end compatible with the pMAL-2 polylinker at the 3' end.
2. pMAL-c2 and/or pMAL-p2, available from New England Biolabs (Beverly, MA) (*see* Note 1).
3. Competent *E. coli* TB1 (or an equivalent α-complementing strain of *E. coli*).
4. LB plates containing 100 μg/mL ampicillin, with and without 80 μg/mL X-gal.

2.2. Pilot Experiment and Affinity Purification

1. Rich medium plus glucose and ampicillin (per liter): 10 g tryptone, 5 g yeast extract, 5 g NaCl, 2 g glucose; autoclave; add sterile ampicillin to 100 μg/mL.
2. $0.1M$ IPTG stock: 1.41 g IPTG hemidioxane adduct (isopropyl-β-D-thio-galactoside hemidioxane adduct; mol wt 282.4) or 1.19 g IPTG (isopropyl-β-D-thiogalactoside; mol wt 238.3). Add H_2O to 50 mL, filter sterilize, store at 4°C. Dioxane adduct solution stable for 6 mo at 4°C; dioxane-free solution is light-sensitive, and therefore considerably less stable.
3. Column buffer: 20 mM Tris-HCl, 200 mM NaCl, 1 mM EDTA, pH 7.4. Optional components: 1 mM sodium azide, and 10 mM β-mercaptoethanol or 1 mM DTT (*see* Note 2). Store at room temperature. The conditions under which MBP fusions will bind to the column are flexible, and the column buffer can be modified without adversely effecting the affinity purification. Buffers other than Tris-HCl that are compatible include MOPS, HEPES, and phosphate, at pH values around 7. MBP binds to amylose primarily by hydrogen bonding, so higher ionic strength increases its affinity. Nonionic detergents such as Triton X-100 and Tween-20 have been seen to interfere with the affinity of some fusions.
4. Sonicator.
5. 30 mM Tris-HCl, 20% sucrose, pH 8.0.
6. 5 mM MgSO$_4$.
7. 2.5 × 10 cm column.
8. Centricon, Centriprep or stirred cell concentrator (Amicon), or equivalent.
9. Amylose resin, available from New England Biolabs (Beverley, MA).

2.3. Regenerating the Amylose Resin

Use 0.1% SDS.

2.4. Denaturing the Fusion Protein

Use 20 mM Tris-HCl, 6M guanidine hydrochloride, pH 7.4.

2.5. Separating the Protein of Interest from MBP after Factor Xa Cleavage

1. 1 × 10 cm column.
2. 20 mM Tris-HCl, 25 mM NaCl, pH 8.0 (Section 3.7.1.).
3. 20 mM Tris-HCl, 0.5M NaCl, pH 8.0 (Section 3.7.1.).
4. DEAE-Sepharose or Q-Sepharose, or equivalent (Section 3.7.1.).
5. Hydroxyapatite resin (Section 3.7.2.).
6. 0.5M Sodium phosphate buffer, pH 7.2 (stock): Prepare 0.5M NaH$_2$PO$_4$ (69.0 g NaH$_2$PO$_4$ · H$_2$O, to 1 L with H$_2$O), and 0.5M Na$_2$HPO$_4$ (134.0 g Na$_2$HPO$_4$ · 7H$_2$O, to 1 L with H$_2$O). Mix 117 mL 0.5M NaH$_2$PO$_4$ with 0.5M Na$_2$HPO$_4$ 383 mL. Store at room temperature (Section 3.7.2.).

3. Methods

3.1. Construction of the Fusion Plasmid

This section gives an example of subcloning a gene of interest into a pMAL vector. The DNA fragment in step 1 can be made by PCR or by oligo-directed mutagenesis to create a restriction site at the 5' and/or 3' end of the gene. The vector to insert ratio in step 10 assumes an insert size of about 1.2 kb. There are many alternative ways to subclone a gene of interest, including direct cloning into one of the other polylinker sites, for example cloning a λgt11 *Eco*RI fragment into the *Eco*RI site. Cloning downstream of the *Xmn*I site creates a fusion that, after cleavage with factor Xa, has vector-encoded amino acids at the N-terminus.

1. Construct a 5'-blunt 3'-sticky-ended fragment containing the gene of interest, for example by PCR. The restriction site for creating the sticky end should be one of the enzymes that cuts in the pMAL-2 polylinker (except for *Xmn*I).
2. Digest 0.5 µg pMAL-2 plasmid DNA in 20 µL with *Xmn*I. Adjust the buffer to the conditions for the second enzyme (e.g., add NaCl to the recommended level), and cleave with the second enzyme.
3. Add EDTA to 20 mM to the restriction digest.
4. Add an equal volume of a 1:1 phenol/chloroform mixture to the restriction digests, mix, and remove the aqueous (top) phase and place in a fresh tube. Repeat with chloroform alone.
5. Add 10 µg glycogen or tRNA to the digest as carrier. Add 1/9th volume 3M sodium acetate, mix, and then add two volumes ethanol. Incubate at room temperature 10 min.

6. Microcentrifuge for 15 min. Pour off the supernatant, rinse the pellet with 70% ethanol, and allow to dry.
7. Resuspend the sample in 25 μL of 10 m*M* Tris-HCl, 1 m*M* EDTA, pH 8.0.
8. Mix 2 μL vector digest (40 ng) with 20 ng insert DNA. Add water to 17 μL, then incubate the DNA mixture at 45°C for 5 min. Cool on ice, then add 2 μL 10X ligase buffer and 1 μL T4 ligase (e.g., New England Biolabs #202; ~400 U). Incubate at 16°C for 2 h to overnight.
9. Heat at 65°C for 5 min; cool on ice.
10. Mix the ligation reaction with 25 μL competent TB1 (or any *lacZ*a-complementing strain) *(11)* and incubate on ice for 5 min. Heat to 42°C for 2 min.
11. Add 0.1 mL LB and incubate at 37°C for 20 min. Spread on an LB plate containing 100 μg/mL ampicillin (do not plate on IPTG; *see below*). Incubate overnight at 37°C. Pick colonies with a sterile toothpick onto a master LB amp plate and an LB amp plate containing 80 μg/mL X-gal and 0.1 m*M* IPTG . Incubate at 37°C for 8–16 h. Determine the Lac phenotype on the X-gal plate and recover the "white" clones from the corresponding patch on the master plate (*see* Note 3).
12. Screen for the presence of inserts in one or both of the following ways:
    a. Prepare miniprep DNA *(12)*. Digest with an appropriate restriction endonuclease to determine the presence and orientation of the insert *(13)*.
       i. Grow a 5 mL culture in LB amp broth to 2×10^8 cells/mL (A_{600} of ~0.5).
       ii. Take a 1 mL sample. Microcentrifuge for 2 min, discard the supernatant and resuspend the cells in 50 μL SDS-PAGE sample buffer.
       iii. Add IPTG to the remaining culture to 0.3 m*M*, for example 15 μL of a 0.1*M* stock solution. Incubate at 37°C with good aeration for 2 h.
       iv. Take a 0.5 mL sample. Microcentrifuge for 2 min, discard the supernatant and resuspend the cells in 100 μL SDS-PAGE sample buffer.
       v. Place samples in a boiling water bath for 5 min. Electrophorese 15 μL of each sample on a 10% SDS-PAGE gel along with a set of protein mol-wt standards *(14,14a)*. Stain the gel with Coomassie brilliant blue *(14,14a,15)*. An induced band should be visible at a position corresponding to the molecular weight of the fusion protein. A band at or around the position of MBP2* (mol wt 42,710 Dalton) indicates either an out of frame fusion or a severe protein degradation problem. These can usually be distinguished by performing a Western blot using the anti-MBP serum *(16,16a)*; even with severe protein degradation, a full length fusion protein can be detected on the Western. The molecular weight of the MBP-β-gal-α fusion is 50,843 Dalton (*see* Notes 4–6).

It can be helpful to perform a control transformation with about 1 ng of the uncut pMAL vector. Whereas most inserts prevent expression of any lacZa fragment, it is possible to get some α-fragment activity in clones with inserts. In this case, a difference in the shade of blue can usually be seen by comparing to transformants containing the vector alone. Because of the strength of the P_{tac}

promoter, transformants taken from a plate containing IPTG can contain mutant plasmids that have either lost part or all of the fusion gene, or no longer express it at high levels (*see* Notes 7 and 8).

3.2. Pilot Experiment

This experiment is designed to determine the behavior of a particular MBP fusion protein. This protocol results in five (pMAL-c2) or six (pMAL-p2) samples: uninduced and induced cells, a total cell crude extract, a suspension of the insoluble material from the crude extract, a fraction containing protein that binds to the amylose resin, and (for pMAL-p2 constructions) a periplasmic fraction prepared by the cold osmotic shock procedure *(17)* (Fig. 2).

1. Inoculate 80 mL rich broth plus glucose and amp with 0.8 mL of an overnight culture of cells containing the fusion plasmid (*see* Note 9).
2. Grow at 37°C with good aeration to 2×10^8 cells/mL (A_{600} of ~0.5). Take a sample of 1 mL and microcentrifuge for 2 min (sample 1: uninduced cells). Discard supernatant and resuspend the cells in 50 μL SDS-PAGE sample buffer. Vortex and freeze at –20°C.
3. Add IPTG (isopropylthiogalactoside) to the remaining culture to give a final concentration of 0.3 mM, e.g., 0.24 mL of a 0.1M stock in H_2O. Continue incubation at 37°C for 2 h (*see* Note 10). Take a 0.5-mL sample and microcentrifuge for two min (sample 2: induced cells). Discard supernatant and resuspend the cells in 100 μL SDS-PAGE sample buffer. Vortex to resuspend cells and freeze at –20°C.

   Additional time points at 1 and 3 h can be helpful in trying to decide when to harvest the cells for a large scale prep.
4. Divide the remaining culture into two aliquots. Harvest the cells by centrifugation at 4000g for 10 min. Discard the supernatants and resuspend one pellet in 5 mL of column buffer, to make sample A. For pMAL-p2 constructions, resuspend the other pellet in 15 mL 30 mM Tris-HCl, 20% sucrose, pH 8.0, to make sample B (8 mL/0.1 g cells wet weight).

Sample A (All Constructions)

5A. Freeze the cells in column buffer in a dry ice-ethanol bath (or overnight at –20°C; –20°C is more effective than –70°C, but takes longer). Thaw in cold water.
6A. Place the cells in an ice-water bath and sonicate in short pulses of 15 s or less. Monitor the release of protein using the Bradford assay *(18)*, adding 10 μL of the sonicate to 1.5 mL Bradford reagent and mixing. Continue sonication until the released protein reaches a maximum (usually about 2 min); a standard containing 10 μL of 10 mg/mL BSA in 1.5 mL of Bradford reagent can be helpful as an approximate endpoint.
7A. Centrifuge at 9000g at 4°C for 20 min. Decant the supernatant (crude extract) and save on ice. Resuspend the pellet in 5 mL column buffer. This is a suspension of the insoluble matter. Add 5 μL 2X SDS-PAGE sample buffer to 5 μL of the crude extract and insoluble matter fractions (samples 3 and 4, respectively).

8A. Place ~200 μL of the amylose resin in a microfuge tube and spin briefly in a microcentrifuge. Remove the supernatant by aspiration and discard. Resuspend the resin in 1.5 mL column buffer, then microcentrifuge briefly and discard the supernatant; repeat. Resuspend the resin in 200 μL of column buffer. Mix 50 μL of crude extract with 50 μL of the amylose resin slurry. Incubate for 15 min on ice. Microcentrifuge 1 min, then remove the supernatant and discard. Wash the pellet with 1 mL column buffer, microcentrifuge 1 min, and resuspend the resin in 50 μL SDS-PAGE sample buffer (sample 5: protein bound to amylose).

Sample B (pMAL-p2 Constructions Only)

5B. Add 30 μL 0.5*M* EDTA (1 m*M* final concentration) to the cells in Tris/sucrose and incubate 5–10 min at room temperature with shaking or stirring.

6B. Centrifuge at 8000*g* at 4°C for 10 min, remove all the supernatant, and resuspend the pellet in 15 mL ice-cold 5 m*M* MgSO$_4$.

7B. Shake or stir for 10 min in an ice-water bath.

8B. Centrifuge as in 6B. The supernatant is the cold osmotic shock fluid. Add 5 μL 2X SDS-PAGE sample buffer to 20 μL of the cold osmotic shock fluid (sample 6).

9. Place the samples in a boiling water bath for 5 min. Microcentrifuge for 1 min. Load 20 μL of the of uninduced cells, induced cells, and amylose resin samples (avoid disturbing the pellets), and all of the remaining samples, on a 10% SDS-PAGE gel *(14,14a)* (*see* Notes 4–6).

10. (Optional) Run an identical SDS-PAGE gel(s) after diluting the samples 1:10 in SDS sample buffer. Prepare a Western blot(s) and develop with anti-MBP serum and, if available, serum directed against the protein of interest *(16,16a)*.

Most fusion proteins expressed from pMAL-c2 will be in the soluble fraction and will bind to the amylose resin. If the protein is insoluble, the best course is to modify the conditions of cell growth to attempt to produce soluble fusion. Two changes that have helped in previous cases are changing to a different strain background and growing the cells at a lower temperature *(8,8a)*. Fusion proteins expressed from pMAL-p2 will be directed to the periplasmic space. If the fusion is in the periplasmic fraction, consider using the method described in Section 3.3.2. In this case, another pilot to optimize expression and export may be desirable.

3.3. Affinity Chromatography

Two methods are presented for purifying the fusion protein. In Section 3.3.1., a total cell crude extract is prepared and passed over the amylose column (Fig. 3). This method can be used for any pMAL construct. In Section 3.3.2., a periplasmic fraction is prepared by the cold osmotic shock procedure *(17)*. This method only works for pMAL-p2 constructions where the fusion protein is exported.

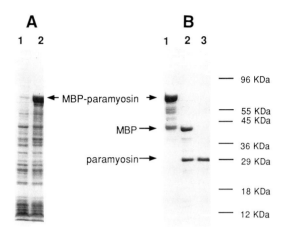

Fig. 3. SDS-polyacrylamide gel stained with Coomassie blue showing fractions from the purification of MBP-paramyosin-ΔSal using pMAL-c. **(A)** Lane 1, uninduced cells; lane 2, induced cells. **(B)** Lane 1, purified protein eluted from amylose column with maltose (this represents about 50% of the fusion protein present in the induced cells); lane 2, purified protein after factor Xa cleavage; lane 3, paramyosin fragment from the flowthrough of a second amylose column, after removal of maltose by the hydroxyapatite procedure.

3.3.1. Method I

1. Inoculate 1 L rich broth plus glucose and ampicillin with 10 mL of an overnight culture of cells containing the fusion plasmid (*see* Note 9).
2. Grow to 2×10^8 cells/mL (A_{600} ~0.5). Add IPTG to a final concentration of 0.3 mM, e.g., 85 mg or 3 mL of a 0.1M stock in H_2O. Incubate the cells at 37°C for 2 h (*see* Note 10).
3. Harvest the cells by centrifugation at 4000g for 20 min and discard the supernatant. Resuspend the cells in 50 mL column buffer (*see* Note 11).
4. Freeze sample in a dry ice-ethanol bath (or overnight at –20°C; –20°C is more effective than –70°C, but takes longer). Thaw in cold water.
5. Place sample in an ice-water bath and sonicate in short pulses of 15 s or less. Monitor the release of protein using the Bradford assay *(18)*, adding 10 μL of the sonicate to 1.5 mL Bradford reagent and mixing. Continue sonication until the released protein reaches a maximum (usually about 2 min sonication time); a standard containing 10 μL of 10 mg/mL BSA can be helpful as an approximate endpoint.
6. Centrifuge at 9,000g for 30 min. Save the supernatant (crude extract). Dilute the crude extract 1:5 with column buffer (*see* Note 12).
7. Pour 15 mL of amylose resin in a 2.5 × 10 cm column (*see* Note 13). Wash the column with 8 column volumes of column buffer.

8. Load the diluted crude extract at a flowrate of 1 mL/min for a 2.5 cm column (flowrate = [10 × (diameter of column in cm)^2]mL/h).
9. Wash with 12 column volumes of column buffer (*see* Note 14).
10. Elute the fusion protein with column buffer plus 10 mM maltose. Collect 10–20 fractions of 3 mL each (fraction size = 1/5th column volume). The fusion protein usually starts to elute within the first five fractions and should be easily detected by UV absorbance at 280 nm or the Bradford protein assay.
11. Pool the protein-containing fractions. If necessary, concentrate to about 1 mg/mL in an Amicon Centricon or Centriprep concentrator, an Amicon stirred-cell concentrator, or the equivalent.

3.3.2. Method II—For Fusion Proteins Exported to the Periplasm (pMAL-p2)

1. Inoculate 1 L rich broth plus glucose and ampicillin with 10 mL of an overnight culture of cells containing the fusion plasmid (*see* Note 9).
2. Grow to 2–4 × 10^8 cells/mL (A_{600} ~0.5). Add IPTG to a final concentration of 0.3 mM, e.g., 85 mg/L or 3 mL of a 0.1M stock in H$_2$O. Incubate the cells at 37°C for 2 h (*see* Note 10).
3. Harvest the cells by centrifugation at 4000g for 20 min and discard the supernatant. Resuspend the cells in 400 mL 30 mM Tris-HCl, 20% sucrose, pH 8.0 (80 mL for each gram of cells wet weight). Add EDTA to 1 mM and incubate for 5–10 min at room temperature with shaking or stirring.
4. Centrifuge at 8000g for 20 min at 4°C, remove all the supernatant, and resuspend the pellet in 400 mL volume ice-cold 5 mM MgSO$_4$. Shake or stir for 10 min in an ice bath.
5. Centrifuge at 8000g for 20 min at 4°C. The supernatant is the cold osmotic shock fluid.
6. Add 8 mL of 1M Tris-HCl, pH 7.4 to the osmotic shock fluid.
7. Continue from Section 3.3.1., step 7.

3.4. Regenerating the Amylose Resin Column

This procedure removes any residue from the crude extract that remains on the column after the affinity purification. Although the column can be washed at 4°C, 0.1% SDS will eventually precipitate at that temperature. It is therefore recommended that the SDS solution be stored at room temperature until needed, and rinsed out of the column promptly. The resin may be reused 3–5 times. On repeated use, trace amounts of amylase in the *E. coli* extract decrease the binding capacity of the column. It is recommended that the column be washed promptly after each use.

1. Wash the column with three column volumes of H$_2$O.
2. Next, wash with three column volumes 0.1% SDS.
3. Wash the column with one column volume of H$_2$O.
4. Finally, equilibrate the column with at least three column volumes of column buffer.

3.5. Cleavage with Factor Xa

1. If necessary, concentrate the fusion protein to at least 1 mg/mL (*see* Note 15).
2. Do a pilot experiment with a small portion of your protein. For example, mix 20 μL fusion protein at 1 mg/mL with 1 μL factor Xa diluted to 200 μg/mL. In a separate tube, place 5 μL fusion protein with no factor Xa (mock digestion). Incubate the tubes at room temperature (*see* Note 16).
3. At 2, 4, 8, and 24 h, take 5 μL of the factor Xa reaction, add 5 μL 2X SDS-PAGE sample buffer, and save at 4°C. Prepare a sample of 5 μL fusion protein plus 5 μL 2X sample buffer (uncut fusion).
4. Place the six samples in a boiling water bath for 5 min and run on an SDS-PAGE gel *(14)* (*see* Notes 4 and 17).
5. Scale the pilot experiment up for the portion of the fusion protein to be cleaved. Save at least a small sample of the uncut fusion as a reference.
6. Check for complete cleavage by SDS-PAGE.

3.6. Denaturing the Fusion Protein

Occasionally, the factor Xa site of a particular fusion is inaccessible to the protease, owing to the three-dimensional structure of the protein. In this case, denaturing the protein can sometimes allow the factor Xa to cleave (*see* Note 18).

1. Either dialyze the fusion against at least 10 volumes 20 mM Tris-HCl, 6M guanidine hydrochloride, pH 7.4 for 4 h, or add guanidine hydrochloride directly to the sample to give a final concentration of 6M.
2. Dialyze against 100 volumes column buffer, 2 times at 4 h each.

During refolding, one has to balance between two objectives. For factor Xa to cleave it must be present before the protein has completely refolded, so removing the denaturant quickly is desirable. However, when the denaturant is removed quickly, some proteins will fail to refold properly and precipitate. Stepwise dialysis against buffer containing decreasing amounts of guanidine hydrochloride can prevent precipitation of the fusion protein; halving the guanidine concentration at each step is convenient, but cases where 0.1M steps are necessary have been reported. However, if the fusion protein is able to refold into a factor Xa-resistant conformation, it may be better to dialyze away the denaturant in one step and take the loss from precipitation in order to maximize the amount of cleavable fusion protein recovered.

3.7. Separating the Protein of Interest from MBP after Factor Xa Cleavage

3.7.1. Method I—DEAE- or Q-Sepharose Ion Exchange Chromatography

This method not only purifies the target protein away from MBP and factor Xa, but also provides an additional purification step for removing trace contaminants. A disadvantage is that occasionally the peak containing the protein of interest overlaps with MBP or factor Xa, giving poor separation (*see* Note 19). The procedure is written for <25 mg, and can be scaled up for larger amounts.

1. Dialyze the fusion protein cleavage mixture vs 20 mM Tris-HCl, 25 mM NaCl, pH 8.0 (2 or 3 changes of 100 volumes at least 2 h each).
2. Wash about 6 mL of Q-Sepharose (or DEAE-Sepharose) in ~20 mL of 10 mM Tris-HCl, 25 mM NaCl, pH 8.0 a couple of times, letting the resin settle and pouring off the supernatant between washes.
3. Pour the resin into a 1 × 10 cm column to give a bed volume of 5 mL (~6–7 cm bed height).
4. Wash the column with 15 mL of the buffer in step 2.
5. Load the fusion protein cleavage mixture onto the column. Collect 2.5 mL fractions of the column flow-through.
6. Wash the column with 3–5 column volumes of the same buffer. Continue collecting 2.5 mL fractions.
7. Start a gradient of 25 mM NaCl to 500 mM NaCl (25 mL each) in 20 mM Tris-HCl, pH 8.0 (*see* Fig. 4). Collect 1-mL fractions.
8. Determine which fractions contain protein by measuring A_{280}, or by the Bradford or Lowry method *(18)*.

The MBP elutes as a sharp peak at 100–150 mM NaCl. Factor Xa elutes at about 400 mM NaCl. The target protein may flow through the column, or it may elute during the gradient. Electrophorese the relevant fractions on an SDS-PAGE gel. Pool the fractions containing the target protein free of MBP and concentrate as desired.

3.7.2. Method II—Removal of Maltose by Hydroxyapatite Chromatography and Domain Separation by Rebinding MBP to Amylose

This method requires 2 steps, but since no dialysis is needed and both columns are step eluted, the procedure is fairly simple. It removes MBP from the cleavage mixture, but not factor Xa or any other trace contaminants. In addition, any MBP that has been denatured or otherwise damaged will not bind to the amylose column. The procedure must be carried out at room temperature to avoid precipitation of the phosphate buffer. It is written for <25 mg, but can be scaled up for larger amounts.

1. Swell 1 g hydroxyapatite in column buffer. Remove fines by allowing the resin to settle and pouring off excess buffer. Add fresh buffer and repeat, twice more.
2. Pour the hydroxyapatite into a 1 × 10 cm column.
3. Load the fusion protein cleavage mixture onto the column.

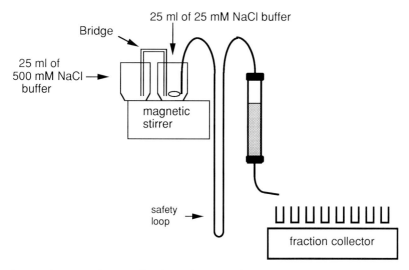

Fig. 4. Diagram of set-up for Q-Sepharose column chromatography is shown.

4. Wash with at least 80 mL of the same buffer (washes away the maltose).
5. Elute with 0.5M sodium phosphate, pH 7.2 (stock solution in Section 2.6.). Collect 2 mL fractions. Assay for protein by A_{280}, Bradford assay or Lowry *(18)*. Most of the protein usually elutes in the first 8 mL.
6. Pour a 15 mL amylose column as described in Section 3.3.1.
7. Load the hydroxyapatite-eluted protein onto the amylose column. Collect the flowthrough as 5-mL fractions. Protein in the flowthrough should be free of MBP and consist primarily of the target protein. Assay for protein by A_{280}, Bradford assay or Lowry. The protein of interest should flow through the column by the seventh fraction.

4. Notes

1. The sequences of the pMAL-c2 and pMAL-p2 are available by mail, fax, or E-mail. For mail or fax versions, call New England Biolabs or their local distributor. The sequences can be obtained via Internet, by anonymous ftp from vent.neb.com, or E-mail a request to riggs@neb.com.
2. DTT or β-mercaptoethanol can be included in the column buffer to prevent interchain disulfide bond formation upon lysis (disulfide bonds usually do not form intracellularly in *E. coli*).
3. An alternative way to use the blue-white screen is to replica plate on LB amp and LB amp X-gal IPTG. One can also divide the ligation mixture into two aliquots, and plate one of them on 80 μg/mL X-gal, 0.3 mM IPTG to observe the percentage of clones with inserts, and if it is acceptably high, screen the transformants on the LB amp plate directly by preparing plasmid DNA.

4. The molecular weight of the MBP-β-galactosidase alpha fragment fusion produced by the vector without an insert is 50.8 kDa. The MBP2* marker supplied by New England Biolabs, as well as the MBP-portion of fusion proteins, is 42.5 kDa. The MBP2*-paramyosinΔSal supplied by New England Biolabs as a control for factor Xa cleavage is 70.2 kDa. Factor Xa is 42.4 kDa, and consists of two disulfide-linked chains. The heavy chain is 26.7 kDa, but runs at about 30 kDa on SDS-PAGE. The light chain is 15.7 kDa, and runs at about 20 kDa.

5. Sometimes SDS-PAGE does not reveal a band the size expected of the fusion protein, but instead a band the size of MBP is seen. There are two likely explanations for this result. If the protein of interest is in the wrong translational reading frame, an MBP2*-sized band will be produced by translational termination at the first in-frame stop codon. If the protein of interest is very unstable, an MBP2*-sized breakdown product is usually produced (MBP is a very stable protein). The best way to distinguish between these possibilities is to run a Western using anti-MBP antiserum. If proteolysis is occurring, at least a small amount of full-length fusion can almost always be detected. DNA sequencing of the fusion junction using the *malE* primer (New England Biolabs) will confirm a reading frame problem. If the problem is proteolysis, protease deficient strains of *E. coli* can sometimes help.

6. If a fusion protein appears to be contaminated with many bands that run between the full-length fusion and MBP on SDS-PAGE, most of the time these bands are not contaminants but rather truncated proteins produced by proteolysis of the fusion protein or (rarely) premature translation termination of the fusion. Occasionally multimers of a fusion protein have been observed, usually due to disulfide bonds that were not adequately reduced before loading the gel. Bands lower than ~39 kDa should not appear, since these fragments do not contain the MBP amylose-binding function. This interpretation can be confirmed by doing a Western with anti-MBP serum (New England Biolabs). Problems with proteolysis of the fusion can sometimes be alleviated by using protease deficient strains of *E. coli*. In addition, some proteins are unstable in the cytoplasm because they do not fold properly, especially proteins with disulfide bonds. Sometimes switching to pMAL-p2, that attempts to export the fusion to the periplasm, leads to proper folding and thus a more stable fusion protein. Other proteins behave the opposite way, i.e., they are much more stable when expressed in the cytoplasm, from pMAL-c2, than when they are exported to the periplasm.

7. If the transformants contain plasmids that are related to the vector but do not contain the full insert, it may indicate that the gene of interest is toxic to *E. coli*. The expression system on the pMAL vectors, $lacI^q/P_{tac}$, has an induction ratio of about 1:50. This means that the basal rate of expression can be up to 1% of the total cellular protein. Sometimes the toxicity of a fusion protein can be reduced by changing the growth conditions, e.g., lowering the temperature or changing the medium. Changing the vector could also help; some fusions are less toxic in pMAL-c2 than in pMAL-p2.

8. Some foreign proteins do not express well in *E coli*. Sometimes this problem can be solved by fusing the foreign protein to the carboxy terminus of another protein

that is expressed well, such as maltose-binding protein. But with some proteins, even a fusion is expressed poorly, probably owing to mRNA instability or a problem with translation. Although no solution to this problem has yet emerged, the affinity purification used in the pMAL system often yields significant amounts of protein (1–5 mg/L of culture) even when the expression level is too low to visualize the protein in the crude extract.

9. During growth of the cells for expression of the fusion protein, glucose in the growth medium is necessary to repress the maltose genes on the chromosome of the *E. coli* host. Besides the chromosomal *malE* gene, one of these is an amylase which can degrade the amylose on the affinity resin.

10. The period of time and the temperature to use during expression depends on several factors (stability of the protein, host strain, and so on), and variations can be tried to find optimum conditions for expression. This is especially true for constructs in pMAL-p2. In this case, partial induction may lead to higher yields, since protein export in *E. coli* may not be able to keep up with full level P_{tac} expression.

11. For many unstable proteins, most of the degradation happens during harvest and cell breakage. Therefore, it is best to harvest the cells quickly and keep them chilled. The EDTA in the lysis buffer is to help inhibit proteases that have a Ca^{2+} cofactor. Addition of PMSF (phenyl methyl-sulfonyl fluoride) and/or other protease inhibitors may help in some cases.

12. The dilution of the crude extract is intended to reduce the protein concentration to about 2.5 mg/mL. If the crude extract is less concentrated, less dilution is required. A good rule of thumb is that 1 g wet weight of cells gives about 120 mg protein.

13. The amount of resin to use depends on the amount of fusion protein produced. The resin binds about 3 mg/mL bed volume, so a column of about 15 mL should be sufficient for a yield of up to 45 mg fusion protein/L culture. A 50 mL syringe plugged with silanized glass wool can be substituted for the 2.5 cm column, but the glass wool should cover the bottom of the syringe (not just in the tip) so the column will have an acceptable flowrate.

14. It is possible to allow the column to wash overnight, if the column has a safety loop to prevent it from running dry (*see* Fig. 4). In this case, it is better to restart the column with elution buffer (step 10), rather than continuing the wash. Avoid loading the column overnight.

15. Factor Xa will work in the column buffer plus maltose used to elute the fusion protein. In addition, Factor Xa will work in a variety of other buffers, with NaCl concentrations from 0–500 mM and pH values around 7–8.

16. Factor Xa cleavage is usually carried out at a w/w ratio of 1% the amount of fusion protein (e.g., 1 mg factor Xa for a reaction containing 100 mg fusion protein). The reaction mixture can be incubated for 3 h to several days, at room temperature or 4°C. Depending on the particular fusion protein, the amount of factor Xa can be adjusted within the range of 0.1–5.0%, to get an acceptable rate of cleavage.

17. Factor Xa will cleave at noncanonical sites in some proteins; for some fusions, there is a correlation between instability of the protein of interest in *E. coli* and

cleavage at additional sites (unpublished observations). Presumably this cleavage activity depends on the three dimensional conformation of the fusion protein.

18. Some fusion proteins are resistant to cleavage with factor Xa, presumably because the site is not accessible to the protease. One strategy that sometimes helps is to denature the fusion to render the factor Xa site accessible to cleavage (*see* Section 3.6. and ref. *9*). Another strategy is to include small amounts of SDS (0.005–0.05%) in the reaction. Presumably this works by relaxing the structure of the fusion enough to allow for cleavage *(19)*. The window of SDS concentrations that work can be small, so a pilot titration with different SDS concentrations is necessary.

19. If the protein of interest is not separated from MBP using Q-Sepharose chromatography, other chromatography resins can be tried. CM-Sepharose (phosphate buffer at pH 7.0; MBP flows through) and gel filtration using Sephadex G-10C have been used successfully.

References

1. Guan, C., Li, P., Riggs, P. D., and Inouye, H. (1987) Vectors that facilitate the expression and purification of foreign peptides in *Escherichia coli* by fusion to maltose-binding protein. *Gene* **67**, 21–30.

2. Maina, C. V., Riggs, P. D., Grandea, A. G., III, Slatko, B. E., Moran, L. S., Tagliamonte, J. A., McReynolds, L. A., and Guan, C. (1988) A vector to express and purify foreign proteins in *Escherichia coli* by fusion to, and separation from maltose binding protein. *Gene* **74**, 365–373.

3. Amann, E. and Brosius, J. (1985) "ATG vectors" for regulated high-level expression of cloned genes in *Escherichia coli*. *Gene* **40**, 183–190.

4. Duplay, P., Bedouelle, H., Fowler, A., Zabin, I., Saurin, W., and Hofnung, M (1984) Sequences of the *malE* gene and of its product, the maltose-binding protein of *Escherichia coli* K12. *J. Biol. Chem.* **259**, 10,606–10,613.

5. Kellerman, O. K. and Ferenci, T. (1982) Maltose binding protein from *E. coli* *Methods in Enzymology* **90**, 459–463.

6. Johnston, T. C., Thompson, R. B., and Baldwin, T. O. (1986) Nucleotide sequence of the *luxB* gene of *Vibrio harveyi* and the complete amino acid sequence of the beta subunit of bacterial luciferase. *J. Biol. Chem.* **261**, 4805–4811.

7. Yanisch-Perron, C., Vieira, J., and Messing, J. (1985) Improved M13 phage cloning vectors and host strains, nucleotide sequences of the M13mp18 and pUC1S vectors. *Gene* **33**, 103–119.

8. Lauritzen, C., Tüchsen, E., Hansin, P. E., and Skovgaard, O. (1991) BPTI and N-terminal extended analogs generated by Factor Xa cleavage and cathepsin C trimming of a fusion protein expressed in *Escherichia coli*. *Protein Expressior and Purification* **2**, 372–378.

8a. Takagi, H., Morinaga, Y., Tsuchiya, M., Ikemura, G., and Inouye, M. (1988) Control of folding of proteins secreted by a high expression secretion vector, pIN-III *ompA*, 16-fold increase in production of active subtilisin E in *Escherichia coli Bio/technology* **6**, 948–950.

9. Nagai, K. and Thøgersen, H. C. (1984) Generation of b–globin by sequence-specific proteolysis of a hybrid protein produced in *Escherichia coli. Nature* **309,** 810–812.

10. Nagai, K. and Thøgersen, H. C. (1987) Synthesis and sequence-specific proteolysis of hybrid proteins produced in *Escherichia coli. Methods in Enzymol.* **153,** 461–481.

11. Sambrook, J., Fritsch, E. F., and Maniatis, T. (1989) Preparation and transformation of competent *E. coli,* in *Molecular Cloning, A Laboratory Manual.* Cold Spring Harbor Laboratory , Cold Spring Harbor, NY, pp. 1.74–1.84.

12. Sambrook, J., Fritsch, E. F., and Maniatis, T. (1989) Small-scale preparations of plasmid DNA, in *Molecular Cloning, A Laboratory Manual.* Cold Spring Harbor Laboratory, Cold Spring Harbor, NY, pp. 1.25–1.28.

13. Bloch, K. D. (1989) Digestion of DNA with restriction endonucleases, in *Current Protocols in Molecular Biology* (Ausebel, F. M. et al., eds.), Greene Publishing & Wiley-Interscience, New York, pp 3.1.1–3.2.5.

14. Sambrook, J., Fritsch, E. F., and Maniatis, T. (1989) SDS-polyacrylamide gel electrophoresis of proteins, in *Molecular Cloning, A Laboratory Manual.* Cold Spring Harbor Laboratory, Cold Spring Harbor, NY, pp. 18.47–18.55.

14a.Smith, J. A. (1989) One-dimensional gel electrophoresis of proteins, in *Current Protocols in Molecular Biology* (Ausebel, F. M. et al., eds.), Greene Publishing & Wiley-Interscience, New York, pp. 10.2.1–10.2.7.

15. Sasse, J. (1989) Detection of proteins in gels, in *Current Protocols in Molecular Biology* (Ausebel, F. M. et al., eds.), Greene Publishing & Wiley-Interscience, New York, pp. 10.6.1–10.6.3.

16. Sambrook, J., Fritsch, E. F., and Maniatis, T. (1989) Transfer of proteins from SDS-polyacrylamide gels to solid supports, immunological detection of immobilized proteins (Western blotting), in *Molecular Cloning, A Laboratory Manual* Cold Spring Harbor Laboratory, Cold Spring Harbor, NY, pp. 18.60–18.75.

16a.Winston, S. E., Fuller, S. A., and Hurrell, J. G. R. (1989) Western Blotting, in *Current Protocols in Molecular Biology* (Ausebel, F. M. et al., eds.), Greene Publishing & Wiley-Interscience, New York, pp. 10.8.1–10.8.6.

17. Neu, H. C. and Heppel, L. A. (1965) The release of enzymes from *Escherichia coli* by osmotic shock and during the formation of spheroplasts. *J. Biol. Chem.* **240,** 3685–3692.

18. Smith, J. A. (1989) Quantitation of proteins, in *Current Protocols in Molecular Biology* (Ausebel, F. M. et al., eds.), Greene Publishing & Wiley-Interscience, New York, pp. 10.1.1–10.1.3.

19. Ellinger, S., Mach, M., Korn, K and Jahn, G. (1991) Cleavage and purification of prokaryotically expressed HIV gag and env fusion proteins for detection of HIV antibodies in the ELISA. *Virology* **180,** 811–813.

10

Detection and Isolation of Recombinant Proteins Based on Binding Affinity of Reporter: Protein A

Stefan Ståhl, Per-Åke Nygren, and Mathias Uhlén

1. Introduction

The strong and specific interaction between the Fc part of IgG and staphylococcal protein A (SpA) was utilized by Uhlén and coworkers *(1)* to create the first described system allowing affinity purification of expressed gene products. To date, a multitude of proteins have been produced as fusions to the IgG-binding domains of staphylococcal protein A, in several different hosts such as gram-positive and gram-negative bacteria *(2,3)*, yeast *(4)*, CHO cells *(5)*, insect cells *(6)* and plants *(7)*.

SpA is an immunoglobulin-binding surface receptor found on the gram-positive bacterium *Staphylococcus aureus*. The SpA, which has found extensive use in immunological and biotechnological research *(8–13)*, binds to the constant (Fc) part of certain immunoglobulins, but the exact biological significance of this property is not clear *(14)*. Functional and structural analysis of SpA has revealed the presence of a signal peptide, S, that is processed during secretion; five highly homologous domains: E, D, A, B, and C, capable of binding to IgG *(15)*; and a cell wall attaching structure, designated XM *(16–18)* (Fig. 1). Here, X represents a charged and repetitive region, postulated to interact with the peptidoglycan cell wall *(16)*, whereas M is a region common for gram-positive cell surface bound receptors containing an $LPX_{aa}TGX_{aa}$ motif, linked to a C-terminal hydrophobic region ending with a charged tail *(17,18)*. It has been demonstrated that all three regions are required for cell surface anchoring *(17)* and it has been suggested that the cell wall sorting is accompanied by proteolytic cleavage at the SpA C-terminus and covalent linking of the surface receptor to the cell wall *(18)*. SpA binds to IgG from most mammalian species, including man. Of the four subclasses of human IgG, SpA binds to IgG1, IgG2, and IgG4 but shows only weak interaction with IgG3 *(19)*.

From: *Methods in Molecular Biology, vol. 63: Recombinant Protein Protocols: Detection and Isolation* Edited by: R. Tuan Humana Press Inc., Totowa, NJ

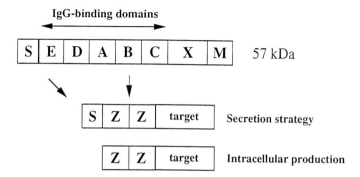

Fig. 1. Staphylococcal protein A and the 7 kDa B-domain-analogue Z, represented in the most widely used divalent form, with a schematic illustration how ZZ is normally fused to a target protein that is to be produced by secreted or intracellular production. In expression vectors designed for secretion of the gene fusion products, the signal sequence S is fused to ZZ to direct produced fusion proteins out from the host cell. The signal peptide is processed during the translocation process.

Several characteristics of the IgG-binding domains of SpA have made them suitable as fusion partners for the production of recombinant proteins: SpA is proteolytically stable itself and the stability of a gene product can be increased in vivo by fusion to SpA *(20,21)*. The structure of the IgG-binding domains, each being a three-helix-bundle *(22–25)* appears to be favorable for independent folding of SpA and the fused product since the N- and C-terminus of each IgG-binding domain is solvent exposed *(24,25)*. It has been demonstrated as feasible to introduce sequences accessible for site-specific cleavage of SpA fusion proteins in order to release the target gene product *(26)*. The high solubility of SpA enables the production of soluble fusion proteins to very high concentrations within the *Escherichia coli* cell *(27–29)*.

2. The Expression Systems

A number of expression vectors, based on gene fusions to the SpA, often in the form of the synthetic divalent SpA analogue ZZ *(30)*, or all five IgG-binding domains, have been developed *(1,9,30)*. The promoter and secretion signal of SpA has been shown to be functional also in the Gram-negative *E. coli (1)*. Protein A fusions expressed in *E. coli* can thus be efficiently secreted to the periplasm of the bacteria as well as to the culture medium *(31–35)* from where they can be easily purified by IgG affinity chromatography *(36)*. To create a "second generation" affinity tag, the synthetic IgG-binding domain, Z, was designed, based on domain B of SpA (Fig. 1) *(30)*. This synthetic domain lacks the methionine residue present in domains E, D, and A, making it resistant to CNBr cleavage. In addition, an NG dipeptide sequence present in all domains was changed to NA by altering the glycine codon to a codon for alanine, making the Z-domain resistant to cleavage with hydroxylamine *(30)*. Analysis of the interaction between IgG and Z domains, polymerized to different multiplicities demonstrated that the dimer, ZZ (Fig. 1), was the optimal fusion partner giving maximal IgG-binding *(37)* and efficient secretion *(30)*.

As discussed earlier (Chapter 5), there is no single strategy applicable on recombinant gene products to achieve secretion. Instead, the inherent properties of the target protein very much decide which strategy will give best results. However, the majority of proteins produced as SpA fusions have been shown to be soluble and efficiently secreted to the periplasmic space and culture medium of *E. coli (10,33,35)*. This may at least partially be owing to the high solubility of the ZZ domains which most likely increases the overall solubility of the fusion proteins. There are a number of advantages connected with a secretion strategy. First, the gene product will be less exposed to cytoplasmic proteases that might enable production of labile proteins *(38)*. Second, disulfide bond formation that is enhanced in the nonreducing environment outside the cytoplasm could improve accurate folding *(33,39)*. Third, the recovery of the recombinant protein is very much simplified since a large degree of the purification has been achieved through the secretion *(33,35)*. Recently, Hansson and coworkers *(35)* described the *E. coli* production of a fusion protein ZZ-M5, a candidate malaria subunit vaccine, using a secretion strategy. More than 65% of the recombinant gene product was secreted to the culture medium, from where it could be recovered in a single step by expanded bed anion exchange adsorption. Since ZZ-M5 was to be used in a preclinical malaria vaccine trial in *Aotus* monkeys *(40)*, a polishing step was included in which the IgG-binding capacity of the fusion protein was employed. After affinity chromatography on IgG-Sepharose, contaminating DNA and endotoxin levels were well below the demands set by regulatory authorities. The overall yield of the process, performed in pilot scale, exceeded 90%, resulting in 550 mg product per L culture *(35)*.

In the vectors designed for secretion, the expressed SpA-fusion products are in most examples transcribed from the SpA-promoter shown to be constitutive and efficiently recognized in *E. coli (1,9)*. However, since all gene products cannot be secreted, owing to inherent properties such as hydrophobic sequences and transmembrane regions, vectors for intracellular production of SpA fusions have also been constructed in which the transcription is under the control of inducible promoters, such as *lac*UV5 *(9)*, λ_{PR} *(9)*, *trp (41,42)*, *trc (43)*, or T7 *(44)* enabling a more controlled expression. The intracellularly produced and still soluble fusion proteins are released by sonication *(9)* or high pressure homogenization *(41)* before affinity purification on IgG-Sepharose. However, even if the proteins precipitate, they can be recovered by affinity chromatography after solubilization (*see* Section 4.1.).

2.1. Expression Vectors

The most extensively used vector for secreted production of SpA fusions is plasmid pEZZmp18 *(45)*, which can be purchased from Pharmacia (Uppsala, Sweden). For intracellular production, the *trp* promoter-regulated pRIT44 *(41)* and T7 promoter-regulated *(46)* pT7ZZ vector systems *(44)*, the latter available in three reading frames, have found extensive use for high level production of various gene products in the cytoplasm of *E. coli*. These vectors can be obtained from the authors of this chapter.

3. Affinity Chromatography Using IgG-Sepharose

The affinity chromatography purification of SpA fusions has been shown to be extremely specific, resulting in a highly purified product after a single-step procedure. The yield in the chromatography step has been demonstrated to be very close to 100% *(33,35,36)*, and the fast kinetics of the binding allows sample loading onto the IgG-Sepharose columns with relatively high flowrates. It has also been demonstrated that the columns can be regenerated for repeated use up to 100 times *(8)*. The human polyclonal IgG used as ligand can be replaced by recombinant Fc fragments, avoiding the use of a human serum protein in the purification protocol *(47)*. Also alternative matrices to Sepharose can be used. For example, IgG-coated paramagnetic beads (Dynal, Oslo, Norway) could be suitable for certain applications.

3.1. Protocol for the Affinity Purification

1.
   a. If a secretion strategy is employed, collect samples containing the SpA-fusion products from the periplasm after an osmotic shock procedure *(48)*, or directly from the culture medium after sedimentation of the cells by centrifugation (10,000*g*).
   b. For intracellularly produced proteins, collect the samples from the supernatant of centrifugated (20,000*g*) cell desintegrates. If the product is precipitated in inclusion bodies, procedures to solubilize the precipitated proteins would be necessary (*see* Section 4.1.).
2. Load the samples directly onto IgG-Sepharose (Pharmacia) columns, previously equilibrated with a washing buffer (50 mM Tris-HCl, pH 7.4, 0.15M NaCl, 0.05% Tween-20). To avoid clogging of the columns, samples should preferably be filtered (0.45 μm) prior to loading.
3. Wash the columns with 5 mM ammonium acetate (pH 6.0) to lower the buffer capacity and remove salt before elution with 0.5M acetic acid, pH 3.3.

4. Fractions are collected, and the protein content is suitably estimated by $A_{280\,nm}$-measurements.

4. Affinity Purification of SpA-Fusions from Inclusion Bodies

Proteins of low solubility and proteins containing hydrophobic transmembrane regions, are extremely difficult to secrete *(43)*, and they often precipitate intracellularly into so-called inclusion or refractile bodies. For such proteins, an intracellular production strategy has to be used. Intracellular expression has become an increasingly attractive alternative owing to recent advances for in vitro renaturation of recombinant proteins from intracellular precipitates *(49)*. Production by the inclusion body strategy has the main advantages that the recombinant product normally is protected from proteolysis and that it can be produced in large quantities. Recently, it was demonstrated that affinity purification indeed can be useful also for the recovery of proteins with a strong tendency to precipitate during renaturation from inclusion bodies following standard protocols *(42)*. Murby and coworkers *(42)* developed an alternative recovery scheme (Fig. 2) for the production of ZZ fusions to various fragments of the fusion glycoprotein (F) from the human respiratory syncytial virus (RSV). Earlier attempts to produce these labile and precipitation-proned polypeptides in *E. coli* had failed, but several different ZZ-F fusions could be produced by this novel strategy and recovered as full-length products with substantial yields (20–50 mg/L). Since it was demonstrated that the IgG-Sepharose was resistant to 0.5M guanidine hydrochloride, efficient recovery from inclusion bodies of the ZZ-F fusions could be achieved by affinity chromatography on IgG-Sepharose in the presence of the chaotropic agent throughout the purification process *(42)*. In contrast, exclusion of guanidine hydrochloride during the procedure resulted in precipitation of the fusion proteins on the affinity column. The described strategy has so far been successfully evaluated for a number of proteins of low solubility and should be of interest for efficient recovery of other heterologous proteins that form inclusion bodies when expressed in a bacterial host.

4.1. Renaturation and Recovery of Proteins by a Modified Affinity Purification Scheme

1. Pellet insoluble material after sonication by centrifugation and recover precipitated intracellular proteins by an initial solubilization in 7M guanidine hydrochloride (Gua-HCl, Sigma, St. Louis, MO) and 25 mM Tris-HCl, pH 8.0 followed by incubation at 37°C for 2 h. When cysteines are present in the fusion proteins, include 10 mM dithiothreitol (DTT) or 100 mM β-mercaptoethanol.
2. Centrifuge the solubilization mixture and pipet the supernatant slowly into 100 mL of renaturation buffer containing 1M GuaHCl, 25 mM Tris-HCl, pH 8.0, 150 mM NaCl, and 0.05 % Triton X100.

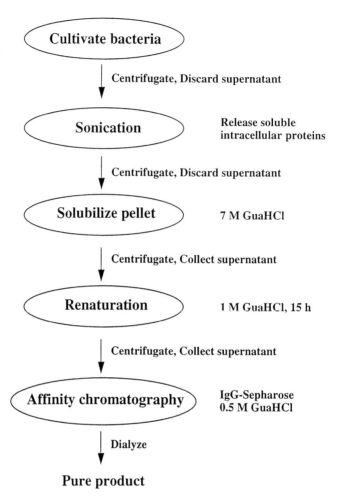

Fig. 2. Schematic presentation of the purification scheme developed by Murby and coworkers *(42)*, used for recovery by affinity chromatography on IgG-Sepharose of SpA fusion proteins, in cases where the fused target product has a strong tendency to precipitate.

3. Incubate the mixture at 4°C under slow stirring for 15 h.
4. Centrifuge and filter the renaturation mixture (0.45 μm) and apply to an affinity column containing 5 mL IgG-Sepharose (Pharmacia, Uppsala, Sweden) at 4°C at a low flowrate (0.5 mL/min). The IgG-Sepharose should previously be pulsed separately with TSTG buffer (50 mM Tris-HCl, pH 8.0, 200 mM NaCl, 0.05% Tween-20, 0.5 mM EDTA, and 0.5M GuaHCl) and 0.3M HAc, pH 3.3 containing 0.5M GuaHCl.

5. After sample loading, wash the column with 100 mL TSTG and 25 mL 5 m*M* ammonium acetate, pH 5.5 containing 0.5*M* GuaHCl.
6. Elute bound proteins with 20 mL 0.3*M* acetic acid, pH 3.3 with 0.5*M* GuaHCl.
7. Remove the chaotropic agent by dialysis against 2 L of 0.3*M* acetic acid, pH 3.3, twice.

5. Detection of Expressed SpA Fusion Proteins

There are several alternative methods to analyze the presence of protein A fusion proteins in a sample. One simple and yet very sensitive method takes advantage of the ability of SpA or ZZ to bind the Fc part of the IgG-molecule, and is very suitable for a modified immunoblotting procedure after protein analysis using polyacrylamide electrophoresis *(50)*.

5.1. PAP-Staining of Protein A Fusion Proteins

1. After protein analysis by SDS-PAGE, electrophoretically transfer proteins by electroblotting to nitrocellulose filters.
2. Block unspecific binding to the filter by incubation in 0.25% gelatin in PBS (50 m*M* phosphate, pH 7.1, 0.9% NaCl) for 30 min at 37°C before treatment with a soluble complex of rabbit antihorseradish peroxidase-IgG and horseradish peroxidase, PAP (Dakopatts, Copenhagen, Denmark), diluted 1:1000 in 0.25% gelatin in PBS for 30 min at 37°C.
3. Wash the filter twice in PBS and once in 50 m*M* Tris-HCl, pH 8.0.
4. The presence of horseradish peroxidase is detected by adding H_2O_2 and 3,3'-diaminobenzidine in 50 m*M* Tris-HCl, pH 8.0. By this procedure, SpA-containing protein bands are stained brown.

6. Competitive Elution of SpA Fusions for Mild Recovery

One of the drawbacks with the protein A system has earlier been the need for elution by low pH from the affinity column. The elution of fusion proteins by pH 3.0, used routinely, has for some products been destructive, yielding biologically inactive products. By a recently presented concept, the low pH-elution can be circumvented by competitive elution based on an engineered competitor protein which efficiently can be removed from the eluate mix *(51)*. The principle for the competitive elution strategy described by Nilsson and coworkers *(51)* is outlined in Fig. 3. The target protein is produced as a fusion to a monovalent Z-domain of SpA. A sample (e.g., crude cell lysate) containing the recombinant fusion Z-target is passed through an IgG-Sepharose column. The recombinant fusion protein is thus captured, enabling extensive washing to remove contaminants (Fig. 3A). A stoichiometric excess of an engineered bifunctional competitor fusion protein ZZ-ABP, where ABP is a second affinity tail (*see* Section 7.) derived from streptococcal protein G *(52)* capable of selective binding ($K_{aff} = 1.4 \times 10^9 M^{-1}$) to a human serum albumin (HSA). The

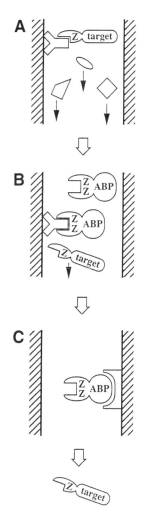

Fig. 3. General concept for the competitive-elution strategy.

competitor ZZ-ABP having a more than 10X higher affinity to human IgG than the monovalent Z, is used for competitive elution of the Z-fusion at neutral pH (Fig. 3B). Specific removal of ZZ-ABP from the effluent mix is accomplished by a passage through a second HSA affinity column *(52)*, to which the ZZ-ABP fusion is bound (Fig. 3C). The described strategy has been used to produce the Klenow fragment of *E. coli* DNA polymerase I as a Z-Klenow fusion, and the recovered product retained full enzymatic activity when recovered by the competitive elution strategy *(51)*.

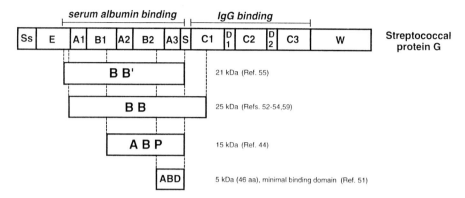

Fig. 4. A schematic illustration of streptococcal protein G (SpG) and a presentation of some existing variants of the ABP affinity tag. SpG is a bifunctional receptor (63 kDa) with the regions responsible for the two binding affinities structurally separated. SpG binds to serum albumins of different mammalian species, including human, mouse, and rat *(53)*. Three serum albumin binding motifs have been suggested *(56)* each being 46 amino acids in size. One of these postulated minimal motifs has been produced and tested for its serum albumin binding *(51)*, and is considered to be a suitable affinity tag, in a mono- or divalent fashion.

7. A Related Affinity Fusion Tag
Based on Streptococcal Protein G

A second affinity tag system, that has a number of features in common with the SpA systems but also some additional advantageous properties, is based on the serum albumin binding region of streptococcal protein G (SpG, Fig. 4) *(52,53)*. The affinity tag, denoted BB (or in recent publications ABP, for albumin binding protein) *(54,55)*, is proteolytically stable and highly soluble. A minimal binding motif (46 amino acids) has been postulated *(51,56)*. Expression vectors for both intracellular *(42–44)* and secreted production *(57)* of ABP-fusions have been developed. Fusions to the ABP-region can be efficiently affinity purified on human serum albumin (HSA)-Sepharose in a single-step procedure *(52,57)* in analogy with the SpA-IgG-system. The SpG-derived expression systems have found several applications. For example, the strong interaction between serum albumin and the ABP region has been shown to increase the in vivo circulation half-life of peptides when administered iv as ABP fusions *(58)*. Furthermore, a dual affinity fusion concept was developed. This strategy, in which proteins of interest were fused between the IgG-binding ZZ and the albumin binding ABP, has resulted in a considerable stabilization of numerous mammalian proteins that were unstable when produced as single fusions *(38,59)*. In addition, consecutive affinity purifications on IgG-

(using the ZZ fusion partner) and HSA-Sepharose (using the ABP fusion partner) yield highly purified full length fusion proteins *(38,59)*. Furthermore, ABP seems to have immunopotentiating properties when used as a carrier protein genetically fused to the immunogen used for immunization *(43,60)*. To date, it is not fully elucidated whether this capacity is a result of strong T-cell epitopes or related to the serum albumin binding activity giving prolonged in vivo half-life.

8. New Ligands Based on the Z Domain

The Z domain has several features including proteolytic stability, high solubility, small size, and compact and robust structure which makes it ideal as fusion partner for the production of recombinant proteins. The residues responsible for the Fc-binding activity have been identified from the crystal structure of the complex between its parent B-domain and human IgG Fc, showing that these residues are situated on the outer surfaces of two of the helices making up the domain structure and do not contribute to the packing of the core *(22)*. The Fc-binding surface covers an area of approx 800 $Å^2$, similar in size to the surfaces involved in many antigen-antibody interactions. Taken together, these data suggest that if random mutagenesis of this binding surface was performed, novel domains could be obtained with the possibility of finding variants capable of binding molecules other than Fc of IgG. In addition, these domains would theoretically share the overall stability and α-helical structure of the wild-type domain. Recently, an attempt to create such ligands with completely new affinities was initiated employing phage display technology *(61)* *(see also* Chapter 5 in companion volume). A genetic library was created in which 13 surface residues of the Z domain have been randomly substituted to all 20 possible amino acids. This Z library encoding approximately 10^{10} Z variants will, after genetic fusion to gene III of filamentous phage M13, be subjected to affinity-selection (panning) against a panel of different molecules for investigation as a source of novel binders *(61)*. Such novel binders might constitute a "next generation" of ligands or "artificial antibodies."

9. Conclusions

SpA and the SpA-derived Z domain have been used as affinity tags for the production and purification of numerous different peptides and proteins. Hundreds of such fusion proteins have been described in publications *(8–13)*. The use of SpA fusions is most widespread for research purposes, but the ZZ expression system can also be adapted for large-scale production. An example of large-scale production of insulin-like growth factor I (IGF-I) has been published *(33)* and production levels of 800 mg/L of excreted ZZ-IGF-I has been described *(62)*. When a secretion strategy can be used, the scale-up is simplified and a single-step production scheme in pilot scale have been presented *(35)*.

The SpA expression systems have been used extensively, often under the control of the SpA promoter/signal sequences, to produce various proteins of interest in biotechnology, medicine and immunology. Applications where the use of the SpA fusion system has proven to be advantageous include: the production and in vitro renaturation of peptide hormones *(20,26–29)*, immobilization and production of enzymes *(1,36,51)*, production of single chain antibodies *(63)*, expression of chimeric gene products to be used in different detection and recovery systems *(64,65)*, and the generation of immunogens for subunit vaccine research *(12,42,66–69)*.

Acknowledgments

All former and present members at the Department of Biochemistry and Biotechnology that have been involved in the development and evaluations of the SpA-expression systems are gratefully acknowledged. In particular we express our gratitude to B. Nilsson, T. Moks, L. Abrahmsén, and B. Jansson. We also thank M. Murby and J. Nilsson who have assisted in preparing protocols for this chapter.

References

1. Uhlén, M., Nilsson, B., Guss, B., Lindberg, M., Gatenbeck, S., and Philipson, L. (1983) Gene fusion vectors based on the gene for staphylococcal protein A. *Gene* **23,** 369–378.
2. Hammarberg, B., Moks, T., Tally, M., Elmblad, A., Holmgren, E., Murby, M., Nilsson, B., Josephson, S., and Uhlén, M. (1990) Differential stability of recombinant human insulin-like growth factor II in *Escherichia coli* and *Staphylococcal aureus. J. Bacteriol.* **14,** 423–438.
3. Vasantha, N. and Thompson, L. D. (1986) Fusions of pro region of subtilisin to staphylococcal protein A and its secretion by *Bacillus subtilis. Gene* **49,** 23–28.
4. Stirling, D. A., Petrle, A., Pulford, D. J., Paterson, D. T. W., and Stark, M. J. R. (1992) Protein A-calmodulin fusions: a novel approach for investigating calmodulin function in yeast. *Molec. Microbiol.* **6,** 703–713.
5. Maeda, Y., Veda, H., Hara, T., Kazami, J., Kawano, G., Suzuki, E., and Nagamune, T. (1996) Expression of a bifunctional chimeric protein A-Vargula hilgendorff: luciferase in mammalian cells. *BioTechniques* **20,** 116–121.
6. Andersons, D., Engström, Å., Josephson, S., Hansson, L., and Steiner, H. (1991) Biologically active and amidated cecropin produced in a baculovirus expression system from a fusion construct containing the antibody-binding part of protein A. *Biochem. J.* **280,** 219–224.
7. Hightower, R., Baden, C., Penzes, E., Lund, P., and Dunsmuir, P. (1991) Expression of antifreeze proteins in transgenic plants. *Plant. Mol. Biol.* **17,** 1013–1021.
8. Uhlén, M. and Moks, T. (1990) Gene fusions for purpose of expression: an introduction. *Methods Enzymol.* **185,** 129–143.

9. Nilsson, B. and Abrahmsén, L. (1990) Fusions to staphylococcal protein A. *Methods Enzymol.* **185,** 144–161.
10. Uhlén, M., Forsberg, G., Moks, T., Hartmanis, M., and Nilsson, B. (1992) Fusion proteins in biotechnology. *Curr. Opin. Biotechnol.* **3,** 363–369.
11. Ståhl, S., Nygren, P.-Å., Sjölander, A., and Uhlén, M. (1993) Engineered bacterial receptors in immunology. *Curr. Opin. Immunol.* **5,** 272–277.
12. Sjölander, A., Ståhl, S., and Perlmann, P. (1993) Bacterial expression systems based on protein A and protein G designed for the production of immunogens: applications to *Plasmodium falciparum* malaria antigens. *Immunomethods* **2,** 79–92.
13. Nygren, P.-Å., Ståhl, S., and Uhlén, M. (1994) Engineering proteins to facilitate bioprocessing. *Trends Biotechnol.* **12,** 184–188.
14. Langone, J. J. (1982) Protein A of *Staphylococcus aureus* and related immunoglobulin binding receptors produced by streptococci and pneumococci. *Adv. Immunol.* **32,** 157–252.
15. Moks, T., Abrahmsén, L., Nilsson, B., Hellman, U., Sjöquist, J., and Uhlén, M. (1986) Staphylococcal protein A consists of five IgG-binding domains. *Eur. J. Biochem.* **156,** 637–643.
16. Guss, B., Uhlén, M., Nilsson, B., Lindberg, M., Sjöquist, J., and Sjödahl, J. (1984) Region X, the cell-wall-attachment part of staphylococcal protein A. *Eur. J. Biochem.* **138,** 413–420.
17. Schneewind, O., Model, P., and Fischetti, V. A. (1992) Sorting of protein A to the staphylococcal cell wall. *Cell* **70,** 267–281.
18. Schneewind, O., Mihaylova-Petkov, D., and Model, P. (1993) Cell wall sorting signals in surface proteins of Gram-positive bacteria. *EMBO J.* **12,** 4803–4811.
19. Eliasson, M., Andersson, R., Olsson, A., Wigzell, H., and Uhlén, M. (1989) Differential IgG-binding characteristics of staphylococcal protein A, streptococcal protein G and a chimeric protein AG. *J. Immunol.* **142,** 575–581.
20. Moks, T., Abrahmsén, L., Holmgren, E., Bilich, M., Olsson, A., Uhlén, M., Pohl, G., Sterky, C., Hultberg, H., Josephson, S., Holmgren, A., Jörnvall, H., and Nilsson, B. (1987) Expression of human insulin-like growth factor I in bacteria: use of optimized gene fusion vectors to facilitate protein purification. *Biochemistry* **26,** 5239–5244.
21. Nilsson, B., Holmgren, E., Josephson, S., Gatenbeck, S., Philipson, L., and Uhlén, M. (1985) Efficient secretion and purification of human insulin-like growth factor I with a gene fusion vector in *Staphylococci.* *Nucl. Acids Res.* **13,** 1151–1162.
22. Deisenhofer, J. (1981) Crystallographic refinement and atomic model of a human Fc fragment and its complex with fragment B of protein A from *Staphylococcus aureus* at 2.9 and 2.8 Å resolution. *Biochemistry* **20,** 2361–2370.
23. Torigoe, H., Shimada, I., Saito, A., Sato, M., and Arata, Y. (1990) Sequential ^{1}H NMR assignments and secondary structure of the B domain of staphylococcal protein A: Structural changes between the free B domain in solution and the Fc-bound B domain in crystal. *Biochemistry* **29,** 8787–8793.

24. Gouda, H., Torigoe, H., Saito, A., Sato, M., Arata, Y., and Shimada, I. (1992) Three-dimensional solution structure of the B domain of staphylococcal protein A: comparison of the solution and crystal structure. *Biochemistry* **31**, 9665–9672.

25. Lyons, B. A., Tashiro, M., Cedergren, L., Nilsson, B., and Montelione, G. (1993) An improved strategy for determining resonance assignments for isotypically enriched proteins and its application to an engineered domain of staphylococcal protein A *Biochemistry* **32**, 7839–7845.

26. Forsberg, G., Baastrup, B., Rondahl, H., Holmgren, E., Pohl, G., Hartmanis, M., and Lake, M. (1992) An evaluation of different cleavage methods for recombinant fusion proteins, applied on des(1-3)insulin-like growth factor I. *J. Prot. Chem.* **11**, 201–211.

27. Samuelsson, E., Wadensten, H., Hartmanis, M., Moks, T., and Uhlén, M. (1991) Facilitated *in vitro* refolding of human recombinant insulin-like growth factor I using a solubilizing fusion partner. *Bio/Technol.* **9**, 363–366.

28. Samuelsson, E., Moks, T., Nilsson, B., and Uhlén, M. (1994) Enhanced *in vitro* refolding of insulin-like growth factor I using a solubilizing fusion partner. *Biochemistry* **33**, 4207–4211.

29. Forsberg, G., Samuelsson, E., Wadensten, H., Moks, T., and Hartmanis, M. (1992) Refolding of human recombinant insulin-like growth factor II (IGF-II) *in vitro* using a solubilizing affinity handle, in *Techniques in Protein Chemistry* (Angeletti, R. H., ed.), Academic Press, San Diego, CA, pp. 329–336.

30. Nilsson, B., Moks, T., Jansson, B., Abrahmsén, L., Elmblad, A., Holmgren, E., Henrichson, C., Jones, T. A., and Uhlén, M. (1987) A synthetic IgG-binding domain based on staphylococcal protein A. *Prot. Eng.* **1**, 107–113.

31. Abrahmsén, L., Moks, T., Nilsson, B., Hellman, U., and Uhlén, M. (1985) Analysis of signals for secretion in the staphyloccal protein A gene. *EMBO J.* **4**, 3901–3906.

32. Abrahmsén, L., Moks, T., Nilsson, B., and Uhlén, M. (1986) Secretion of heterologous gene products to the culture medium of *Escherichia coli. Nucl. Acids Res.* **14**, 7487–7500.

33. Moks, T., Abrahmsén, L., Österlöf, B., Josephson, S., Östling, M., Enfors, S.-O., Persson, I., Nilsson, B., and Uhlén, M. (1987) Large-scale affinity purification of human insulin-like growth factor I from culture medium of *Escherichia coli. Bio/Technol.* **5**, 379–382.

34. Ståhl, S., Sjölander, A., Hansson, M., Nygren, P.-Å., and Uhlén, M. (1990) A general strategy for polymerization, assembly and expression of epitope-carrying peptides applied on the *Plasmodium falciparum* antigen Pf155/RESA. *Gene* **89**, 187–193.

35. Hansson, M., Ståhl, S., Hjorth, R., Uhlén, M., and Moks, T. (1994) Single-step recovery of a secreted recombinant protein by expanded bed adsorption. *Bio/Technol.* **12**, 285–288.

36. Nilsson, B., Abrahmsén, L., and Uhlén, M. (1985) Immobilization and purification of enzymes with staphylococcal protein A gene fusion vectors. *EMBO J.* **4**, 1075–1080.

37. Ljungquist, C., Jansson, B., Moks, T., and Uhlén, M. (1989) Thiol-directed immobilization of recombinant IgG-binding receptors. *Eur. J. Biochem.* **186,** 557–561.
38. Murby, M., Cedergren, L., Nilsson, J., Nygren, P.-Å., Hammarberg, B., Nilsson, B., Enfors, S.-O., and Uhlén, M. (1991) Stabilization of recombinant proteins from proteolytic degradation in *Escherichia coli* using a dual affinity fusion strategy. *Biotechnol. Appl. Biochem.* **14,** 336–346.
39. Carter, P., Kelley, R. F., Rodrigues, M. L., Snedecor, B., Covarrubias, M., Velligan, M. D., Wong, W. L. T., Rowland, A. M., Kotts, C. E., Carver, M. E., Yang, M., Bourell, J.H., Shepard, H. M., and Henner, D. (1992) High level *Escherichia coli* expression and production of a bivalent humanized antibody fragment. *Bio/Technol.* **10,** 163–167.
40. Berzins, K., Adams, S., Broderson, J. R., Chizzolini, C., Hansson, M., Lövgren, K., Millet, P., Morris, C. L., Perlmann, H., Perlmann, P., Sjölander, A., Ståhl, S., Sullivan, J. S., Troye-Blomberg, M., Wåhlin-Flyg, B., and Collins, W. E. (1995) Immunogenicity in *Aotus* monkeys of ISCOM formulated repeat sequences from the *Plasmodium falciparum* asexual blood stage antigen Pf155/RESA. *Vaccine Res.* **4,** 121–133.
41. Köhler, K., Ljungquist, C., Kondo, A., Veide, A., and Nilsson, B. (1991) Engineering proteins to enhance their partitioning coefficients in Aqeous two-phase systems. *Bio/Technol.* **9,** 642–646.
42. Murby, M., Nguyen, T. N., Binz, H., Uhlén, M., and Ståhl, S. (1994) Production and recovery of recombinant proteins of low solubility, in *Separations for Biotechnology 3* (Pyle, D. L., ed.), Bookcraft, Bath, UK, pp. 336–344.
43. Sjölander, A., Ståhl, S., Lövgren, K., Hansson, M., Cavelier, L., Walles, A., Helmby, H., Wåhlin, B., Morein, B., Uhlén, M., Berzins, K., Perlmann, P., and Wahlgren, M. (1993) *Plasmodium falciparum:* The immune response in rabbits to the clustered asparagine-rich protein (CARP) after immunization in Freund's adjuvant or immunostimulating complexes (ISCOMs). *Exp. Parasitol.* **76,** 134–145.
44. Larsson, M., Brundell, E., Nordfors, L., Höög, C., Uhlén, M., and Ståhl, S. (1996) A general bacterial expression system for functional analysis of cDNA-encoded proteins. *Prot. Exp. Purif.,* in press.
45. Löwenadler, B., Jansson, B., Paleus, S., Holmgren, E., Nilsson, B., Moks, T., Palm, G., Josephson, S., Philipson, L., and Uhlén, M. (1987) A gene fusion system for generating antibodies against short peptides. *Gene* **58,** 87–97.
46. Studier, F. W., Rosenberg, A. H., Dunn, J. J., and Dubendorff, J. W. (1990) Use of T7 RNA polymerase to direct expression of cloned genes. *Methods Enzymol.* **185,** 60–89.
47. Nygren, P.-Å. and Uhlén, M. Unpublished data.
48. Nossal, N. G. and Heppel, L. A. (1966) The release of enzymes by osmotic shock from *Escherichia coli* in exponential phase. *J. Biol. Chem.* **241,** 3055–3062.
49. Rudolph, R. (1995) Successful protein folding on an industrial scale, in *Principles and Practice of Protein Folding* (Cleland, J. L. and Craik, C. S., eds.), John Wiley and Sons, New York, pp. 283–298.
50. Jansson, B., Palmcrantz, C., Uhlén, M., and Nilsson, B. (1990) A dual-affinity gene fusion system to express small recombinant proteins in a soluble form: expression and characterization of protein A deletion mutants. *Prot. Eng.* **2,** 555–561.

51. Nilsson, J., Nilsson, P., Williams, Y., Pettersson, L., Uhlén, M., and Nygren, P.-Å. (1994) Competitive elution of protein A fusion proteins allows specific recovery under mild conditions. *Eur. J. Biochem.* **224,** 103–108.

52. Nygren, P.-Å., Eliasson, M., Palmcrantz, E., Abrahmsén, L., and Uhlén, M. (1988) Analysis and use of the serum albumin binding domains of streptococcal protein G. *J. Mol. Recognit.* **1,** 69–74.

53. Nygren, P.-Å., Ljungquist, C., Trömborg, H., Nustad, K., and Uhlén, M. (1990) Species-dependent binding of serum albumins to the streptococcal receptor protein G. *Eur. J. Biochem.* **193,** 143–148.

54. Murby, M., Samuelsson, E., Nguyen, T., Mignard, L., Power, U., Binz, H., Uhlén, M., and Ståhl, S. (1995) Hydrophobicity engineering to increase solubility and stability of a recombinant protein from respiratory syncytial virus. *Eur. J. Biochem.* **230,** 38–44.

55. Samuelson, P., Hansson, M., Ahlborg, N., Andréoni, C., Götz, F., Bächi, T., Nguyen, T. N., Binz, H., Uhlén, M., and Ståhl, S. (1995) Cell-surface display of recombinant proteins on *Staphylococcus carnosus. J. Bacteriol.* **177,** 1470–1476.

56. Nygren, P.-Å. (1992) Characterization and use of the serum albumin binding region of streptococcal protein G. Doctoral Thesis (ISBN 91-7170-087-0), Royal Institute of Technology, Stockholm, Sweden.

57. Ståhl, S., Sjölander, A., Nygren, P.-Å., Berzins, K., Perlmann, P., and Uhlén, M. (1989) A dual expression system for the generation, analysis and purification of antibodies to a repeated sequence of the *Plasmodium falciparum* malaria antigen Pf155/RESA. *J. Immunol. Meth.* **124,** 43–52.

58. Nygren, P.-Å., Flodby, P., Andersson, R., Wigzell, H., and Uhlén, M. (1991) *In vivo* stabilization of a human recombinant CD4 derivative by fusion to a serum-albumin-binding receptor, in *Vaccines 91* (Chanock, R. M., Ginsberg, H. S., Brown, F., and Lerner, R. A., eds.), Cold Spring Harbor Laboratory Press, Cold Spring Harbor, NY, pp. 363–368.

59. Hammarberg, B., Nygren, P.-Å., Holmgren, E., Elmblad, A., Tally, M., Hellman, U., Moks, T., and Uhlén, M. (1989) Dual affinity fusion approach and its use to express recombinant insulin-like growth factor II. *Proc. Natl. Acad. Sci. USA* **86,** 4367–4371.

60. Sjölander, A., Andersson, R., Nygren, P.-Å., and Ståhl, S. (1995) Unpublished data.

61. Nord, K., Nilsson, J., Nilsson, B., Uhlén, M., and Nygren, P.-Å. (1995) A combinatorial library of an a-helical bacterial receptor domain. *Prot. Eng.* **6,** 601–608.

62. Josephson, S. and Bishop, R. (1988) Secretion of peptides from *E. coli*: a production system for the pharmaceutical industry. *Trends Biotechnol.* **6,** 218–224.

63. Tai, M. S., Mudgett-Hunter, M., Levinson, D., Wu, G. M., Haber, E., Oppermann, H., and Huston, J. S. (1990) A bifunctional fusion protein containing Fc-binding fragment B of staphylococcal protein A amino terminal to antidigoxin single-chain Fv. *Biochemistry* **29,** 8024–8030.

64. Sano, T., Smith, C., and Cantor, C. R. (1992) Immuno-PCR: very sensitive antigen detection by means of specific antibody-DNA conjugates. *Science* **258,** 120–122.

65. Ljungquist, C., Lundeberg, J., Rasmussen, A.-M., Hornes, E., and Uhlén, M. (1993) Immobilization and recovery of fusion proteins and B-lymphocyte cells using magnetic separation. *DNA Cell Biol.* **12,** 191–197.
66. Sjölander, A., Ståhl, S., Nygren, P.-Å., Åslund, L., Ahlborg, N., Wåhlin, B., Scherf, A., Berzins, K., Uhlén, M., and Perlmann, P. (1990) Immunogenicity and antigenicity in rabbits of a repeated sequence of the *Plasmodium falciparum* antigen Pf155/RESA fused to two IgG-binding domains of stahylococcal protein A. *Infect. Immun.* **58,** 854–859.
67. Löwenadler, B., Svennerholm, A.-M., Gidlund, M., Holmgren, E., Krook, K., Svanholm, C., Ulff, S., and Josephson, S. (1990) Enhanced immunogenicity of recombinant peptide fusions containing multiple copies of a heterologous T helper epitope. *Eur. J. Immunol.* **20,** 1541–1545.
68. Lepage, P., Heckel, C., Humbert, S., Ståhl, S., and Rautmann, G. (1993) Recombinant technology as alternative to chemical peptide synthesis: Expression and characterisation of HIV1-REV recombinant peptides. *Anal. Biochem.* **213,** 40–48.
69. Ahlborg, N., Iqbal, J., Hansson, M., Uhlén, M., Perlmann, P., Ståhl, S., and Berzins, K. (1995) Immunogens containing antigen Pf332 sequences induce *Plasmodium falciparum* reactive antibodies inhibiting parasite growth but not cytoadherence. *Parasite Immunol.* **17,** 341–352.

11

Expression and Purification of Recombinant Streptavidin-Containing Chimeric Proteins

Takeshi Sano, Cassandra L. Smith, and Charles R. Cantor

1. Introduction

Streptavidin, a protein produced by *Streptomyces avidinii*, binds a water-soluble vitamin, D-biotin (vitamin H), with remarkably high affinity *(1,2)*. The dissociation constant of the streptavidin-biotin complex is approx $10^{-15}M$; the binding of streptavidin to biotin is one of the strongest noncovalent interactions found in biological systems. The extremely tight and specific biotin-binding ability of streptavidin has made this protein a very powerful biological tool for a variety of biological and biomedical analyses *(3,4)*. The ability of biotin to be incorporated easily into various biological materials has also expanded the application of the streptavidin-biotin technology to a wider range of biological systems.

Several years ago, the streptavidin gene was cloned from a genomic library of *S. avidinii (5)*. The isolation of the streptavidin gene allowed the development of an efficient bacterial expression system for streptavidin *(6)*, which is based on the bacteriophage T7 expression system *(7)*. The establishment of the expression and purification methods for recombinant streptavidin allowed the design and production of streptavidin-containing chimeric proteins. Such streptavidin-containing chimeras, if successfully produced, should have extremely tight and specific biotin-binding ability derived from the streptavidin moiety; this should provide the fused partner protein with highly specific, strong biological recognition capability that would allow, for example, specific conjugation, labeling, and targeting of the fused partner protein to any biological material containing biotin. Thus, such streptavidin-containing chimeras could serve as new, powerful biological tools in a variety of molecular

From: *Methods in Molecular Biology, vol. 63: Recombinant Protein Protocols: Detection and Isolation* Edited by: R. Tuan Humana Press Inc., Totowa, NJ

and cellular analyses, thereby further expanding the efficacy and versatility of the streptavidin-biotin technology.

This article describes general methods for expressing recombinant streptavidin-partner protein chimeras in *Escherichia coli* and purifying the expressed chimeric proteins. Although these methods have been used successfully for a few streptavidin-containing chimera constructs, some conditions and procedures may need to be modified or optimized for a particular partner protein fused to streptavidin to maximize the production of functional chimera molecules.

2. Materials

2.1. Construction of a Gene Fusion of Streptavidin with a Partner Protein

Construction of a gene fusion of streptavidin with a partner protein is carried out by using standard molecular cloning methods and procedures. *See* general molecular cloning manuals (e.g., refs. *8* and *9*) for necessary materials.

2.2. Expression of a Streptavidin-Partner Protein Chimera in E. coli

1. Erlenmeyer flasks or equivalent, suitable for culture of *E. coli*.
2. A shaking incubator at 37°C, suitable for culture of *E. coli*.
3. A spectrophotometer or equivalent, suitable for monitoring the growth of *E. coli*.
4. Host *E. coli* lysogens, BL21(DE3)(pLysS) and BL21(DE3)(pLysE) *(7)*. These lysogen strains are commercially available from Novagen, Madison, WI.
5. M9 minimal medium: 6.0 g Na_2HPO_4, 3.0 g KH_2PO_4, 0.5 g NaCl, 1.0 g NH_4Cl/L), supplemented with 1 m*M* $MgSO_4$, 0.2% glucose, 1.5 μ*M* thiamine (vitamin B1), 0.5% Casamino acids, 8.2 μ*M* (2 μg/mL) biotin, 150 μg/mL ampicillin, and 25 μg/mL chloramphenicol. Supplementary materials are added to M9 minimal medium at the time of use from stock solutions: 1.0*M* $MgSO_4$; 20% glucose; 15 m*M* thiamine; 20% Casamino acids; 3.25 mM (800 μg/mL) biotin in 90% ethanol; 25 mg/mL ampicillin; and 34 mg/mL chloramphenicol in ethanol.
6. 100 m*M* Isopropyl-β-D-thiogalactopyranoside (IPTG) in water.
7. *See* refs. *8* and *9* for other materials commonly used for culture of *E. coli*.

2.3. Purification of an Expressed Streptavidin-Partner Protein Chimera

1. A centrifuge and centrifuge tubes that can be used at a centrifugal force of 18,000*g* or greater.
2. 100 m*M* NaCl, 10 m*M* Tris-HCl (pH 8.0), 1 m*M* EDTA.
3. 2 m*M* EDTA, 30 m*M* Tris-HCl (pH 8.0), 0.1% Triton X-100.
4. 1.0*M* $MgSO_4$.

5. 10 mg/mL DNase I (deoxyribonuclease I) in 150 mM NaCl, 50% glycerol, stored at –20°C.
6. 10 mg/mL RNase A (ribonuclease A) in 15 mM NaCl, 10 mM Tris-HCl, pH 7.5, 50% glycerol, stored at –20°C.
7. 7M Guanidine hydrochloride, pH 1.5.
8. 0.2M Ammonium acetate (pH 6.0), 0.02% NaN$_3$.
9. 1.0M NaCl, 50 mM sodium carbonate (pH 11.0), 0.02% NaN$_3$.
10. 2-Iminobiotin agarose. 2-Iminobiotin agarose is commercially available from several manufacturers, including Sigma, St. Louis, MO, and Pierce, Rockford, IL. The resin is packed in a column of an appropriate size. The resin can be used at least several times without appreciable loss of the binding capacity for streptavidin. After each use, the resin is washed with 6M urea, 50 mM ammonium acetate, pH 4.0, 0.02% NaN$_3$ to remove any biological materials. Then, the resin is equilibrated with a neutral solution, such as 0.5M NaCl, 20 mM Tris-HCl, pH 7.5, 0.02% NaN$_3$, and stored at 4°C.
11. 50 mM ammonium acetate, pH 4.0, 0.02% NaN$_3$ containing urea at a concentration of 50 mM–6M. The urea concentration can be varied dependent of the stability of a partner protein in urea. If the partner protein is unstable in urea, 0.5M NaCl, 50 mM ammonium acetate, pH 4.0, 0.02% NaN$_3$ without urea can be used with reduced recoveries.
12. Dialysis membrane tubing. Any commonly used dialysis membrane tubing, such as regenerated natural cellulose, with a molecular mass cut-off of 12 kDa or smaller can be used.

3. Methods

3.1. Construction of a Gene Fusion of Streptavidin with a Partner Protein

Expression vectors have been designed that are useful for the construction of a gene fusion of streptavidin with a partner protein (Fig. 1) *(10)*. These fusion vectors, pTSA-18F and pTSA-19F, carry a truncated streptavidin gene, encoding only the essential region of streptavidin (amino acid residues 16–133), with a translation initiation codon (ATG) under the control of the $\Phi10$ promoter of bacteriophage $T\Phi$, followed by the $T\Phi$ transcription terminator *(7)*. A poly-linker region, derived from pUC19, is placed downstream of the coding sequence for streptavidin, so that a gene fusion encoding a chimera consisting of the N-terminal streptavidin moiety and the C-terminal partner protein moiety can be easily constructed. Alternatively, the 3'-terminus of a coding sequence for a partner protein can also be fused to the 5'-terminus of the truncated streptavidin gene to produce a chimeric protein, consisting of the N-terminal partner protein moiety and the C-terminal streptavidin moiety, by inserting the partner protein gene in frame into a unique *Nde*I site located at the translation initiation site.

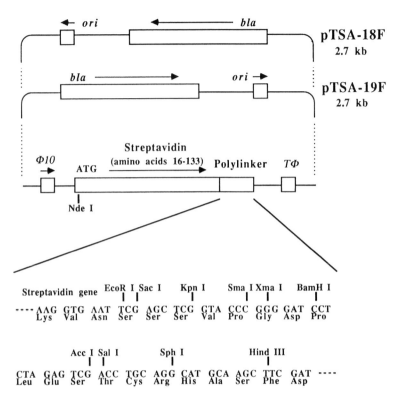

Fig. 1. Expression vectors, pTSA-18F and pTSA-19F, for streptavidin-containing chimeric proteins. These vectors carry a coding sequence for only the essential region of streptavidin (amino acid residues 16–133) with a translation initiation codon (ATG) under the control of the *Φ10* promoter of bacteriophage T7. A polylinker region, derived from pUC19, is placed downstream of the coding sequence. This polylinker region greatly facilitates the construction of gene fusions of streptavidin with partner proteins. (Reproduced with permission from ref. *10*; copyright, Academic, Orlando, FL, 1991.)

Standard molecular cloning methods and procedures are used for construction of a gene fusion of streptavidin with a partner protein *(8,9)*. The resulting expression vector should be stably maintained in *E. coli* strains commonly used for cloning, such as DH5α, unless they carry the T7 RNA polymerase gene or its fragment.

The expression vector encoding a streptavidin-partner protein chimera is isolated from host *E. coli* cells by using standard plasmid minipreparation methods *(8,9)*. The isolated expression vector is used to transform *E. coli* lysogens, BL21(DE3)(pLysS) and BL21(DE3)(pLysE), which carry the T7

RNA polymerase gene under the *lac*UV5 promoter in the chromosome. These lysogen strains also contain an additional plasmid, pLysS or pLysE, which carries the T7 lysozyme gene under the *tet* promoter and the chloramphenicol acetyltransferase (CAT) gene. Constitutive production of T7 lysozyme, which is a natural inhibitor for T7 RNA polymerase and is used to reduce the basal T7 RNA polymerase activity in an uninduced state *(7)*, is essential to maintain toxic genes, such as the streptavidin gene, stably in host BL21(DE3) lysogens.

3.2. Expression
of a Streptavidin-Partner Protein Chimera in E. coli

Expression of a gene fusion encoding a streptavidin-partner protein chimera is carried out by using the T7 expression system *(7)*, which allows efficient production of various streptavidin mutants and streptavidin-containing chimeras. The procedure described below is written for a culture volume of 100 mL and can be scaled up as needed.

1. BL21(DE3)(pLysS) or BL21(DE3)(plLysE) carrying an expression vector encoding a streptavidin-partner protein chimera is grown overnight at 37°C with vigorous shaking in 5 mL of M9 minimal medium supplemented with 1 mM MgSO$_4$, 0.2% glucose, 1.5 μM thiamine, 0.5% Casamino acids, 8.2 μM biotin, 150 μg/mL ampicillin, and 25 μg/mL chloramphenicol.
2. Approximately 2 mL of fresh culture of BL21(DE3)(pLysS) or BL21(DE3) (pLyse) carrying the expression vector is added to 100 mL of M9 minimal medium supplemented with 1 mM MgSO$_4$, 0.2% glucose, 1.5 μM thiamine, 0.5% Casamino acids, 8.2 μM biotin, 150 μg/mL ampicillin, and 25 μg/mL chloramphenicol. Cells are grown at 37°C with vigorous shaking.
3. When the absorbance of the culture reaches 1.0 for cells carrying pLysS or 0.6 for those carrying pLyse, 100 mM isopropyl-β-D-thiogalactopyranoside (IPTG) is added to a final concentration of 0.4–0.5 mM to induce the expression of the T7 RNA polymerase gene placed under the *lac*UV5 promoter. After induction, cells are incubated at 37°C with vigorous shaking.

Expression of the encoded streptavidin-partner protein is analyzed by SDS-PAGE (sodium dodecyl sulfate-polyacrylamide gel electrophoresis) *(11)* of total cell protein (*see* Fig. 2). A part of the culture (e.g., 500 μL) is collected periodically in a microcentrifuge tube, and cells are precipitated by brief centrifugation. The supernatant is discarded, and the cell pellet is dissolved in a sample solution for SDS-PAGE, such as 3% SDS, 100 mM Tris-Cl (pH 6.8), 30% glycerol, 0.01% bromophenol blue. The dissolved sample is heated in boiling water for 3 min, and applied to a polyacrylamide gel of an appropriate polyacrylamide concentration. Total cell protein from approx 100 μL of culture is applied to each lane, but appropriate amounts need to be found empiri-

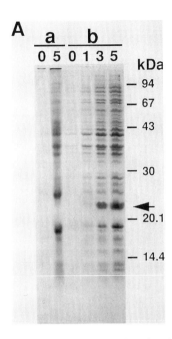

Fig. 2. Expression of a streptavidin-metallothionein chimera in *E. coli* Total cell protein of *E. coli* BL21(DE3)(pLysE) carrying an expression vector, pTSAMT-2, encoding a streptavidin-metallothionein chimera was analyzed by SDS-PAGE using a 15% polyacrylamide gel. Proteins were stained with Coomassie brilliant blue R-250. Lanes a, BL21(DE3)(pLysE) without the expression vector; b, BL21(DE3)(pLysE) (pTSAMT-2). The expressed streptavidin-metallothionein chimera is indicated by an arrow. The number above each lane is the time in hours after induction. Each lane contains the total cell protein from 167 μL of culture, except for the lane at 5 h for lanes a, where 83 μL of culture was used. (Reproduced with permission from ref. *14*.)

cally according to the growth of host cells and gel electrophoresis apparatus used. Proteins are stained with Coomassie brilliant blue R-250. Usually, expressed chimera is the dominant species in total cell protein and thus can be easily identified.

Generally, expression of a streptavidin-partner protein chimera reaches a maximum at 2–3 h after induction for BL21(DE3)(pLysS) and at around 4 h after induction for the equivalent strain carrying pLysE. Expression efficiencies are generally higher with BL21(DE3)(pLysE) than with BL21(DE3) (pLysS). However, degradation of expressed chimera by cellular proteinases is generally greater with BL21(DE3)(pLysE). Thus, for a particular streptavidin-partner protein chimera, both strains should be first tested to see which strain has higher expression efficiency and provides a higher yield of purified, functional chimera molecules.

3.3. Purification of an Expressed Streptavidin-Partner Protein Chimera

A streptavidin-partner protein chimera, expressed in *E. coli* by using the T7 expression system, generally forms inclusion bodies. Thus, denaturation of the expressed chimera, followed by renaturation, is needed to obtain functional chimera molecules. Most steps are performed at 4°C or on ice to prevent the denaturation and proteolysis of the chimera molecule at higher temperatures, though the streptavidin moiety itself is extremely stable once it is folded. The final purification step is affinity chromatography using 2-iminobiotin as a ligand *(12)*. 2-Iminobiotin binds to streptavidin reversibly in a pH-dependent manner; this avoids the need for harsh conditions to dissociate bound streptavidin-partner protein chimeras from affinity resins. If specific purification methods are available for the partner protein, e.g., affinity chromatography using a ligand of the partner protein, they should be used in combination with 2-iminobiotin affinity chromatography to obtain chimera molecules in which both the streptavidin and the partner protein moieties are functional. The procedure, described below, is written for purification of a streptavidin-partner protein chimera expressed in 100 mL of culture.

1. *E. coli* host cells, which have been incubated for an appropriate time after induction, are centrifuged at 2900g for 20 min, and the supernatant is discarded.
2. (Optional) The cell pellet is washed briefly with 20 mL of 100 mM NaCl, 10 mM Tris-HCl, pH 8.0, 1 mM EDTA, followed by centrifugation as above.
3. The cell pellet is suspended in approx 10 mL of 2 mM EDTA, 30 mM Tris-HCl, pH 8.0, 0.1% Triton X-100 with vigorous shaking to lyse host cells. This cell lysate can generally be stored at −70°C until used.
4. (Optional) The cell lysate is subjected to a few freeze-thaw cycles to achieve complete cell lysis. Because of the presence of T7 lysozyme in host cells, lysis is usually efficient even without freeze-thaw cycles.
5. To the cell lysate, 1.0M MgSO$_4$, 10 mg/mL DNase I, and 10 mg/mL RNase A are added to final concentrations of 12 mM, 10 μg/mL, and 10 μg/mL, respectively. The mixture is incubated at room temperature for 20–30 min. During incubation, the viscosity of the cell lysate is significantly reduced due to the digestion of chromosomal DNA derived from *E. coli* host cells.
6. The cell lysate is centrifuged at 39,000g for 20 min, and the supernatant (soluble fraction) is discarded. Generally, expressed streptavidin-partner protein chimeras are in the precipitate (insoluble) fraction, as they form inclusion bodies in host cells.
7. The precipitate (insoluble) fraction is dissolved in 5 mL of 7M guanidine hydrochloride (pH 1.5) with vigorous shaking. If needed, the suspension can be heated

at, for example, 37°C to facilitate solubilization. If the partner protein is unstable in $7M$ guanidine hydrochloride at pH 1.5, the conditions may be changed; for example, the concentration of guanidine hydrochloride is reduced and the pH is raised to neutral. However, it is important that all insoluble materials are fully dissolved.

8. (Optional) Dissolved fractions are dialyzed against the same solution, i.e., $7M$ guanidine hydrochloride (pH 1.5), for several hours to remove cellular biotin, if any. This step slightly increases the final yield.

9. The dialyzed crude protein fraction is diluted with $7M$ guanidine hydrochloride (pH 1.5) to 100 mL (= the original culture volume).

10. The diluted crude protein solution is dialyzed against $0.2M$ ammonium acetate (pH 6.0), 0.02% NaN$_3$ to remove guanidine hydrochloride and renature the chimera. It is of key importance that guanidine hydrochloride is removed very slowly to maximize the final yield. Fast removal of guanidine hydrochloride causes the chimera to form aggregates; this reduces the final yield of the functional chimera.

11. The dialyzed fraction is centrifuged at 39,000g for 20 min, and the supernatant is collected. If needed, the dialyzed fraction can also be filtered through a 0.45-μm filter, such as Millipore's (Bedford, MA) Durapore membrane, to remove insoluble material.

12. The supernatant (soluble) fraction is dialyzed against $1.0M$ NaCl, 50 mM sodium carbonate (pH 11.0), 0.02% NaN$_3$. Make sure that the final pH of the dialyzed fraction has reached 10.5 or higher before proceeding to the next step. If not, the pH should be adjusted with NaOH. Alternatively, solid NaCl and sodium carbonate are added directly to the supernatant fraction and dissolved to final concentrations of approximately $1.0M$ and 50 mM, respectively. Then, the pH of the crude protein solution is adjusted to 11.0 with NaOH. There should be no insoluble materials present in the crude protein solution at pH 11.0. However, if any, they should be removed by centrifugation or filtration.

13. The crude protein solution at pH 11.0 is applied to a 2-iminobiotin agarose column (1.2×1.5 cm; bed volume, 1.7 mL), which has been equilibrated with $1.0M$ NaCl, 50 mM sodium carbonate (pH 11.0), 0.02% NaN$_3$. Unbound fraction is collected and reapplied to the column to ensure that all functional proteins are bound to 2-iminobiotin.

14. The column is washed with 17 mL (10 bed volumes) of $1.0M$ NaCl, 50 mM sodium carbonate (pH 11.0), 0.02% NaN$_3$ to remove unbound materials.

15. Bound material, i.e., the streptavidin-partner protein chimera, is eluted with 7 mL (4 bed volumes) of 50 mM sodium acetate (pH 4.0), 0.02% NaN$_3$, containing urea at a concentration of 50 mM–$6M$. The addition of urea enhances the recovery of the chimera from the 2-iminobiotin agarose column. In principle, the higher the concentration of urea is, the sharper the elution is, thus requiring smaller volumes of the elution solution. If the partner protein is stable in urea, higher concentrations of urea should be used to minimize the dilution of purified chimera. If no urea or a relatively low concentration of urea is used, larger volumes of the elution solution are needed. If the partner protein is unstable in urea, $0.5M$

NaCl, 50 mM sodium acetate (pH 4.0), 0.02% NaN$_3$ may be used as an eluate with reduced recoveries.

16. The eluted fraction, which contains the streptavidin-partner protein chimera, is dialyzed against any desired solution to remove urea. If insoluble materials are formed during dialysis, they should be removed by centrifugation or filtration.

4. Notes

1. The methods described above have been used successfully for a few streptavidin-containing chimera constructs *(13,14)*. However, the conditions may need to be adjusted or optimized for a particular streptavidin-partner protein chimera to maximize the production and final yield of functional chimera molecules.

2. Rich media, such as LB medium, can also be used for expression of a streptavidin-partner protein chimera and could give higher expression efficiency than with M9 minimal medium. In our experience, however, the use of rich media often causes cellular degradation of the partner protein moiety of an expressed chimera, thereby reducing the final yield. The streptavidin moiety is generally resistant to cellular proteinases, even in a denatured form. Thus, rich media should be used for expression if the stability of a partner protein against cellular proteolysis is sufficiently high.

3. During purification, each fraction should be analyzed by SDS-PAGE to find the optimal purification procedure for a particular streptavidin-partner protein chimera.

4. If degradation of chimera molecules during purification is observed, proteinase inhibitors, such as phenylmethylsulfonyl fluoride, pepstatin, leupeptin, and E-64, are added to all solutions used for purification. Because the streptavidin moiety is extremely stable against proteolysis when it is folded, the degradation of the chimera is likely to occur within the partner protein moiety. Such degradation of the chimera can be analyzed by Western blot analysis using anti-streptavidin after SDS-PAGE. Degradation products of the expressed chimera are detected by labeling the streptavidin moiety with anti-streptavidin. The determination of the size of each labeled protein component allows one to locate the sites in the partner protein moiety where proteolysis occurred.

5. The biotin-binding ability of purified chimera can be determined by a gel filtration method *(15)* using a desalting column, such as PD-10 columns (Pharmacia, Piscataway, NJ), and radiolabeled biotin, such as D-[*carbonyl-*^{14}C]biotin (Amersham, Arlington Heights, IL) and D-[8,9-^{3}H]biotin (Amersham and DuPont NEN, Boston, MA). Alternatively, the biotin-binding ability can also be determined semiquantitatively using nonradioactive methods, such as enzyme-binding assays *(16)*. Usually, the purified chimera has almost full biotin-binding ability, as it has been purified by using its biotin-binding ability.

6. The purified chimera is generally tetrameric; the subunit association of the chimera is determined by the streptavidin moiety whose subunit association is exceptionally stable. However, the subunit association of a particular streptavidin-partner protein chimera should be determined experimentally by using, for example, gel filtration chromatography.

Acknowledgments

The work described here was supported by Grant CA39782 from the National Cancer Institute, National Institutes of Health, and Grant DF-FG01-93ER61656 from the US Department of Energy.

References

1. Chaiet, L. and Wolf, F. J. (1964) The properties of streptavidin, a biotin-binding protein produced by *Streptomycetes. Arch. Biochem. Biophys.* **106**, 1–5.
2. Green, N. M. (1990) Avidin and streptavidin. *Methods Enzymol.* **184**, 51–67.
3. Wilchek, M. and Bayer, E. A. (1990) Avidin-biotin technology. *Methods Enzymol.* **184**, 5–13.
4. Wilchek, M. and Bayer, E. A. (1990) Applications of avidin-biotin technology: literature survey. *Methods Enzymol.* **184**, 14–45.
5. Argaraña, C. E., Kuntz, I. D., Birken, S., Axel, R., and Cantor, C. R. (1986) Molecular cloning and nucleotide sequence of the streptavidin gene. *Nucleic Acids Res.* **14**, 1871–1881.
6. Sano, T. and Cantor, C. R. (1990) Expression of a cloned streptavidin gene in *Escherichia coli. Proc. Natl. Acad. Sci. USA* **87**, 142–146.
7. Studier, F. W., Rosenberg, A. H., Dunn, J. J., and Dubendorff, J. W. (1990) Use of T7 RNA polymerase to direct expression of cloned genes. *Methods Enzymol.* **185**, 60–89.
8. Sambrook, J., Fritsch, E. F., and Maniatis, T. (1989) *Molecular Cloning: A Laboratory Manual*, 2nd ed., Cold Spring Harbor Laboratory, Plainview, NY.
9. Ausubel, F. M., Brent, R., Kingston, R. D., Moore, D. D., Seidman, J. G., Smith, J. A., and Struhl, K. (eds.) (1992) *Short Protocols in Molecular Biology*, 2nd ed., Wiley, New York, NY.
10. Sano, T. and Cantor, C. R. (1991) Expression vectors for streptavidin-containing chimeric proteins. *Biochem. Biophys. Res. Commun.* **176**, 571–577.
11. Laemmli, U. K. (1970) Cleavage of structural proteins during the assembly of the head of bacteriophage T4. *Nature* **227**, 680–685.
12. Hofmann, K., Wood, S. W., Brinton, C. C., Montibeller, J. A., and Finn, F. M. (1980) Iminobiotin affinity columns and their application to retrieval of streptavidin. *Proc. Natl. Acad. Sci. USA* **77**, 4666–4668.
13. Sano, T. and Cantor, C. R. (1991) A streptavidin-protein A chimera that allows one-step production of a variety of specific antibody conjugates. *Bio/Technology* **9**, 1378–1381.
14. Sano, T., Glazer, A. N., and Cantor, C. R. (1992) A streptavidin-metallothionein chimera that allows specific labeling of biological materials with many different heavy metal ions. *Proc. Natl. Acad. Sci. USA* **89**, 1534–1538.
15. Wei, R.-D. (1970) Assay of avidin. *Methods Enzymol.* **18A**, 424–427.
16. Bayer, E. A., Ben-Hur, H., and Wilchek, M. (1990) Colorimetric enzyme assays for avidin and biotin. *Methods Enzymol.* **184**, 217–223.

12

Bacterial Expression, Purification, and Potential Use of His-Tagged GAL4 Fusion Proteins

M. Lienhard Schmitz and Patrick A. Baeuerle

1. Introduction

Fusion of proteins to DNA-binding domains or transactivation domains has become an extremely powerful approach to study the transcriptional properties of proteins and to characterize and clone interacting proteins *(1,2)*. The yeast transcriptional activator protein GAL4 contains the sequences necessary for DNA-binding, dimerization and nuclear localization within its N-terminal 147 amino acids *(3)*. A dimer of this fragment can bind specifically to a 17-bp DNA recognition sequence, which displays an imperfect dyad symmetry *(4)*. It seems that higher eukaryotic cells do not possess proteins with GAL4-specific binding activity. Recombinant proteins between GAL4 and fused protein sequences of interest can be used to investigate two types of questions. First, potential protein/protein interactions can be detected by using an electrophoretic mobility shift assay (EMSA). This approach allows the discrimination of radioactively labeled unbound DNA, complexes between the GAL4 fusion protein and its cognate DNA, and more slowly migrating complexes containing additional protein(s) binding to the protein sequences fused to the GAL4 protein. Second, the potential transcriptional activity of regulatory proteins can be assayed by fusing them to the DNA-binding domain of GAL4 and testing their transcriptional activity on well-characterized promoters bearing Gal4 binding sites in a cell-free assay (for review, *see* ref. *1*). This strategy has the advantage that transcription extracts have not to be depleted of the transcription factor of interest. It also allows the direct comparison of transactivation domains from different transcription factors regardless of the effects probably resulting from their DNA-binding domains.

From: *Methods in Molecular Biology, vol. 63: Recombinant Protein Protocols: Detection and Isolation* Edited by: R. Tuan Humana Press Inc., Totowa, NJ

To facilitate fusion protein expression two vectors, that combine high-level expression of GAL4 proteins in *E. coli* with a simple purification of the recombinant proteins, were designed. Both vectors allow the inducible expression of N-terminally His-tagged GAL4 proteins and differ in the length of the GAL4 DNA-binding domain *(5,6)*. The hexahistidine sequence, that allows efficient purification of the bacterially produced protein on nickel nitrilotriacetic acid (Ni-NTA) columns, is attached to the N-terminal end of the protein. This hexahistidine sequence does not interfere with DNA-binding of GAL4, since the first eight amino acids of GAL4 display considerable conformational mobility as determined by NMR spectroscopy *(7)*. Only the amino acid residues 9–40 were found to form a well-defined cluster containing two Zn(II) ions coordinated through six cysteins such that the metals share two (Cys 11 and 28) of the ligands *(8)*. The histidine-tagged GAL4 protein purified from *E. coli* can bind to its cognate DNA with the same affinity as the untagged version *(5)*. Competition experiments showed that the tagged GAL4 protein also retained its DNA-binding specificity *(6)*. The six adjacent histidines provide an elegant and simple method for rapid protein purification based on the selective affinity of proteins with polyhistidines for a Ni-metal chelate adsorbent *(9,10)*. This binding occurs in physiological buffers that are suitable for the purification of soluble proteins present in the cytoplasm or periplasm as well as in denaturing buffers, which are used for proteins of limited solubility present in inclusion bodies. Furthermore, the His-tagged GAL4 protein can be purified in the presence of detergents, which do not interfere with the binding of the hexahistidine sequence to Ni-NTA columns. Figure 1 illustrates the binding of the His-residues to the immobilized Ni metal ions.

This interaction between the His residues and the Ni^{2+} ions is reversible, and the attached GAL4 protein can be eluted either with increasing amounts of imidazole or by lowering the pH.

The GAL4 protein alone is contained as a soluble protein in the cytoplasm of the bacterial cells and can be easily purified as will be described below. Most of the protein sequences that are attached to the GAL4 DNA-binding domain do not alter the solubility of the protein, such as the transcription activation domains of NF-κB p65, GAL4, and the AH domain *(5,6)*. However, fusion of some protein sequences, such as transcription factor Sp1 to GAL4 rendered the GAL4 fusion protein insoluble and required purification in denaturing buffers *(5)*. It is therefore necessary to determine for each bacterially produced fusion protein the optimal purification scheme and to optimize other parameters such as time and temperature of expression. The purification of soluble proteins is described below. If the fusion protein is not contained in

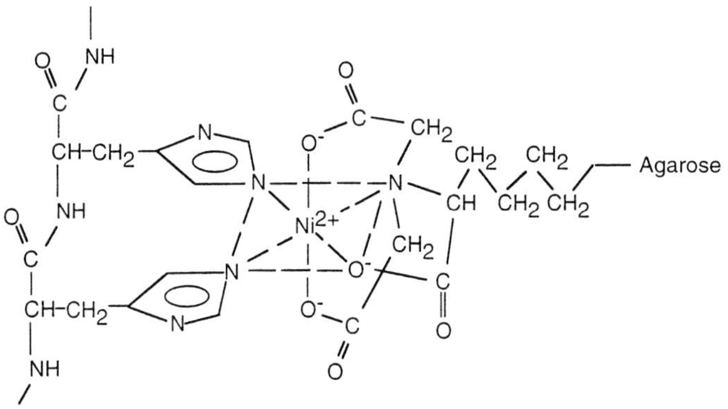

Fig. 1. Schematic model for the binding of the NI-NTA resin to neighboring 6× His residues is shown.

this fraction it will be necessary to purify it under denaturing conditions. Once the optimal conditions for the purification of the fusion protein are worked out, it can be expressed on a preparative scale.

2. Materials

All of the equipment used including pipets, glassware, and plasticware should be clean and autoclaved. The stock solutions for all reagents were prepared exactly as described in *(11)*, unless otherwise indicated.

2.1. Expression of GAL4 in E. coli

1. LB-Medium: 5 g yeast extract, 10 g bactotrypton, 5 g NaCl/L of medium, autoclave.
2. Antibiotics: 1000X stock solution of ampicillin (50 mg dissolved in 1 mL H_2O, sterilize by filtration and store at $-20°C$). 1000X stock solution of kanamycin-sulfate (25 mg dissolved in 1 mL H_2O, sterilize by filtration and store at $-20°C$).
3. IPTG: Prepare a $1M$ stock solution by dissolving 11.9 g of isopropyl-β-D-thiogalactopyranoside (IPTG) in 50 mL of H_2O. Aliquots thereof should be stored at $-20°C$.
4. Bacterial strains and plasmids: Principally any *E. coli* strain bearing the $lacI^q$ gene can be used. The *E. coli* strains M15 *(12)* and SG13009 *(13)* are both recommended for high-level expression of the His-tagged GAL4 proteins. These strains carry the pREP4 plasmid, which contains the $lacI^q$ gene and confers resistance to kanamycin. The inducible expression of His-tagged GAL4 proteins in *E. coli* can be achieved either with the plasmid pRJR1-containing amino acids 1–93 of GAL4 *(5)* or with pHisGal, which comprises amino acids 1–147 of GAL4 *(6)*. Both plasmids contain a multiple cloning site at the C-terminus of the GAL4 DNA-binding domain, allowing the con-

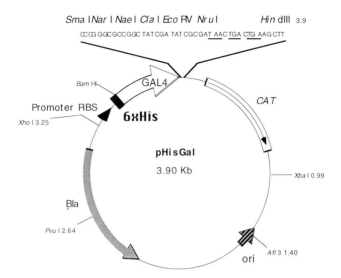

Fig. 2. The promoter/ribosome binding site of phisGal is shown by an arrow. The hexahistidine sequence is represented by a black box upstream from the first amino acids of GAL4. The coding regions for GAL4, β-lactamase (bla) and chloramphenicol acetyltransferase (cat) are highlighted with different shadings. The positions of the origin of replication (ori) and the unique restriction sites are indicated.

venient in-frame insertion of the sequence of interest. The structure of the phisGal plasmid is displayed in Fig. 2.

2.2. Purification of His-tagged GAL4 Proteins

1. Lysis buffer: 50 mM Nah$_2$PO$_4$, pH 8.0 (adjust pH with NaOH) 300 mM NaCl, 1 mM phenylmethylsulfonyl fluoride (PMSF), 5 mM 2-mercaptoethanol, 0.1% Tween-20. PMSF and 2-mercaptoethanol are toxic substances. The buffer should be autoclaved and PMSF, 2-mercaptoethanol and Tween-20 are added just prior to use. Do not use the Ni^{2+} reducing agents dithiothreitol (DTT) or dithioerythrit (DTE).

2. Denaturing buffer A: 6M guanidine hydrochloride, 0.1M Nah$_2$PO$_4$, 0.01M Tris (adjust pH to 8.0 with NaOH), 5 mM 2-mercaptoethanol.

3. Denaturing buffer B: 8M urea, 0.1M Nah$_2$PO$_4$, 0.01M Tris (pH adjusted to 8.0 with NaOH immediately prior to use), 5 mM 2-mercaptoethanol.

4. Denaturing buffer C: 8M urea, 0.1M Nah$_2$PO$_4$, 0.01M Tris (pH adjusted to 6.3 with HCl immediately prior to use), 5 mM 2-mercaptoethanol.

5. Denaturing buffer D: corresponds exactly to denaturing buffer C, except that the pH is adjusted to 5.9 with HCl immediately prior to use.

6. Denaturing buffer E: corresponds exactly to denaturing buffer C, except that the pH is adjusted to 4.5 with HCl immediately prior to use. It is not neces-

sary to autoclave denaturing buffers A to E. Avoid direct contact of the buffers with skin.

7. BC200: 20 mM Tris-HCl, pH 7.9, 200 mM KCl, 2 mM EDTA, 20% (v/v) glycerol, 10 μM ZnCl$_2$, 1 mM 2-mercaptoethanol, 1 mM PMSF. PMSF and 2-mercaptoethanol are added just prior to use.

8. BC400: corresponds exactly to BC200, except that it contains 400 mM KCl.

9. Ni-NTA agarose, siliconized glass wool, columns of different sizes, Heparin agarose.

3. Methods
3.1. Cloning and Bacterial Transformation

The protein sequence of interest can be inserted into the polylinker of Phisgal in the chosen reading frame by conventional cloning methods *(11)*. Transform 1–20 ng of the generated plasmid in the bacterial strains SG13009 or M15, both of which are recommendable for expression of the His-tagged GAL4 fusion proteins. These bacterial strains are already resistant to kanamycin and can easily be made competent for high-voltage electroporation or for CaCl$_2$ transformation *(11)*. Plate various amounts of the transformed bacteria on LB agar plates containing 25 μg/mL kanamycin and 50 μg/mL ampicillin. Within the next day, single colonies are visible. They contain the plasmid pREP4 and the plasmid encoding the His-tagged GAL4 protein and can be picked for further analysis.

3.2. Bacterial Expression of His-Tagged GAL4 Proteins

For a test expression, two single colonies of the potentially correct clones are inoculated and tested for their ability to produce the His-tagged GAL4 protein. In our experience it is not necessary to pick a larger number of clones. As a positive control for a soluble protein, the DNA binding domain of GAL4 alone should be included in the initial analytical purification. Each of the clones is grown in a shaker (210 rpm) as an overnight culture in 3 mL LB containing 25 μg/mL kanamycin and 50 μg/mL ampicillin at 37°C to stationary phase. One milliliter of each of the cultures are added to a flask containing 20 mL of LB (kan/amp) prewarmed to 37°C. For optimal growth of the cells and good expression of the GAL4 fusion protein, it is necessary to provide good aeration of the cultures. Therefore the volume of the vessel should be at least four times larger than the volume of the medium it contains. The protocol below is given for the analytical range but can be upscaled without alterations for the preparative range.

1. Determine the rate of growth by measuring the OD_{600} of the bacterial cultures after another 90 min on a shaker (210 rpm).

2. When an OD_{600} of 0.7 is reached, the protein expression is induced by the addition of IPTG to a final concentration of 1 mM. At the same time, ZnCl$_2$ is added

to a final concentration of 10 μ*M*, because the GAL4 DNA-binding domain requires two Zn(II) ions per molecule. For the test expression it is recommended to analyze one bacterial culture not induced by IPTG as an additional control.

3. Expression is allowed for 4 h, which should be ample time for the expression of most GAL4 fusion proteins. The optimal expression time has to be determined individually for each fusion protein.

4. Transfer the bacteria from the vessels into centrifuge tubes. Harvest the cells by centrifugation at 4000*g* for 15 min. Carefully remove the supernatant. At this stage the cell pellets can be stored at –20°C.

The analytical purification described in the next section is suitable for the His-tagged GAL4 proteins present in the soluble fraction. If the protein is not present in this fraction it has to be purified in denaturing buffers.

3.3. Purification of the Soluble GAL4 Fusion Proteins on Ni-NTA-Agarose

This protocol is suitable for the analytical scale, the instructions written in parentheses are given for the 1-L scale.

1. Dissolve the bacterial pellet in 200 μL (6 mL) of cold lysis buffer. All subsequent steps are carried out with cold solutions on ice. Break the cells by three cycles of freeze-thawing and subsequent sonication for 30 s. Cool the cells during the sonication and strictly avoid the generation of foam. (Cells from the 1-L culture should be sonifyed for 2 min. Subsequently, 60 mg of lysozyme are added and dissolved by carefully agitating the tube. After 1 h of incubation on ice, the cells are sonified again for 2 min. The cells can be lysed also with alternative techniques such as glass beads or a french press.) Add 0.5 m*M* of PMSF to the protein solution and carefully agitate the tube. If necessary, additional proteinase inhibitors may be used (*see* Section 4.1.). The viscosity is reduced by passing the solution three times through a needle that is attached to a syringe.

2. Equilibrate the Ni-NTA-agarose with lysis buffer just prior to use. This is done by transferring the Ni-NTA-agarose to an Eppendorf tube and adding lysis buffer. After centrifugation for 1 min at 6000*g* the supernatant is removed and the procedure is repeated twice. The Ni-NTA-agarose is now ready for use.

3. Centrifuge the protein extract at 4°C with 12,000*g* for 30 min. Transfer the supernatant to a fresh tube and add 20 μL (350 μL) of freshly equilibrated Ni-NTA-agarose. Incubate the tube containing the protein extract and the Ni-NTA-agarose on a spinning wheel at 4°C for 1 h.

4. Spin the tube for 2 min at 6000*g* in a cooled centrifuge. Take off the supernatant and add 1 mL (40 mL) of fresh lysis buffer. Agitate the tube in order to completely resuspend the Ni-NTA-agarose pellet. Spin the tube again and repeat the washing procedure four times. (For purification in the preparative scale, the Ni-NTA-agarose can alternatively be transferred to a column and washed with a flowrate of approx 2 mL/min. In an additional step, wash the Ni-NTA-agarose with 30 mL of BC200 buffer.) To reduce the background proteins subsequently

add 30 µL lysis buffer containing 1 mM imidazole (5 mL BC200 containing 1 mM imidazole) to the Ni-NTA-agarose pellet containing the His-tagged GAL4 protein. After 5 min incubation on ice, centrifuge the tube for 2 min at 6000g, take off the supernatant, and repeat the procedure once. Collect these two fractions as a control.

5. The His-tagged GAL4 fusion protein is eluted by adding 30 µL lysis buffer containing 100 mM imidazole (5 mL BC200 containing 100 mM imidazole) to the Ni-NTA-agarose pellet. After incubation for 5 min and a subsequent centrifugation, the supernatant is collected and this procedure is repeated once. The Ni-NTA-agarose can be reused up to five times for the same recombinant protein. Analyze the fractions including controls by reducing SDS-polyacrylamide gel electrophoresis (SDS-PAGE). If the protein is in the soluble fraction, it should now be at least 90% pure and can be further purified using heparin-agarose chromatography. Otherwise purify the fusion proteins in denaturing buffers as described in the following section.

3.4. Purification of Insoluble GAL4 Fusion Proteins on Ni-NTA-Agarose

This protocol is suitable for the analytical scale, the instructions written in parentheses are given for the 1-L scale.

1. Dissolve the bacterial pellet in 200 µL (6 mL) of cold denaturing buffer A. Shake cells at room temperature on a spinning wheel for 1 h. (During lysis, cells from the 1 L culture can be stirred using a magnetic bar [150 rpm]).
2. The Ni-NTA-agarose is prepared by washing it sequentially with 20 vol of denaturing buffers, A, B, C, and F, respectively. After a final wash in 20 vol of denaturing buffer A, the resin is ready for use.
3. Centrifuge the protein extract at 4°C with 12,000g for 30 min. Carefully transfer the supernatant to a fresh tube and add 20 µL (350 µL) of freshly equilibrated Ni-NTA-agarose. Incubate the tube containing the protein extract and the Ni-NTA-agarose on a spinning wheel at 4°C for 1 h.
4. Spin the tube for 2 min at 6000g in a cooled centrifuge. Take off the supernatant, and add 1 mL (40 mL) of fresh denaturing buffer A. Agitate the tube in order to completely resuspend the Ni-NTA-agarose pellet. Centrifuge the tube again, discard the supernatant and wash with denaturing buffer B. (Again, as already described for the purification of soluble GAL4 fusion proteins, the Ni-NTA-agarose can be transferred to a column and washed with the various buffers at a flowrate of approx 2 mL/min.) Wash with 1 mL (40 mL) of denaturing buffer C. Most of the *E. coli* proteins are washed out during this step. Collect a fraction thereof as a control.
5. Elute and collect the recombinant protein by adding 200 µL (3.5 mL) of denaturing buffer D followed by incubation with 100 µL (1.5 mL) of denaturing buffer E. The proteins contained in the various fractions can now be analyzed by SDS-PAGE.

The insoluble GAL4 fusion proteins are eluted by denaturing buffers D or E and are unfolded in the presence of $8M$ urea. The fusion proteins can subsequently be renatured by dialyzing them sequentially in BC200 buffer containing $6M$, $4M$, $2M$, and no urea. To allow complete and proper refolding of the proteins, the buffers should not be changed more often than every other hour. Alternatively, it is possible to renature the fusion protein when it is still attached to the Ni-NTA-agarose. In this case the BC200 buffers containing decreasing amounts of urea are passed over the Ni-NTA-agarose column after it has been washed in denaturing buffer C. The refolded protein is eluted in a 200 mM imidazole buffer of pH 7.0.

3.5. Further Purification and Characterization of the Fusion Proteins

The DNA-binding property of GAL4 is usually maintained when heterologous sequences are attached to its C-terminal part. This property of GAL4 can be exploited to further purify the GAL4 fusion protein to homogeneity. Either specific or nonspecific DNA or the polyanion heparin are suitable resins for the purification of DNA-binding proteins such as GAL. Use of DNA affinity chromatography has the advantage that only properly folded, active protein is purified.

1. Equilibrate a heparin agarose column (200 µL resin/mg protein) with six bed volumes of BC200, and subsequently pass the GAL4 fusion protein contained in BC200 over the column. Reload the flowthrough once and wash the column with 15 bed volumes of BC200 at a flowrate of 2 mL/min.
2. Elute the GAL4 protein with three bed volumes of BC400. Except for the aliquot that is used for analysis by SDS-PAGE, it is recommendable to add 1 mg/mL of BSA to the eluate in order to stabilize the proteins prior to aliquoting and freezing them at –80°C. The proteins are characterized further by examining their DNA-binding activity in EMSAs *(5,6)*.

4. Notes

The described protocols work very well for the expression and purification of the DNA-binding domain of the yeast GAL4 protein. There are some potential problems that may arise owing to the properties of the attached sequence. One is that no or little full-length protein is produced.

1. This might be owing to proteolytic degradation, a problem that can be overcome by the addition of further protease inhibitors to the various buffers.
2. The use of special *E. coli* strains devoid of proteases or the reduction of expression time and temperature may be helpful to reduce proteoloysis.
3. Many proteases are activated during harvesting and cell breakage. It may therefore become necessary to purify a particular fusion protein in denaturing buffers.

4. Alternatively, it is possible that the attached protein sequence is toxic for *E. coli*, a property sometimes observed with hydrophobic sequences. If the reduction of full-length fusion proteins can be attributed to the presence of rare codons in the coding sequence, they must be replaced in order to prevent premature termination of translation.

References

1. Ptashne, M. (1988) How eucaryotic transcriptional activators work. *Nature* **335**, 683–689.
2. Fields, S. and Song, O. K. (1989) A novel genetic system to detect protein–protein interactions. *Nature* **340**, 245–246.
3. Carey, M., Kakidani, H., Leatherwood, J., Mostashari, F., and Ptashne, M. (1989) An amino-terminal fragment of GAL4 binds DNA as a dimer. *J. Mol. Biol.* **209**, 423–432.
4. Marmorstein, R., Carey, M., Ptashne, M., and Harrison, S. C. (1992) DNA recognition by GAL4: structure of a protein-DNA complex. *Nature* **356**, 408–414.
5. Reece, R. J., Rickles, R. J., and Ptashne, M. (1993) Overproduction and single-step purification of GAL4 fusion proteins from *Escherichia coli*. *Gene* **126**, 105–107.
6. Schmitz, M. L. and Baeuerle, P. A. (1994) A vector, phisGal, allowing bacterial production of proteins fused to a hexahistidine-tagged GAL4 DNA-binding domain. *Biotechniques* **17**, 714–718.
7. Baleja, J. D., Marmorstein, R., Harrison, S. C., and Wagner, G. (1992) Solution structure of the DNA-binding domain of Cd_2-GAL4 from *S. cerevisiae*. *Nature* **356**, 450–453.
8. Kraulis, P. J., Raine, A. R. C., Gadhavi, P. L., and Laue, E. D. (1992) Structure of the DNA-binding domain of zinc GAL4. *Nature* **356**, 448–450.
9. Hochuli, E., Döbeli, H., and Schacher, A. (1987) New metal chelate adsorbents selective for proteins and peptide containing neighbouring histidine residues. *J. Chromatography* **411**, 177–184.
10. Hochuli, E. (1990) Purification of recombinant proteins with metal chelate adsorbent, in *Genetic Engineering, Principle and Methods, vol. 12* (Setlow, J. K., ed.), Plenum Press, New York, pp. 87–98.
11. Sambrook, J, Fritsch, E. F., and Maniatis, T. (1989) *Molecular cloning, Second Edition*, Cold Spring Harbor Laboratory, Cold Spring Harbor, New York.
12. Villarejio, M. R. and Zabin, I. (1974) β-galactosidase from termination and deletion mutant strains. *J. Bacteriol.* **110**, 171–178.
13. Gottesman, S., Halpern, E., and Trisler, P. (1981) Role of sulA and sulB in filamentation by Lon mutants of Escherichia coli K-12. *J. Bacteriol.* **148**, 265–273.

13

Detection of Expressed Recombinant Protein Based on Multidrug Resistance: P-Glycoprotein

Ursula A. Germann

1. Introduction

The simultaneous resistance of cells to multiple structurally and functionally unrelated cytotoxic agents is known as multidrug resistance (MDR). The phenomenon of MDR was first observed in the clinic in tumors of patients undergoing chemotherapy. Multidrug resistance was recognized as a major factor contributing to the failure of chemotherapeutic treatment of cancer *(1)*. Multidrug resistance also occurs in cultured cells selected for resistance to anticancer drugs *(2,3)*. Exposure of cultured cells to a single cytotoxic agent, e.g., Adriamycin (or doxorubicin) will enhance their resistance to anthracyclins and related agents (daunorubicin, idarubicin, mitoxantrone), as well as to *Vinca* alkaloids (e.g., vincristine, vinblastine), epipodophyllotoxins (VP-16 [etoposide] and VM-26 [teniposide]), and other anticancer drugs (e.g., actinomycin D, mitomycin C, and topotecan) *(4,5)*. Numerous cell lines have been established as in vitro model systems to demonstrate the clinical relevance of MDR, to elucidate its molecular basis and mechanism(s), and to design therapeutically applicable strategies to circumvent and overcome MDR *(6)*. Frequently, MDR is due to reduced intracellular accumulation of drugs resulting from overexpression of the MDR *(mdr)* gene product P-glycoprotein (pgp), also known as the multidrug transporter. P-glycoprotein is thought to act as an energy-dependent drug efflux pump at the cell surface *(4,5)*. Increased levels of P-glycoprotein correlate with increased MDR. Efforts directed at circumventing or overcoming MDR in the clinic are focused on inhibition or modulation of P-glycoprotein function or *MDR*1 gene expression. Approaches include rational design and screening for P-glycoprotein inhibitory compounds (so-called MDR modulators, MDR reversing agents, or chemosensitizers), inhibitory antibodies, or antisense oligodeoxynucleotides *(7)*.

From: *Methods in Molecular Biology, vol. 63: Recombinant Protein Protocols: Detection and Isolation* Edited by: R. Tuan Humana Press Inc., Totowa, NJ

Within the past decade, the human *MDR*1 gene, and its rodent homologs, the mouse *mdr*1 (or *mdr*1b) and *mdr*3 (or *mdr*1a) genes, the hamster *pgp*1 and *pgp*2 genes, and the rat *pgp*1 *and pgp*2 (or *mdr*1b) genes have been cloned from highly multidrug resistant cultured cells *(4,5)*. Upon transfection or retroviral transduction, all these cDNAs endow drug-sensitive host cells with MDR and, thus, are useful as selectable markers for gene transfer experiments *(8-10)* as described by Kane (*see* Chapter 27 in companion volume). Potential applications include the cotransfection of the *MDR*1 gene and an unselectable gene present in two different vectors *(11)*, or present in the same vector under the control of two different promoters *(12)*, or as a transcriptional fusion under the same promoter by using the internal ribosomal entry sites (IRES) from picornaviruses for cap-independent translation of the second gene *(13,14)*, or as a translational fusion to produce a hybrid protein *(15,16)*. This chapter focuses on approaches for the detection of P-glycoprotein in DNA-transfected or retrovirally transduced, cultured cells after drug selection (*see* Chapter 27 in companion volume).

P-glycoproteins are high molecular weight (apparent mol wt 145,000–180,000) single chain phosphoglycoproteins that are expressed within the plasma membranes *(4,5)*. Pulse-chase labeling experiments have revealed that P-glycoprotein is synthesized as a nonglycosylated precursor (apparent mol wt 130,000–140,000) that is slowly ($t_{1/2}$ = 1–2 h in human cells, $t_{1/2}$ = 20–30 min in mouse cells) processed to its mature form *(17,18)*. The metabolic half-life is 48–72 h for human P-glycoprotein and approximately 18 h for the mouse isoforms *(17,18)*. Based on hydropathy analysis of the amino acid sequence deduced from an *mdr* cDNA, a topological model for P-glycoprotein was postulated that predicts the polypeptide chain to contain 12 membrane-spanning regions (Fig. 1) *(19,20)*. According to this model, P-glycoprotein is composed of two similar halves, each of which contains a transmembrane domain (six membrane-spanning segments) and a nucleotide binding domain. The orientation of the molecule is such that both the amino-terminus and the carboxy-terminus are located on the cytoplasmic side of the plasma membranes. Moreover, the two predicted ATP binding sites are located intracellularly which is in agreement with their suggested function to fulfill the energy requirements of the drug transport process *(4,5)*. The predominant glycosylation sites of human P-glycoprotein are located in the first extracytoplasmic loop, but distinct isoforms may differ in the degree and sites of glycosylation *(4)*. This hypothetical 12-transmembrane model of P-glycoprotein, although not proven, is consistent with cellular epitope localization data obtained with various antibodies that specifically bind to the amino- or carboxy-terminal ends, near the two ATP binding sites, and within the first and fourth extracytoplasmic loop (Fig. 1) *(21–24)*.

Specific antibodies are useful to explore the structure and function of *mdr* gene products. They also represent sensitive reagents to detect P-glycoprotein

MRK16, 4E3

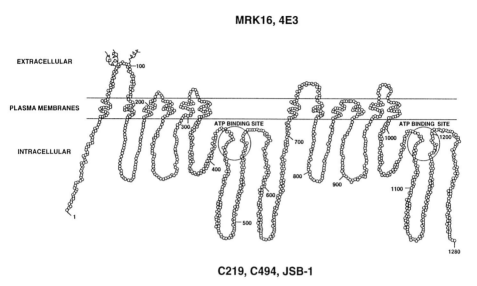

C219, C494, JSB-1

Fig. 1. Topological model of the human *MDR*1 gene product and recognition of P-glycoprotein by commercially available Mabs. A 12-transmembrane domain model is predicted by a computer-assisted hydropathy profile analysis of the amino acid sequence of P-glycoprotein, six membrane spanning segments are present in each half of the molecule *(19,20).* Two predicted ATP-binding sites are circled. Wiggly lines represent putative N-linked carbohydrates. The MAbs MRK16 and 4E3 crossreact with extracellular epitopes, whereas the epitopes for C219, C494, and JSB-1 are located intracellularly (*see* text for details). Adapted with permission from ref. *4,* by Annual Reviews.

in *mdr*-transfected or retrovirally transduced host cells. As summarized in Table 1, different commercially available mouse MAbs have been generated as tools for immunodetection of P-glycoprotein. C219 was prepared by using SDS-solubilized plasma membranes of multidrug-resistant Chinese hamster ovary and human leukemic cells as immunogen *(21).* C219 detects a cytoplasmic sequence (VQXALD where X is A, or E, or V) that is located within the two ATP binding domains of P-glycoprotein *(24).* This region is highly conserved among all known P-glycoprotein isoforms from different species. Thus, C219 is universally useful for detection of different P-glycoproteins *(24)* in *mdr*-transfected cells by Western blot analysis, immunoprecipitation, fluorescence-activated cell sorting (FACS) analysis, or immunocytochemistry. C494 was obtained with the same immunogen as C219 *(21)* and recognizes the amino acid sequence PNTLEGN present within the carboxy-terminal ATP binding domain of the human *MDR*1 and the hamster *pgp*1 gene product *(24).* JSB-1 was established after immunization of mice with the colchicine-

Table 1
Commercially Available P-Glycoprotein-Specific MAbs

Antibody	Isotype	Investigator	Commercial Source
C219	IgG2a	*21*	Signet Laboratories
C494	IgG2a	*21*	Signet Laboratories
JSB-1	IgG1	*25*	Signet Laboratories
			Boehringer Mannheim
MRK16	IgG2a	*26*	Signet Laboratories
			Kamiya Biomedical
4E3	IgG2a	*27*	Signet Laboratories

resistant Chinese hamster ovary cell line CHrC5 *(25)*. It recognizes a highly conserved intracellular epitope of P-glycoprotein which is different from the epitope for C219 *(25)*. Both C494 and JSB-1 are useful for immunoprecipitation and Western blot analysis of P-glycoprotein in *mdr*-transfected cells. Various mouse MAbs, including MRK16 and 4E3, crossreact specifically with the external domain of the human *MDR1* gene product *(26,27)*. Intact multidrug resistant human myelogenous leukemia K562/ADM cells were used as immunogen to generate MRK16 *(26)*, which recognizes epitopes present in the first and fourth predicted extracytoplasmic loop of human P-glycoprotein *(24)*. 4E3 was established by immunizing mice with human squamous lung cancer SW-1573-500 or uterine adenocarcinoma ME180/Dox500 cells *(27)*. MRK16 and 4E3 are excellent reagents for detection of human P-glycoprotein at the cell surface of human *MDR1*-transfected cells by FACS analysis or immunocytometry. They can also be used for immunoprecipitation of human P-glycoprotein, but not for Western blot analysis. Moreover, they are useful for magnetic affinity cell sorting of *MDR1*-transfectants *(28,29)*.

In the following, three basic protocols are described for detection of P-glycoprotein in *mdr*-transfected, drug-selected cells. All three protocols make use of the mouse MAb C219 for specificity of detection.

1. P-glycoprotein is metabolically labeled with ^{35}S-methionine, a whole cell extract is prepared, and P-glycoprotein is detected by immunoprecipitation, SDS polyacrylamide gel electrophoresis, and autofluorography.
2. P-glycoprotein present at the surface of viable multidrug resistant cells is photoaffinity labeled using ^{3}H-azidopine, a triton extract is prepared, and labeled P-glycoprotein is detected after immunoprecipitation by SDS polyacrylamide gel electrophoresis, and subsequent fluorography.
3. Crude membranes are prepared from P-glycoprotein expressing cells, proteins are size fractionated by SDS polyacrylamide gel electrophoresis, and P-glycoprotein is detected by Western blot analysis.

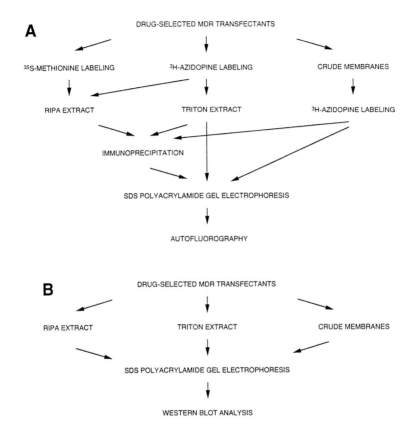

Fig. 2. Flowchart diagram of different experimental strategies for detection of P-glycoprotein. **(A)** Detection of labeled P-glycoprotein. **(B)** Detection of nonlabeled P-glycoprotein by Western blot analysis.

As shown in the flowchart diagram presented in Fig. 2, however, strategies can be devised that deviate from these three suggested routes. The protocols described in the following have been intentionally limited to procedures using commercially available reagents (*see* Notes 1–5).

2. Materials
2.1. Labeling of P-Glycoprotein
2.1.1. Metabolic Labeling Using ^{35}S-Methionine

1. ^{35}S-methionine (e.g., 1000 Ci/mM, Amersham, Arlington Heights, IL).
2. Complete cell growth medium.
3. Methionine-free cell growth medium.

2.1.2. ³H-Azidopine Photoaffinity Labeling

1. Dulbecco's phosphate-buffered saline (DPBS) without Mg^{2+}/Ca^{2+}: 200 mg/L KCl, 8000 mg/L NaCl, 2160 mg/L $Na_2HPO_4 \cdot 7H_2O$, 200 mg/L KH_2PO_4, sterile filter or autoclave, store at room temperature.
2. ³H-azidopine (40–50 Ci/mM, Amersham), store in the dark at –20°C.
3. Vinblastine stock solution: 10 mg/mL vinblastine (Sigma, St. Louis, MO) in dimethyl sulfoxide (DMSO), store in the dark (wrapped in aluminum foil) at –20°C. A dilution series may be prepared for competition assays. Alternatively, other agents that interact with P-glycoprotein may be used, including *Vinca* alkaloids, anthracyclines, etoposoide *(30)*.

2.2. Extraction of P-Glycoprotein from Tissue Culture Cells

2.2.1. Radioimmunoprecipitation Extracts

1. STE: 20 mM Tris-HCl, pH 7.4, 0.15M NaCl, 1 mM ethylenediaminetetraacetic acid (EDTA), sterile filter or autoclave, store at 4°C.
2. 10X Tris-buffered saline (TBS): 200 mM Tris-HCl, pH 7.2, 1.5M NaCl , sterile filter or autoclave, store at room temperature.
3. EDTA stock solution: 250 mM EDTA, pH 8.0, sterile filter or autoclave, store at room temperature.
4. 1X Radioimmunoprecipitation (RIPA): 1X TBS, pH 7.2, 1% (v/v) Triton-X-100, 1% (w/v) sodium deoxycholate, 0.1% (w/v) sodium dodecyl sulfate (SDS), 1 mM EDTA, 1% (v/v) aprotinin (Sigma), store at 4°C.

2.2.2. Triton Extracts

1. DPBS without Mg^{2+}/Ca^{2+}: (*see* Section 2.1.2.) store at room temperature, some at 4°C.
2. Dithiothreitol (DTT) stock solution: 1M DTT in H_2O, store in aliquots at –20°C.
3. DNase stock solution: 1 mg/mL DNase (Sigma) in 50% (v/v) glycerol, 150 mM NaCl, store at –20°C.
4. Triton elution buffer (TEB): 10 mM Tris HCl, pH 8.0, 0.1% (v/v) Triton-X-100, 10 mM $MgSO_4$, 2 mM $CaCl_2$, 10 µg/mL DNase, 1 mM DTT, 1% (v/v) aprotinin (Sigma). TEB can be prepared without DNase, DTT, and aprotinin, and stored at 4°C. DNase, DTT, and aprotinin are added immediately prior to use.

2.2.3. Preparation of Crude Membranes

1. DPBSAp: DPBS without Mg^{2+}/Ca^{2+} (*see* Section 2.1.2.) supplemented with 1% (v/v) aprotinin (Sigma), store at 4°C.
2. Phenylmethylsulfonylfluoride (PMSF) stock solution: 100 mM PMSF in isopropanol. Store in aliquots at –20°C.
3. Hypotonic lysis (HTL): 10 mM Tris-HCl, pH 8.0, 10 mM NaCl, 1 mM $MgCl_2$, 1 mM PMSF, 1 % (v/v) aprotinin (Sigma). HTL prepared without PMSF and aprotinin can be stored at 4°C; PMSF and aprotinin are added immediately prior to use.

4. Trypan blue stain (e.g., GibcoBRL, Gaithersburg, MD).
5. TSNa: 10 mM Tris-HCl, pH 7.5, 250 mM sucrose, 50 mM NaCl, 1% (v/v) aprotinin (Sigma), prepare fresh and prechill at 4°C.
6. Protein assay kit (e.g., BCA protein assay kit, Pierce, Rockford, IL).

2.3. Immunoprecipitation of P-Glycoprotein

1. 1X RIPA: (*see* Section 2.2.1.) In some cases, 2X RIPA is useful for dilution of samples.
2. Normal rabbit serum (Organon Teknika-Capel, Durham, NC).
3. *Staphylococcus aureus*: Zysorbin (Zymed, South San Francisco, CA), use according to manufacturers condition (10% suspension in DPBS w/o Mg^{2+}/Ca^{2+}, store in aliquots at –20°C), spin prior to use for 1 min at 15,000g, resuspend in 1X RIPA, spin for 1 min at 15,000g, and resuspend pellet again in 1X RIPA to final concentration of 10%.
4. 1% (w/v) Ovalbumin in H$_2$O.
5. 20% (w/v) Trichloroacetic acid (TCA) in H$_2$O.
6. 1M NaOH.
7. 0.4M Acetic acid.
8. Aquasol scintillation fluid (DuPont NEN, Wilmington, DE).
9. C219 affinity-purified MAb, 0.1 mg/mL (Signet Laboratories, Dedham, MA).
10. Protein A sepharose (PAS): (Pharmacia, Uppsala, Sweden) Swell and wash beds according to instructions of the manufacturer, prepare 50% (w/v) solution in 1X RIPA. Prepare fresh, or store at 4°C in presence of 0.1% sodium azide, but wash once with 1X RIPA prior to use.
11. Rabbit antimouse IgG (Organon Teknika-Capel, Durham, NC).
12. 1X RIPA/KCl: 1X TBS, pH 7.2, 1% (v/v) Triton-X-100, 1% (w/v) sodium deoxycholate, 0.1% (w/v) SDS, 1% (v/v) aprotinin (Sigma), 2.5M KCl, store at 4°C.
13. Elution buffer (EB): 100 mM Tris-HCl, pH 7.5, 2% (w/v) SDS, 5% (v/v) β-mercaptoethanol.
14. Sucrose: 50% (w/v) sucrose, 0.02% (w/v) sodium azide, store at room temperature.
15. Acetone; keep at –20°C.
16. Gel fix: 40% (v/v) methanol, 10% (v/v) acetic acid.
17. Amplify (Amersham) or other reagent for gel fluorography.

2.4. Western Blot Analysis and Immunodetection of P-Glycoprotein

1. Towbin transfer buffer: 25 mM Tris-base, 192 mM glycine, 20 % (v/v) methanol, pH 8.3.
2. DPBS w/o Mg^{2+}/Ca^{2+} (*see* Section 2.1.2.).
3. DPBST: DPBS w/o Mg^{2+}/Ca^{2+} with 0.1% (v/v) Tween-20 (Bio-Rad, Hercules, CA); store at room temperature.
4. Blotto: 10% (w/v) Carnation nonfat dry milk in DPBS w/o Mg^{2+}/Ca^{2+}, store at 4°C for a few days.

5. Primary antibody solution: 0.2 µg/mL C219 mouse Mab (Signet Laboratories) in Blotto, dilute freshly prior to use. The primary antibody solution can be stored at 4°C in the presence of 0.02% sodium azide and reused, if desired.

6. Secondary antibody solution: antimouse IgG-horseradish peroxidase (Amersham) diluted 1:1000 in Blotto, prepare immediately before use.

7. Enhanced chemoluminescence (ECL) Western blotting detection system (Amersham), store at 4°C.

8. 3,3'-diaminobenzidine tetrahydrochloride (DAB) stock solution: 40 mg/mL DAB (Bio-Rad) in H_2O; store at –20°C.

9. $NiCl_2$ stock solution: 80 mg/mL in H_2O, store at room temperature.

10. 30% H_2O_2 (Aldrich, Milwaukee, WI); store at 4°C.

11. DAB staining solution: 100 mM Tris-HCl, pH 8.0, 0.8 mg/mL DAB, 0.4 mg/mL $NiCl_2$, 0.01% H_2O_2; use immediately after addition of H_2O_2.

3. Methods
3.1. Labeling of P-Glycoprotein
3.1.1. Metabolic Labeling Using ^{35}S-Methionine

[35]S-methionine is commonly used to metabolically label proteins during biosynthesis. According to the amino acid sequence deduced from the *MDR*1 cDNA sequence, human P-glycoprotein contains 32 methionine residues. As mentioned in the introduction, it has a relatively long half-life. Thus, to allow for efficient labeling of P-glycoprotein, drug-selected *MDR*1-transfectants are grown in medium containing ^{35}S-methionine for an extended period of time, i.e., 16 h (*see* Note 6).

1. Plate cells that grow as monolayer at approx 4–6 × 10[6] cells/100-mm dish in regular growth medium approx 4–6 h prior to use. The cell density should be subconfluent.

2. Remove the growth medium, add 5–10 mL methionine-free growth medium to wash the cells briefly, aspirate again, and add 5 mL methionine-free growth medium. Add ^{35}S-methionine to a final concentration of 1 mCi/dish and incubate cells at 37°C, 5% CO_2 for 16 h.

3. Harvest the labeled cells, wash cells, prepare cell extract as described in section 3.2.1., and perform radioimmunoprecipitation (*see* Section 3.3.).

3.1.2. ^{3}H-Azidopine Photoaffinity Labeling

This section describes a method for photoaffinity labeling of P-glycoprotein. Among various photoactivatable reagents, arylazides are most widely used. They form covalent bonds to the polypeptide acceptor site(s) upon UV irradiation via nitrene intermediates *(31)*. Several photoaffinity probes are known to label P-glycoprotein specifically, including different analogs of cytotoxic drugs and different analogs of MDR modulators

(30,32). ^{3}H-Azidopine is a commercially available, relatively affordable dihydropyridine calcium channel blocker that photolabels P-glycoprotein at two major sites, one in each half of the molecule *(33,34).* ^{3}H-Azidopine photoaffinity labeling of P-glycoprotein is inhibited by an excess of vinblastine, doxorubicin, actinomycin D, and many MDR modulators *(30).* Thus, competition assays with increasing concentrations of nonradioactive anticancer drugs or MDR modulators can be performed to identify the photolabeled protein as P-glycoprotein (*see* Note 7).

1. Harvest tissue culture cells. Trypsinize cells that grow as monolayers, dilute with serum-containing growth medium, and centrifuge at 500g for 10 min at room temperature. Wash the cell pellet 3 times with DPBS w/o Mg^{2+}/Ca^{2+} and resuspend in DPBS w/o Mg^{2+}/Ca^{2+} at a density of 1 × 10^7 cells/mL.
2. Add 1 µCi (i.e., 1 µL) ^{3}H-azidopine/10^6 cells (final concentration is approximately 0.2 µ*M*). ^{3}H-azidopine is supplied as an ethanol solution and will rapidly evaporate, thus, work as quickly as possible and keep solution chilled. Divide the cell suspension into 100 µL aliquots in 1.5-mL polypropylene microtubes. For competition assays, add 1 µL of vinblastine solution (or other compound, concentration can be varied, *see* Note 7) to the appropriate tubes. Add 1 µL DMSO to control tube.
3. Incubate cells for 1 h in the dark (wrapped in aluminum foil) at room temperature with gentle agitation (e.g., rocking).
4. Pellet the cells by low speed centrifugation at 1000g for 1 min, remove the supernatant which contains excess radioactive label, and resuspend the cell pellet in 100 µL DPBS w/o Mg^{2+}/Ca^{2+}.
5. Open the tubes, place in rack on ice to keep samples cold during crosslinking reaction, and irradiate at a distance of approximately 5 cm for 20 min with UV lamp using two, 15-W, self-filtering, longwave UV tubes (e.g., Blak-Ray UV Bench/display lamp, UVP, VWR Scientific).
6. Collect the cells by centrifugation as above and remove the supernatant.
7. The photolabeled cell pellet can be extracted immediately, or frozen on dry ice and stored at −70°C until use. Usually triton extracts (*see* Section 2.2.2.) are prepared which are used for immunoprecipitation (*see* Section 3.3.3.) after dilution with an equal volume of 2X RIPA buffer and adjustment of the total volume to 500 µL with 1X RIPA. The ^{3}H-azidopine labeled cell pellets can also be extracted according with RIPA buffer (*see* Section 3.2.1.). In transfectants that express high levels of P-glycoprotein due to multistep selection in increasing concentrations of cytotoxic agent (*see* Chapter 27 in companion volume), ^{3}H-azidopine-labeled P-glycoprotein may be detectable without immunoprecipitation by directly analyzing triton extracts by SDS polyacrylamide gel electrophoresis, followed by autofluorography (dilute samples with 2X loading buffer for SDS polyacrylamide gel electrophoresis, then proceed to step 8 in Section 3.3.3.). To concentrate the proteins in triton extracts before loading on an SDS polyacrylamide gel, perform an acetone precipitation (*see* steps 6,7 in Section 3.3.3.).

3.2. Extraction of P-Glycoprotein from Tissue Culture Cells

3.2.1. Radioimmunoprecipitation Extracts

Cell extracts for radioimmunoprecipitation are usually prepared by detergent lysis. The protocol described here is for cells growing as a monolayer (*see* Section 3.1.1.). The protein extract is clarified by centrifugation at 100,000g. Nucleic acids, which make the sample viscous, are pelleted in this step (*see* Notes 6 and 8).

1. After metabolic labeling, monolayer cells are rinsed 2X with 5 mL ice-cold STE buffer, scraped into STE buffer and centrifuged at 500g for 5 min at 4°C.
2. Dissolve the cell pellet in 1 mL ice cold RIPA buffer and let sit on ice for 10 min.
3. Perform ultracentrifugation at 100,000g for 30 min at 4°C. P-glycoprotein will be in the supernatant.
4. Immunoprecipitate radiolabeled P-glycoprotein and quantitate (*see* Section 3.3.).

3.2.2. Triton Extracts

Compared to the protocol for whole cell extracts described in Section 3.2.1., this method uses mildly denaturating conditions. Although the overall yield of extracted proteins is somewhat smaller than with RIPA buffer, P-glycoprotein is efficiently extracted with 0.1% (v/v) Triton-X-100. The cells are lysed by repeated freeze-thawing and mild sonication. DNAse is present in the triton elution buffer to decrease the viscosity of the samples. The protein extract is cleared by centrifugation at 15,000g (*see* Notes 6 and 8).

1. Resuspend cells in 1.5 mL polypropylene microtubes at 1×10^7 cells/mL in TEB by pipeting up and down.
2. Freeze-thaw cells 3X, i.e., freeze cells on dry ice, thaw at room temperature or at 37°C (in water bath or heating block).
3. Sonicate samples for 30 s with Teqmar sonic disruptor (*see* Note 8).
4. Incubate cell extracts at 37°C for 10 min to complete DNAse reaction.
5. Spin at 15,000g for 5 min, P-glycoprotein will be in the supernatant.
6. If desired, the protein in the supernatants can be concentrated by acetone precipitation (*see* steps 6 and 7 in Section 3.3.3.).

3.2.3. Preparation of Crude Membranes

P-glycoprotein is a polytopic transmembrane protein present at the cell surface. Alternatively to whole cell extracts, crude membranes can be prepared to enhance the relative P-glycoprotein content in a protein extract and, thus, facilitate detection of P-glycoprotein. Briefly, a preparation of crude membranes involves hypotonic lysis and homogenization of cells, low speed centrifugation to separate unbroken cells and nuclei, and high speed centrifugation of the low-speed supernatant to pellet the crude membrane fraction which contains plasma membranes, as well as various intracellular membrane fractions. All the steps for the crude membrane preparation need to be carried out in the cold; i.e., buffers and equipment are prechilled to 4°C and sample(s) are kept on ice (*see* Notes 6 and 9).

1. Grow approx 5×10^7 monolayer cells. Ideally, monolayer cells are near confluency at time of harvesting. Aspirate the medium and harvest cells by scraping into DPBSAp. Spin cells at 500g for 10 min, remove supernatant, resuspend the cell pellet in DPBSAp by pipeting up and down, and wash twice. Determine total cell number before third centrifugation with Coulter counter or hematocytometer.
2. Resuspend cells at 1×10^7 cells/mL in HTL (i.e., 5 mL). Spin immediately at 500g for 10 min. As the cell pellet will be very soft, carefully remove the supernatant and resuspend cells again in the same volume HTL. Incubate on ice for 20 min.
3. Transfer the cells to a Dounce homogenizer and perform 20 strokes with tight-fitting pestle. A staining with trypan blue can be performed with a few µL cell homogenate to ascertain that at least 90% of cells are broken. If so, add an equal volume of TsNa to the cell lysate.
4. Transfer the diluted cell lysate to a centrifuge tube and spin at 500g for 10 min. Resuspend the cell pellet in 1–2 mL TsNa and keep a small aliquot for protein assay (make a note of the volume of this fraction). The rest of the cell pellet can be discarded unless used otherwise.
5. Transfer the supernatant into an ultracentrifuge tube and spin at 100,000g for 1 h. The high-speed supernatant will contain the cytosolic fraction. A small aliquot should be kept for protein assay (make a note of the volume of this fraction), the remainder can be discarded unless used otherwise.
6. Resuspend the high-speed pellet containing the crude membranes in TsNa, use approx 1/10 volume of HTL volume from step 2 (i.e., 500 µL TsNa). Membrane pellets are generally difficult to resuspend, it helps to use a syringe with an 18-gage needle (1.5 in. long) that is bent at a 90° angle. Resuspended crude membranes can be stored frozen at –70°C. Repeated freeze-thawing of crude membranes should be avoided, and storage in aliquots is recommended. If membranes aggregate upon thawing, again use a syringe with a bent 18-gage needle to resuspend.
7. Determine protein concentrations in the low-speed pellet, the high-speed pellet, and the high-speed supernatant with protein assay kit according to manufacturer's instructions. Calculate the total amount of protein in each fraction. Typically, the crude membrane fraction will contain approx 15–20% of the total cell protein, i.e., the P-glycoprotein content in crude membranes is approximately fivefold enhanced as compared to whole cell extracts.

3.3. Immunoprecipitation of P-Glycoprotein

Radiolabeled cell extracts are first preadsorbed to normal rabbit serum and heat-inactivated, formalin-fixed *Staphylococcus aureus* containing high amounts of protein A to reduce nonspecific binding to primary antibody, secondary antibody, and protein A sepharose (Section 3.3.1.). A protein precipitation using trichloroacetic acid (TCA) is performed to allow for normalization of radioactivity from different samples (Section 3.3.2.). Equal amounts of radiolabeled protein are then incubated with C219, a P-glycoprotein-specific mouse MAb, and subsequently with protein A sepharose containing rabbit antimouse IgG secondary antibody (Section 3.3.3.). The immune complexes are pelleted by centrifugation and washed. Finally, radiolabeled P-glycoprotein is eluted, concentrated, and analyzed by SDS-polyacrylamide gel electrophoresis followed by autofluorography. If desired, immunoprecipitations can be quantitated (Section 3.3.4., *see* Notes 10 and 11).

3.3.1. Preadsorption of Cell Extracts
to Staphylococcus aureus Protein A

1. Add 0.1 mL normal rabbit serum to 1 mL RIPA extract and incubate on ice for 30 min.
2. Add 0.1 mL *Staphylococcus aureus* suspension and incubate with gentle agitation (e.g., on a rotating wheel) at 4°C for 30 min.
3. Spin at 15,000g for at least 15 min at 4°C and transfer the supernatant to a new microtube for further analysis. Protein concentrations are determined, if desired.

3.3.2. TCA Precipitation of Cell Extracts

To normalize radioactivity of extracts from different cell lines, the ^{35}S-methionine content (cpm/mL) is determined. To this end, three protein aliquots of each sample are precipitated by TCA as follows.

1. To 2 μL, 5 μL, and 10 μL of extract in 1.5-mL microtubes add 0.1 mL ovalbumin and 1.0 mL 20% TCA, incubate for 5 min at 4°C (on ice), spin for 5 min at 15,000g and discard supernatant.
2. Solubilize the pellet in 0.1 mL 1M NaOH, then neutralize with 0.3 mL of 0.4M acetic acid. Transfer the sample to a scintillation vial and count in scintillation fluid (e.g., 12 mL Aquasol).
3. Use equal number of ^{35}S-cpm from different cell lines for immunoprecipitation and adjust volumes to 500 μL with 1X RIPA.

3.3.3. Immunoprecipitation of P-Glycoprotein

All steps are performed in 1.5 mL polypropylene microtubes at 4°C or on ice unless indicated otherwise.

1. Add 1 μg C219 MAb to antigen prepared in 500 μL RIPA and incubate overnight.
2. For each sample, preadsorb 50 μL Protein A sepharose with 5 μL rabbit antimouse IgG secondary antibody for 60 min with rocking or other gentle agitation (e.g., rotating wheel). Spin for 2 min at 1000g and wash twice with 1X RIPA.

3. Resuspend PAS containing secondary antibody in 50 µL 1 x RIPA and add to cell extract. Incubate for 2 h with rocking or other gentle agitation (e.g., rotating wheel).
4. Spin for 2 min and wash once with 1 mL 1X RIPA, once with 1 mL 1X RIPA/ KCl, and once again with 1 mL 1X RIPA.
5. Elute antigen and IgG from sepharose beads into 400 µL EB by incubating for 30 min at room temperature with rocking or other agitation. Spin for 2 min and transfer supernatant into new microtube.
6. Precipitate proteins by adding 1/10 volume, i.e., 45 µL 50% (w/v) sucrose and 1/1 volume, i.e., 450 µL cold acetone. Freeze samples on dry ice or at –70°C.
7. When you are ready for SDS-polyacrylamide gel electrophoresis, thaw samples at –20°C for at least an hour. Spin at 15,000g for 15 min in the cold, remove supernatant and dry pellet briefly (2–5 min) in lyophilizer or SpeedVac. Do not overdry pellets or they will be difficult to resuspend.
8. Resuspend pellets with 1X loading buffer for SDS polyacrylamide gel electrophoresis. Since P-glycoprotein may aggregate at high temperature, do not boil, but incubate at 37°C for 5–10 min before loading on low percentage gel (for best resolution in high-mol-wt range, use gel containing 6–8% polyacrylamide). After electrophoresis fix proteins in at least 5 gel volumes of gel fix for at least 30 min. Subsequently soak gel for fluorography in Amplify for 20–30 min according to the instructions of the manufacturer. Dry gel and expose to X-ray film (e.g., Kodak XAR5 film) at –70°C.

3.3.4. Quantitation of Immunoprecipitation

There are several different ways to quantitate immunoprecipitations which are described in the following.

1. Using a phosphoimager or similar instrument (e.g., Bio-Rad GS-363 Molecular Imager system).
2. The signals on the X-ray film can be quantitated by densitometry scanning (e.g., PharmaciaLKB UltroScan XL). X-ray films needs to be preflashed, if accurate quantitation is required.
3. The radiolabeled P-glycoprotein band can be excised from the SDS polyacrylamide gel, solubilized in 1 mL 30% H_2O_2/1% NH_4OH overnight at 37°C, and counted in 20 mL Aquasol scintillation fluid.

3.4. Western Blot Analysis

Conventional Western blotting is performed which includes the separation of a mixture of proteins by SDS polyacrylamide gel electrophoresis, followed by electrophoretic transfer of the proteins to a solid support (e.g., nitrocellulose membrane or polyvinylidene difluoride [PVDF] membrane). After blocking of the membrane with nonspecific milk proteins, immunocomplexes are formed with P-glycoprotein using C219 mouse MAb and sheep antimouse Ig that is covalently coupled to horseradish peroxidase (HRP). Finally, the HRP-labeled protein complexes are detected using the enhanced chemoluminescence (ECL) Western blotting detection system. If desired, the Western blot can also be stained with DAB subsequently.

1. Prepare protein sample in 1X loading buffer for SDS polyacrylamide gel electrophoresis. Protein amount required for signal detection depends on the levels of expression of P-glycoprotein. For unknown samples, it is recommended to initially load the maximum amount of sample without overloading the gel. Heat samples at 37°C for 10 min, but do not boil since P-glycoprotein may aggregate. Load and run a low percentage (6–8%) SDS polyacrylamide gel and include a prestained molecular weight marker in a control lane.
2. Tranfer proteins to nitrocellulose (0.45 μ or 0.2 μ) or PVDF membranes using Towbin transfer buffer. Use prechilled transfer buffer (4°C) and apply current of approx 0.7 mA/cm^2 of gel for 1.5–2 h. The prestained markers are used to ascertain the quality of the Western transfer.
3. Incubate the blot in approx 0.1 mL/cm^2 blocking solution at room temperature for 1 h with gentle agitation (e.g., rocking).
4. Discard the blocking solution and immediately add the same volume of primary antibody solution. Incubate overnight at 4°C with gentle agitation (e.g., rocking).
5. Wash blot for at least 30 min at room temperature with DPBST (>1 mL/cm^2). Change DPBST 4–5 times; then wash once with DPBS.
6. Add secondary antibody solution (approx 0.1 mL/cm^2) and incubate for 1 h at room temperature with gentle agitation (e.g., rocking).
7. Wash with DPBST for at least 30 min. Change DPBST 4–5 times.
8. Perform ECL detection according to instructions of manufacturer (Amersham). Briefly drain membrane until no liquid is visible (do not blot dry). To avoid artefacts, this is done best by holding the blot with forceps at one corner, letting the opposite corner touch a Whatman filter paper and airdry the membrane (but be careful not to overdry the blot). Then wrap the semidry membrane with Saran Wrap and immediately expose to X-ray film. Start with an exposure of 1 min, develop film and adapt exposure time according to signals (from 1 s up to 1 h). For quantitative purposes, the signals should be in the linear range of the X-ray film and X-ray film should be preflashed. If signals are too strong even upon exposure of 1 s, prechill film cassette including membrane for 10 min at –20°C, this will slow down signal detection.
9. If ECL detection shows high background, wash blot several times with DPBS and repeat ECL detection.
10. If background is still too high, wash blot several times with DPBS and try staining with DAB solution (approx 0.1 mL/cm^2). Unfortunately it is impossible to suppress background staining completely, thus, it is very important to monitor the course of the peroxidase reaction closely. As soon as desired signals have optimal intensity, rinse the blot briefly with deionized water and transfer it to a tray containing approximately 250 mL DPBS. Photograph the blot as soon as

possible to obtain a permanent record of the experiment, since signals will fade after several hours of exposure to light.

11. Quantitate P-glycoprotein signals (e.g., by densitometry scanning, *see* Section 3.3.4.).

4. Notes

1. P-glycoprotein subjected to SDS polyacrylamide gel electrophoresis always appears as a characteristic broad band with an apparent 145,000–180,000 mol wt. The apparent molecular weight of P-glycoprotein depends on the host cell line, it is generally somewhat lower in mouse than human cell lines presumably due to differences in glycosylation. Protein fusions with P-glycoprotein will have an increased apparent molecular weight, depending on the size of the fusion partner. Again the signals for such hybrid proteins are generally broad.

2. It is very important to include positive and negative controls in all experiments. The KB-3-1 and KB-V1 cell lines are ideal control cell lines which are available from ATCC. KB-3-1 is a drug-sensitive subclone of the human adenocarcinoma cell line HeLa. KB-V1 was derived from KB-3-1 via two steps of mutagenesis with ethyl methanesulfonate (EMS) and multiple steps of drug selection in increasing concentrations of vinblastine *(35)*. The KB-V1 cell line is maintained at 1 µg/mL vinblastine and exhibits a relative resistance of 213 for vinblastine, 422 for Adriamycin, and 171 for colchicine when compared with the KB-3-1 parental cell line *(35)*. The *MDR*1 gene in the KB-V1 cell line is amplified (approx 100 copies are present) *(36)*, and KB-V1 cells express approximately 320-fold higher levels of *MDR*1 mRNA than KB-3-1 cells *(37)*. P-glycoprotein levels expressed in the KB-V1 cell line are estimated to represent approximately 1% of the total plasma membrane protein (M. M. Gottesman, personal communication), thus, P-glycoprotein is readily detectable in KB-V1 cells.

3. Generally it is advisable to use two or more antibodies for detection of P-glycoprotein to corroborate specificity of the reagents. The protocols are described for C219 mouse MAb, but similar protocols can be used for C494 or JSB-1 mouse MAb. It is recommended, however, to titer all antibodies for individual experiments since different lots of MAbs may vary in their quality. Thus, the amounts of antibodies indicated in the protocols have to be considered approximate. Some nonspecific signals may have to be expected depending on the cell line analyzed. C219 may cross-react with muscle myosin *(38)*. C494 has been reported to cross-react with pyruvate carboxylase, a mitochondrial protein of an apparent 130,000 mol wt *(39)*.

   Polyclonal rabbit antisera have been prepared by several laboratories (e.g., anti-P, anti-C, 4007, 4077, anti-PEPG2, anti-PEPG13, *22,40,41)*; these may be useful for both immunoprecipitations and Western blot analysis of P-glycoprotein, although nonspecific background will be somewhat higher than with MAbs. Obviously the protocols described above would have to be adapted to include the appropriate secondary antibodies.

   Alternatively, if human P-glycoprotein is used as a selectable marker, whole cells can be subjected to fluorescence activated cell sorting (FACS) analysis using the mouse MAbs MRK16 or 4E3 and a secondary antibody that carries a fluorescent tag (e.g., fluorescein isothiocyanate (FITC)-conjugated secondary antibody). It is important to include in each analysis an IgG2a isotype-matched negative control primary antibody. MRK16 and 4E3 can also be used to immunoprecipitate P-glycoprotein from metabolically or surface labeled cells, but they require different protocols, e.g., incubation of cells with primary antibody before lysis and/or use of mild detergents for the preparation of cell lysates *(27,42)*. Several other human P-glycoprotein-specific MAbs have been developed by different laboratories, but many of them are not commercially available *(43)*.

4. The sensitivity of various methods for detection of P-glycoprotein has not been determined accurately to date and can only be estimated. Metabolic labeling with ^{35}S-methionine may give somewhat higher sensitivity than photoaffinity labeling with ^{3}H-azidopine, although experiments performed with ^{3}H-azidopine can be exposed to X-ray film for extended periods of time (several months as opposed to several weeks for experiments with ^{35}S-methionine) to increase the sensitivity of detection. ^{3}H-azidopine, however, will label less proteins than ^{35}S-methionine, and the specificity of labeling of P-glycoprotein can be analyzed by competition assays with nonlabeled cytotoxic agents. Western blot analysis of crude membrane preparations has similar sensitivity as immunoprecipitation of radiolabeled cells.

Experiments have been performed with a series of multidrug resistant KB cell lines which have demonstrated that P-glycoprotein can be detected in KB-8-5 cells, which are three- to sixfold resistant to various drugs compared with KB-3-1 cells, but not in KB-8 cells which are only twofold resistant *(44)*. Levels of P-glycoprotein in drug-selected *mdr*-transfectants are generally high enough to allow detection of P-glycoprotein. If problems occur and no P-glycoprotein is detected, it is advised to analyze *MDR*1 mRNA levels. Different methodologies have been established to measure *MDR*1 mRNA levels, including Northern blot analysis, slot/dot blot analysis, RNase protection assay, *in situ* hybridization and reverse transcriptase-polymerase chain reaction (RTPCR), the last one being the most sensitive *(45)*.

5. The protocols described have been optimized for *MDR*1-transfected cells that grow as monolayers such as human HeLa cells or mouse NIH 3T3 cells, but they can also be performed with other cell lines, including cells that grow in suspension. Depending on the size of the cells, the cell number may have to be adapted. Most protocols can be scaled up or down proportionally.

6. The protocols described in this chapter are optimized for detection of human P-glycoprotein and need to be modified for concomitant detection of coexpressed proteins. Depending on the half-life of the protein, metabolic labeling with ^{35}S-methionine may have to be performed for a shorter time period (1–2 h) and in this case, starvation of cells for 1 h in methionine-free growth medium may enhance efficiency of labeling. Proteins containing only few methionine residues, but many cysteine residues can be labeled efficiently with mixtures of ^{35}S-methionine and ^{35}S-cysteine (e.g., Tran^{35}S-label; ICN Radiochemicals). Tran^{35}S-label will also label P-glycoprotein efficiently and is more reasonably priced than pure

^{35}S-methionine. Cell growth medium that is depleted in methionine alone, or both cysteine and methionine can be used. For the analysis of proteins coexpressed with P-glycoprotein, the preparation of cell extracts will have to be adapted, e.g., by including additional protease inhibitors, by using milder or stronger detergents, or by isolating different cell fractions using differential centrifugation, sucrose gradient centrifugation, or percoll density gradient fractionation.

7. Not every cytotoxic agent that represents a substrate for the multidrug transporter will block ^{3}H-azidopine photoaffinity labeling of P-glycoprotein efficiently *(30)*. Vinblastine, actinomycin D, or doxorubicin interfere, but not colchicine *(46)*. Many MDR modulators will also inhibit ^{3}H-azidopine labeling of P-glycoprotein *(30)*. The efficacy of blocking ^{3}H-azidopine labeling of P-glycoprotein depends on the concentration and the P-glycoprotein isoform, so competitive agents need to be titrated. Since ^{3}H-azidopine and competitive agents are incubated with P-glycoprotein for 1 h before crosslinking occurs, the order of adding these reagents is not important. As described in Section 3.1.2., viable cells containing P-glycoprotein at the cell surface are used for ^{3}H-azidopine labeling. Alternatively, crude membranes (50–100 µg protein per sample) prepared from P-glycoprotein-expressing cells can be used. Labeling of whole cells, however, appears more efficient and gives lower background. Step 4 in Section 3.1.2. is optional. The ^{3}H-azidopine labeling protocol can be scaled up several-fold, but it is important to limit the volume in steps 4 and 5 to 100 µL.

8. With regard to the yield of proteins, extractions with RIPA buffer or triton elution buffer (*see* Sections 3.2.1. and 3.2.2.) are similar. The triton extraction protocol has the perhaps marginal advantage that the extracted proteins, if necessary, can be concentrated by using acetone precipitation. For the preparation of triton extracts, the sonication step (step 3) is quite crucial and determines how effectively proteins are extracted. A Teqmar sonic disruptor may not be accessible in every laboratory. Other types of sonicators may be used alternatively, but sonication time should be adjusted (e.g., three times a second for much more effective sonicators that use a microtip, such as Heat Systems Ultrasonic processor XL, or 5 min with much less effective sonication bath, such as Cole-Palmer 8851). The DNAse reaction is often completed before step 4, so the incubation of the sample at 37°C may eventually be skipped or shortened, especially if labile proteins coexpressed with P-glycoprotein are to be analyzed.

9. Usually it is sufficient to isolate crude membranes to detect low levels of P-glycoprotein by Western blot analysis. If required, plasma membranes can be purified by differential centrifugation, or via sucrose or percoll density gradient fractionation. Generally such protocols require, however, a higher number of cells for the experiment.

10. The radioimmunoprecipitation protocol usually gives very low background. If background is too high, bovine serum albumin (BSA) can be used to block nonspecific binding to protein A sepharose. Protein A sepharose is incubated with BSA before rabbit antimouse secondary antibody is bound and/or BSA can be added to the extract (final concentration 10 mg/mL). Occasionally it is helpful to

use several different concentrations of radiolabeled extract to establish the immunoprecipitation protocol. As indicated in Note 3, antibodies may vary from lot to lot, thus, they need to be titered. Occasionally shortening of incubation periods may also help to reduce the background. The method described for eluting proteins from protein A sepharose (step 5 in Section 3.3.3.) involves a relatively large volume of elution buffer and, therefore, is quite quantitative. The subsequent acetone precipitation, however, may cause problems since occasionally the acetone precipitate is difficult to dissolve. Sucrose is added during the acetone precipitation to make the pellet easier to suspend. As an alternative to steps 5 and 6 in Section 3.3.3., immunoprecipitated proteins may be eluted from the protein A sepharose in 1X SDS polyacrylamide gel loading buffer directly.

11. Since ^{3}H-azidopine labels P-glycoprotein more specifically than ^{35}S-methionine, preadsorption of cell extracts to *Staphylococcus aureus* protein A may not be required. When comparing different ^{3}H-azidopine-labeled cell lines, the protein extracts should be normalized for protein content rather than for incorporated radioactivity.

References

1. Gottesman, M. M. (1993) How cancer cells evade chemotherapy—sixteenth Richard and Linda Rosenthal Foundation Award Lecture. *Cancer Res.* **53**, 747–754.
2. Kessel, D., Botterill, V., and Wodinsky, I. (1968) Uptake and retention of daunomycin by mouse leukemic cells as factors in drug response. *Cancer Res.* **28**, 938–941.
3. Biedler, J. L. and Riehm, H. (1970) Cellular resistance to actinomycin D in Chinese hamster cells *in vitro*: cross-resistance, radioautographic, and cytogenetic studies. *Cancer Res.* **30**, 1174–1184.
4. Gottesman, M. M. and Pastan, I. (1993) Biochemistry of multidrug resistance mediated by the multidrug transporter. *Annu. Rev. Biochem.* **62**, 385–427.
5. Germann, U. A. (1993) Molecular analysis of the multidrug transporter. *Cytotechnology* **12**, 33–62.
6. Nielsen, D. and Skovsgaard, T. (1992) P-glycoprotein as multidrug transporter: a critical review of current multidrug resistant cell lines. *Biochim. Biophys. Acta* **1139**, 169–183.
7. Ford, J. M. and Hait, W. N. (1993) Pharmacologic circumvention of multidrug resistance. *Cytotechnology* **12**, 171–212.
8. Gottesman, M. M. and Pastan, I. (1991) The multidrug resistance (MDR1) gene as a selectable marker in gene therapy. *Human Gene Transfer* **219**, 185–191.
9. Kane, S. E. and Gottesman, M. M. (1993) Use of multidrug resistance gene in mammalian expression vectors. *Methods Enzymol.* **217**, 34–47.
10. Gottesman, M. M., Germann, U. A., Aksentijevich, I., Sugimoto, Y., Cardarelli, C. O., and Pastan, I. (1994) Gene transfer of drug resistance genes. Implications for cancer therapy. *Ann. N.Y. Acad. Sci.* **716**, 126–138.
11. Kane, S. E., Troen, B. R., Gal, S., Ueda, K., Pastan, I., and Gottesman, M. M. (1988) Use of a cloned multidrug-resistance gene for co-amplification and over-

production of MEP, a transformation-regulated secreted acid protease. *Mol. Cell. Biol.* **8**, 3316–3321.

12. Kane, S. E., Reinhard, D. H., Fordis, C. M., Pastan, I., and Gottesman, M. M. (1989) A new vector using the human multidrug resistance gene as a selectable marker enables overexpression of foreign genes in eukaryotic cells. *Gene* **84**, 439–446.

13. Aran, J. M., Gottesman, M. M., and Pastan, I. (1994) Drug-selected coexpression of human glucocerebrosidase and P-glycoprotein using a bicistronic vector. *Proc. Natl. Acad. Sci. USA* **91**, 3176–3180.

14. Sugimoto, Y., Aksentijevich, I., Gottesman, M. M., and Pastan, I. (1994) Efficient expression of drug-selectable genes in retroviral vectors under control of an internal ribosome entry site. *Biotechnology* **12**, 694–698.

15. Germann, U. A., Gottesman, M. M., and Pastan, I. (1989) Expression of a multidrug resistance-adenosine deaminase fusion gene. *J. Biol. Chem.* **264**, 7418–7424.

16. Germann, U. A., Chin, K.-V., Pastan, I., and Gottesman, M. M. (1990) Retroviral transfer of a chimeric multidrug resistance-adenosine deaminase gene. *FASEB J.* **4**, 1501–1507.

17. Richert, N. D., Aldwin, L., Nitecki, D., Gottesman, M. M., and Pastan, I. (1988) Stability and covalent modification of P-glycoprotein in multidrug-resistant KB cells. *Biochemistry* **27**, 7607–7613.

18. Greenberger, L. M., Lothstein, L., Williams, S. S., and Horwitz, S. B. (1988) Distinct P-glycoprotein precursors are overproduced in independently isolated drug-resistant cell lines. *Proc. Natl. Acad. Sci. USA* **85**, 3762–3766.

19. Chen, C.-J., Chin, J. E., Ueda, K., Clark, D. P., Pastan, I., Gottesman, M. M., and Roninson, I. B. (1986) Internal duplication and homology with bacterial transport proteins in the *mdr*1 (P-glycoprotein) gene from multidrug-resistant human cells. *Cell* **47**, 381–389.

20. Gros, P., Croop, J., and Housman, D. E. (1986) Mammalian multidrug resistance gene: Complete cDNA sequence indicates strong homology to bacterial transport proteins. *Cell* **47**, 371–380.

21. Kartner, N., Evernden-Porelle, D., Bradley, G., and Ling, V. (1985) Detection of P-glycoprotein in multidrug-resistant cell lines by monoclonal antibodies. *Nature* **316**, 820–823.

22. Yoshimura, A., Kuwazuru, Y., Sumizawa, T., Ichikawa, M., Ikeda, S., Ueda, T., and Akiyama, S.-I. (1989) Cytoplasmic orientation and two-domain structure of the multidrug transporter, P-glycoprotein, demonstrated with sequence-specific antibodies. *J. Biol. Chem.* **264**, 16,282–16,291.

23. Georges, E., Bradley, G., Gariepy, J., and Ling, V. (1990) Detection of P-glycoprotein isoforms by gene-specific monoclonal antibodies. *Proc. Natl. Acad. Sci. USA* **87**, 152–156.

24. Georges, E., Tsuruo, T., and Ling, V. (1993) Topology of P-glycoprotein as determined by epitope mapping of MRK-16 monoclonal antibody. *J. Biol. Chem.* **268**, 1792–1798.

25. Scheper, R. J., Bulte, J. W. M., Brakkee, J. G. P., Quak, J. J., Van Der Schoot, E., Balm, A. J. M., Meijer, C. J. L. M., Broxterman, H. J., Kuiper, C. M., Lankelma, J., and Pinedo, H. M. (1988) Monoclonal antibody JSB-1 detects a highly conserved epitope on the P-glycoprotein associated with multidrug resistance. *Int. J. Cancer* **42**, 389–394.

26. Hamada, H. and Tsuruo, T. (1986) Functional role for the 170- to 180-kDa glycoprotein specific to drug-resistant tumor cells as revealed by monoclonal antibodies. *Proc. Natl. Acad. Sci. USA* **83**, 7785–7789.

27. Arceci, R. J., Stieglitz, K., Bras, J., Schinkel, A., Baas, F., and Croop, J. (1993) Monoclonal antibody to an external epitope of the human mdr1 P-glycoprotein. *Cancer Res.* **53**, 310–317.

28. Padmanabhan, R., Tsuruo, T., Kane, E. S., Willingham, M. C., Howard, B. H., Gottesman, M. M., and Pastan, I. (1991) Magnetic-affinity cell sorting of human multidrug-resistant cells. *J. Natl. Cancer Inst.* **83**, 565–569.

29. Padmanabhan, R., Padmanabhan, R., Howard, T., Gottesman, M. M., and Howard, B. H. (1993) Magnetic affinity cell sorting to isolate transiently transfected cells, multidrug-resistant cells, somatic cell hybrids, and virally infected cells. *Methods Enzymol.* **218**, 637–651.

30. Safa, A. R. (1993) Photoaffinity labeling of P-glycoprotein in multidrug-resistant cells. *Cancer Invest.* **11**, 46–56.

31. Bayley, H. and Knowles, J. R. (1977) Photoaffinity labeling. *Methods Enzymol.* **46**, 69–115.

32. Beck, W. T. and Qian, X.-D. (1992) Photoaffinity substrates for P-glycoprotein. *Biochem. Pharmacol.* **43**, 89–93.

33. Bruggemann, E. P., Germann, U. A., Gottesman, M. M., and Pastan, I. (1989) Two different regions of P-glycoprotein are photoaffinity labeled by azidopine. *J. Biol. Chem.* **264**, 15,483–15,488.

34. Greenberger, L. M., Lisanti, C. J., Silva, J. T., and Horwitz, S. B. (1991) Domain mapping of the photoaffinity drug-binding sites in P-glycoprotein encoded mouse mdr1b. *J. Biol. Chem.* **266**, 20,744–20,751.

35. Shen, D.-W., Cardarelli, C., Hwang, J., Cornwell, M., Richert, N., Ishii, S., Pastan, I., and Gottesman, M. M. (1986) Multiple drug resistant human KB carcinoma cells independently selected for high-level resistance to colchicine, Adriamycin or vinblastine show changes in expression of specific proteins. *J. Biol. Chem.* **261**, 7762–7770.

36. Roninson, I. B., Chin, J. E., Choi, K., Gros, P., Housman, D. E., Fojo, A. T., Shen, D.-W., Gottesman, M. M., and Pastan, I. (1986) Isolation of human mdr DNA sequences amplified in multidrug-resistant KB carcinoma cells. *Proc. Natl. Acad. Sci. USA* **83**, 4538–4552.

37. Shen, D.-W., Fojo, A. T., Chin, J. E., Roninson, I. B., Richert, N., Pastan, I., and Gottesman, M. M. (1986) Human multidrug resistant cell lines: increased mdr1 expression can precede gene amplification. *Science* **232**, 643–645.

38. Thiebaut, F., Tsuruo, T., Hamada, H., Gottesman, M. M., Pastan, I., and Willingham, M. C. (1989) Immunohistochemical localization in normal tissues of

different epitopes in the multidrug transport protein, P170: evidence for localization in brain capillaries and cross-reactivity of one antibody with a muscle protein. *J. Histochem. Cytochem.* **37,** 159–164.

39. Rao, V. V., Anthony, D. C., and Piwnica-Worms, D. (1994) MDR1 gene-specific monoclonal antibody C494 cross-reacts with pyruvate carboxylase. *Cancer Res.* **54,** 1536–1541.

40. Tanaka, S., Currier, S. J., Bruggemann, E. P., Ueda, K., Germann, U. A., Pastan, I., and Gottesman, M. M. (1990) Use of recombinant P-glycoprotein fragments to produce antibodies to the multidrug transporter. *Biochem. Biophys. Res. Commun.* **166,** 180–186.

41. Bruggemann, E. P., Chaudhary, V., Gottesman, M. M., and Pastan, I. (1991) *Pseudomonas* exotoxin fusion proteins are potent immunogens for raising antibodies against P-glycoprotein. *Biotechniques* **10,** 202–209.

42. Mechetner, E. B. and Roninson, I. B. (1992) Efficient Inhibition of P-glycoprotein-mediated multidrug resistance with a monoclonal antibody. *Proc. Natl. Acad. Sci. USA* **89,** 5824–5828.

43. Heike, Y. and Tsuruo, T. (1993) Antibodies in the study of multiple drug resistance. *Cytotechnology* **12,** 91–107.

44. Gottesman, M. M., Goldstein, L. J., Bruggemann, E., Currier, S. J., Galski, H., Cardarelli, C., Thiebaut, F., Willingham, M. C., and Pastan, I. (1989) Molecular diagnosis of multidrug resistance, in *Cancer Cells, Molecular Diagnosis of Human Cancer* (Furth, M. and Greaves, M., eds.), Cold Spring Harbor Laboratory, Cold Spring Harbor, NY, pp.75–80.

45. O'Driscoll, L., Daly, C., Saleh, M., and Clynes, M. (1993) The use of reverse transcriptase-polymerase chain reaction (RT-PCR) to investigate specific gene expression in multidrug-resistant cells. *Cytotechnology* **12,** 289–314.

46. Germann, U. A., Willingham, M. C., Pastan, I., and Gottesman, M. M. (1990) Expression of the human multidrug transporter in insect cells by a recombinant baculovirus. *Biochemistry* **29,** 2295–2303.

14

Detection and Isolation of Recombinant Proteins from Mammalian Cells by Immunoaffinity Chromatography: p53

Jamil Momand and Bahman Sepehrnia

1. Introduction

Recombinant proteins can be purified through immunoaffinity chromatography. The advantage of this technique over conventional chromatography methods is the reduced number of steps required for purification. However, owing to the high avidity of antibodies for their respective antigens, it is often difficult to elute the recombinant protein from the immunoaffinity column without denaturing the antibody or protein antigen. If the antibody binding epitope on the recombinant protein is mapped to a short primary sequence a synthetic epitope peptide can be used to elute the protein from the antibody under mild conditions. This technique has been employed to elute proteins from immunoaffinity columns with the retention of biological function.

This chapter sets out to describe the methods for:

1. The growth of monoclonal antibody (MAb)-secreting hybridomas;
2. The purification and preparation of antibodies for coupling to activated resin;
3. The growth and harvesting of mammalian cells expressing recombinant protein;
4. The application of soluble cellular lysate to the immunoaffinity column; and
5. The elution of the recombinant protein from the column.

Recombinant p53 is used to illustrate this methodology. However, this protocol should be applicable to the purification of any recombinant or native protein that contains a defined epitope of short sequence and for which the antibody is available. A brief background of the p53-specific MAb used for the purification and the source of mammalian expressed recombinant p53 is presented, followed by the detailed protocol.

From: *Methods in Molecular Biology, vol. 63: Recombinant Protein Protocols:*
Detection and Isolation Edited by: R. Tuan Humana Press Inc., Totowa, NJ

P53 was first detected as a cellular protein in transformed cell lines utilizing polyclonal antiserum derived from mice or hamsters infected with SV40 virus *(1,2)*. Spleen cell/myeloma hybridomas were generated from a mouse injected with syngeneic tumor fragments from animals injected with SV40-transformed murine cells *(3)*. Whereas several MAbs derived from the spleen of this mouse recognized SV40 viral encoded proteins, one MAb, PAb421 (originally named L21), was observed to detect a cellular protein called p53. The MAb PAb421 belongs to the IgG2a class and binds amino acid residues 370–378 of murine p53 *(4,5)*. It is ideally suited for purifying recombinant p53 because it recognizes most mutant and wild-type forms of p53 from a variety of different species *(3,6)*.

As with any method, there are limitations to immunoaffinity chromatography. Conveniently, a few studies employing PAb421 can be used to illustrate these limitations. It has been reported that there are some forms of recombinant p53 and native p53 not recognized by PAb421 *(7,8)*. One mechanism that prevents binding is phosphorylation of p53. It has been shown that at least one serine residue within the PAb421 epitope of p53 can be phosphorylated in vivo, and that the phosphorylated p53 is unable to bind PAb421 *(9,10)*. Another mechanism that prevents an antibody from binding its target protein, is alternative splicing of mRNA. Murine p53 mRNA has been shown to undergo an alternative splicing reaction resulting in the replacement of 26 carboxyl terminal amino acid residues with 17 amino acids derived from intron 10 *(11,12)*. The PAb421 epitope lies within the region of p53 that is spliced out. A third mechanism leading to epitope loss is a conformational alteration of p53. Several missense mutations in the central region of p53 have been shown to lead to an abnormal conformation incapable of binding DNA. This altered conformation masks the epitope usually recognized by the MAb called PAb246 *(6)*. The missense mutations giving rise to the altered conformations actually lie outside of the actual epitope. In sum, an antibody may not recognize its protein target due to posttranslational modifications, alternative splicing of its mRNA, or conformational changes. This warrants some caution in using this technique (*see* Note 1). Notwithstanding these potential pitfalls, however, immunoaffinity chromatography remains an effective method to quickly purify recombinant proteins from a variety of sources.

As a source of p53, our laboratory used rat embryo fibroblasts transfected with a plasmid overexpressing mutant murine p53 *(6)*. This cell-line was derived from primary rat embryo fibroblasts transformed by cotransfection of a plasmid expressing activated H-*ras* and a plasmid expressing mutant murine p53. Focus-forming cells were cloned and screened for p53 expression by immunoprecipitation of the soluble cell lysates after metabolically radiolabeling the cells with ^{35}S-methionine. The p53 was separated from bound antibody on a denaturing sodium dodecyl sulfate-polyacrylamide gel (SDS-PAGE) and detected by autoradiography. Recombinant p53 was purified from these cells by constructing a column containing the PAb421 antibody covalently bound to crosslinked agarose *(13)*. P53 was specifically eluted by incubation with a 14 amino acid residue epitope peptide encompassing amino acid residues 370–378 of p53. A scaled-up version of this immunoaffinity chromatographic procedure was used to copurify a 90 kDa polypeptide along with p53. Through protein sequence analysis, it was identified as the product of the *MDM2* proto-oncogene *(14)*. An immunoaffinity column constructed with PAb421 has also been used to purify p53 from *Spodoptera frugiperda* insect cells infected with a recombinant baculovirus expressing human p53 *(15)*. The p53 was shown to bind DNA in a sequence-specific DNA manner and activate transcription in vitro *(16–18)*. These studies demonstrated that immunopurified p53 retained biological activity after purification.

2. Materials

2.1. Columns and Column Resins

1. Precolumn resin for protein A-Sepharose column: SepharoseCL-4B (Sigma, St. Louis, MO).
2. Empty, low-pressure glass column for SepharoseCL-4B resin: 1 × 5 cm Econo-Column (Bio-Rad, Hercules, CA).
3. Resin for purifying PAb421 antibody: protein A-Sepharose CL-4B (Sigma).
4. Empty, low-pressure glass column for protein A-Sepharose CL-4B: 1.5 × 20 cm Econo-Column (Bio-Rad).
5. Precolumn resin for immunoaffinity column: BioGel A-5m 100–200 mesh (Bio-Rad).
6. Empty, low pressure glass column for BioGel A-5m: 1 × 5 cm Econo-Column (Bio-Rad).
7. Resin for immunoaffinity column: Affigel-10 or Affigel-Hz (Bio-Rad).
8. Empty low pressure glass column for immunoaffinity resin: 1.5 × 20 cm Econo-Column (Bio-Rad).

2.2. Buffers

1. Lysis buffer: 50 mM Tris-HCl, pH 8.0, 5 mM EDTA, 150 mM NaCl, 0.5% Nonidet-P40 (v/v).
2. Phosphate-buffered saline (PBS): 137 mM NaCl, 2.7 mM KCl, 4.3 mM Na$_2$HPO$_4$, 1.4 mM KH$_2$PO$_4$.
3. TEN buffer: 50 mM Tris-HCl, pH 7.4, 5 mM EDTA, 150 mM NaCl, 0.5% Nonidet-P40 (v/v).
4. SNNTE buffer: 5% sucrose (w/v), 1% Nonidet-P40 (v/v), 500 mM NaCl, 50 mM Tris-HCl, pH 7.4, 5 mM EDTA.
5. RIPA buffer: 20 mM Tris-HCl, pH 7.4, 150 mM NaCl, 1% Triton X-100 (v/v), 1% deoxycholate (w/v), 0.1% sodium dodecyl sulfate (w/v).
6. TBS buffer: 15 mM Tris-HCl (pH 7.4), 140 mM NaCl, 5 mM KCl, 3 mM MgCl$_2$.

2.3. Other Materials

1. Epitope peptide LKTKKGQSTSRHKK (*see* Note 2).
2. Beckman UltraClear Centrifuge Tubes (25 × 89 mm).
3. Phenylmethylsulfonylfluoride (abbreviated as PMSF): 100 mM stock stored in ethanol at 4°C. This is available from Sigma.
4. Pepstatin (Boehringer Mannheim, Mannheim, Germany): 1 mM stock stored in ethanol at 4°C.
5. E-64 (Boehringer-Mannheim): 1 mM stock stored in water at 4°C.
6. Dulbecco's modified essential media (Irvine Scientific, Santa Ana, CA), stored at 4°C.
7. Heat-inactivated fetal calf serum (Irvine Scientific): heat-inactivated in the laboratory by incubation at 55°C for 30 min, stored at –20°C.
8. Penicillin/streptomycin solution to prevent microbial growth in cultured cells (Irvine Scientific): penicillin G 10,000 U/mL, streptomycin sulfate 10,000 μg/mL in normal saline. This stock solution is diluted 100-fold in the final tissue culture medium.
9. 3-[(3-Cholamidopropyl)-dimethylammonio]-2-hydroxyl-1-propanesulfonate (Sigma): this is a detergent abbreviated as CHAPS.

3. Methods
3.1. Growing MAb-Secreting Hybridomas

1. Maintain PAb421-secreting hybridoma cells (*see* Note 3) at a density of 10^6 cells/mL in media consisting of 90% Dulbecco's modified essential media (DMEM), 10% heat-inactivated fetal calf serum, 100 U/mL penicillin G, 100 μg/mL streptomycin in a humidifying incubator of 5% CO_2 at 37°C. Grow the cells in T-700 flasks (Becton Dickinson, Franklin Lakes, NJ). The cells grow in suspension as well as attached to the plastic surface. Tilt the flasks at a slight angle against the inside wall of the incubator. This provides a large surface area for some of the cells to attach. Unscrew the cap enough to allow incubator gases to equilibrate with the media.
2. Swirl the flasks once a day to allow for proper mixing of the cells with the media. Remove a small aliquot (100 μL) of the cell suspension to quantify the cell density with a Coulter counter. Add media to the growing cells to maintain a density of 10^6 cells/mL.
3. When the flask is filled to capacity (approx 500 mL) and the cell density has reached 10^6 cells/mL, allow the cells to incubate another three days before harvesting. Clear the supernatant by centrifugation at 2300g for 20 min at 4°C in a swinging bucket rotor. Remove the supernatant to fresh storage tubes and maintain at 4°C for a maximum of three days prior to application to the protein-A Sepharose column. For long-term storage, store the hybridoma supernatant at –20°C. Recentrifuge the supernatant just prior to application to the protein A-Sepharose column. The concentration of antibody in the hybridoma supernatant ranges from 1 to 5 μg/mL (*see* Section 3.2.2. for determination of antibody concentration).

3.2. Purification and Preparation of Antibodies for Coupling to Activated Resin.

1. Swell 1.5 g of protein A-Sepharose CL4B at a ratio of 1:14 (w/v) in TEN buffer for at least 1 h at room temperature. Wash the swollen resin (6 mL vol) twice in TEN buffer by gentle mixing. Remove each wash by centrifugation at 1500g for 5 min. Resuspend the washed resin in 10 vol of TEN buffer and pour into a 1.5 × 20 cm Econo column connected to a peristaltic pump located in line postcolumn. Pack the column at a flowrate of 0.5 mL/min. Pack a 1-mL precolumn with SepharoseCL-4B in TEN buffer in a similar fashion. Attach the outlet of precolumn to the inlet of the protein A-Sepharose column with tygon tubing.

2. Apply the hybridoma supernatant to the column at a flowrate of 0.5 mL/min at 4°C. The flowrate can be adjusted with the peristaltic pump. After one pass, wash the protein-A Sepharose column with seven column volumes of 0.1M sodium borate, pH 9.0. Elute the PAb421 antibody with 0.1M sodium citrate, pH 4.0 into prealiquoted glass borosilate tubes (13 × 100 mm) containing 1.5 mL of 1.0M Tris-HCl, pH 8.0. This step neutralizes the PAb421 antibody solution as quickly as possible to avoid denaturation by the acidic eluant. Quantify the amount of PAb421 in each fraction with a Bradford assay (available from Bio-Rad) using manufacturer-supplied sheep IgG as a protein standard (19). Pool the antibody-containing fractions. Run a small aliquot of the pooled fractions (1–3 µg) on 10% SDS-PAGE and stain with Coomassie brilliant blue R (Sigma) to ensure that the antibody heavy and light chains (relative mol wts of approx 55 and 25 kDa, respectively) are present *(20)*. The amount of antibody obtained from 3.7 L of hybridoma supernatant was 56 mg, and the final concentration was 1.8 mg/mL.

3. Concentrate the antibody to 5.4 mg/mL with an Amicon Ultrafiltrator (Amicon, Beverly, MA) using a YT membrane with a mol-wt cutoff of 30,000. Dialyze the concentrated antibody against coupling buffer (Bio-Rad), and couple the antibody to Affigel-10 or Affigel-Hz according to the manufacturer's instructions (Bio-Rad). Determine the amount of antibody covalently bound to the column by measuring the protein concentration of the antibody in the coupling buffer before and after coupling with the Bradford assay. In our hands, approx 73% of the PAb421 coupled to Affigel-Hz resulting in 2.73 mg of antibody coupled per mL of column bed. To verify that antibody did couple to the resin, run the resin on SDS-PAGE and visualize the heavy and light chain of the antibody by Coomassie staining. If Affigel-Hz is employed, the antibody covalently couples to the resin through the Fc domain (via oligosaccharides) so that the Fab domain is free to interact with antigen. Boil the antibody-coupled resin in sample buffer *(20)*. This results in the dissociation of the light chain and heavy chain. The heavy chain, for the most part, remains covalently bound to the resin. Thus, during electrophoresis most of the heavy chain bound to the resin remains in the well of the stacking gel and the light chain migrates to its normal relative size of approx 25 kDa mol wt (*see* Note 4). Store the affinity column as a 50% slurry in lysis buffer supplemented with 0.01% sodium azide.

3.3. Growth and Harvesting Mammalian Cells Expressing Recombinant Protein

1. The source of p53 used for our studies has been transformed A-1 rat cells expressing recombinant murine p53 *(6)*. P53 constitutes 0.1–0.24% of the total soluble protein in these cells and represents the highest level of p53 expression in mammalian cells reported *(13,14)*. Grow the cells in 90% DMEM, 10% heat-inactivated fetal calf serum, 100 U/mL penicillin G, 100 μg/mL streptomycin in a humidifying incubator of 5% CO_2 at 37°C. Maintain cells in 15 cm diameter dishes and split 1:5 or 1:10 when cells form a confluent monolayer.

2. After the cells have reached confluency (approx 5×10^6 cells per plate) discard the media and add approx 3 mL of ice-cold PBS to each plate. Scrape the cell/PBS mixture with a rubber policeman (we use a customized 6 cm wide rubber policeman but commercially-available cell lifters [Costar, Cambridge, MA] will suffice). With a Pasteur pipet, transfer the suspended cells to a 50 mL, graduated, conical polypropylene tube (Becton Dickinson, Franklin Lakes, NJ) prechilled on ice. Perform the next steps at 0–4°C unless otherwise noted. Typically, 10 plates of cells can be combined in one 50-mL tube.

3. Spin the cells in a refrigerated clinical centrifuge at 1000*g* for 5 min. Discard the supernatant and store the cell pellet at –80°C until the next step.

3.4. Application of Soluble Cellular Lysate to the Immunoaffinity Column

1. Add 2 mL of lysis buffer per 15-cm dish of confluent cells to each cell pellet. To inhibit proteolysis of p53 during the purification, supplement the lysis buffer with protease inhibitors at a final concentration of 1 m*M* PMSF, 1 μ*M* E-64, and 1 μ*M* pepstatin (*see* Note 5). Thaw the frozen cell pellet/lysis buffer mixture on ice for 20 min and occasionally vortex to disperse the pellet.

2. Sonicate the resuspended cells at a ratio of one 10-s pulse/mL of suspended cells on a Branson Sonifier (Model 250, Danbury, CT) equipped with a water/ice-filled cuphorn. Set the control setting to three. This gives an output power of 75 W. Submerge the capped 50 mL tube containing the suspended cells into the cuphorn just above the metal horn. Use one hand to hold the tube steady, and the other to control the sonicator output (wear hearing protective earmuffs). Deliver a 10-s pulse for each round of sonication. Wait approx 30 s between each pulse to prevent the lysate from warming during sonication. Care is taken to re-plenish the cuphorn with ice during large preparations to keep the lysate cold during this procedure.

3. Pour the lysate into Beckman UltraClear centrifuge tubes and centrifuge at 100,000*g* for 1 h in a SW28 rotor. Fill the tubes to capacity (approx 38 mL) and balance them to prevent tube failure during the centrifuge run. If not enough lysate is available to completely fill the centrifuge tube it can be topped off with lysis buffer.

4. Immediately transfer the supernatant to an Erlenmeyer flask and apply to the PAb421 column equipped with a 1-mL precolumn at a flowrate of 0.25 mL/min. Ensure that the bead size of the precolumn is the same as the bead size of the affinity column to maintain the proper flowrate. The precolumn serves two purposes: to filter insoluble material not removed by the centrifugation step and to bind nonspecific proteins prior to encountering the PAb421 column. It is important to change the precolumn once every 6 h during the run to prevent clogging. Failure to change the precolumn results in drawing air into the PAb421 column. This may denature the PAb421 antibody. Save a small aliquot of the lysate (0.1–1 mL) before and after each column pass and store at –80°C for recovery analysis.

3.5. Washing the Column and Elution of the Recombinant Protein

1. Remove the resin from the column, place in a 50-mL Falcon tube, and batch wash (*see* Note 6). For high stringency washing, wash the column twice with four volumes of the following buffers:
   a. Lysis buffer.
   b. Lysis buffer, 5 mM CHAPS.
   c. Lysis buffer, 350 mM NaCl.
   d. Lysis buffer, 500 mM LiCl.
   e. 0.12M sodium thiocyanate (pH 7.5), 0.5% Nonidet-P40.
   f. Lysis buffer.
   For each wash step, gently rock the resin/buffer slurry 1–3 s. Remove the wash buffer by centrifugation at 1500g for 3 min. Avoid vigorous mixing as it can damage the column.
2. For most purposes a nonstringent wash will be sufficient and lead to recombinant protein that is 90–98% pure. For nonstringent washing, the resin should be washed with lysis buffer, SNNTE, RIPA, and again with lysis buffer (*see* Note 7).
3. Elute the p53 by adding four volumes of 0.25 mg/mL (w/v) epitope peptide in lysis buffer. Rock the resin/buffer slurry for 1 h and collect the eluted p53 by centrifugation as described in Section 3.5.1. Store the supernatant containing the p53 at –80°C.
4. If the column is saturated after the initial pass of lysate, reapply the flow-through to the column after equilibration with lysis buffer (*see* Note 8).
5. Store the column in lysis buffer supplemented with 0.01% (w/v) sodium azide at 4°C.
6. Assess p53 recovery by immunoblot analysis and assess purity by SDS-PAGE followed by silver staining. Figure 1A is an autoradiograph of an immunoblot monitoring the presence of p53 during purification. Lane 1 represents the level of p53 in the soluble cell lysate prior to purification. Lane 2 represents the level of p53 remaining in the lysate after one pass through the column. A comparison of the p53 level in lane 1 and lane 2 suggests that the column was saturated and that the lysate from the first pass could be used to obtain more p53. Lane 3 shows the level of p53 eluted from the column after incubation with the epitope peptide. Since it appeared that the column was saturated, the lysate from the first pass was reapplied to the column. Lane 4 shows the level of p53 eluted from the column

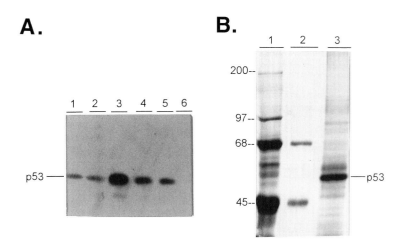

Fig. 1. Purification of p53 by immunoaffinity chromatography. **(A)** Immunoblot analysis of p53 at successive steps of purification. Lane 1, Soluble A-1 cellular lysate (30 μL of 1 mg/mL total protein); lane 2, column flowthrough (30 μL); lane 3, peptide eluted p53 from the column (30 μL); lane 4, peptide eluted p53 from the column after a second pass of the initial flowthrough (30 μL); lane 5, column resin after the second peptide elution (7.5 μL); lane 6, flowthrough of second pass (30 μL). Methods: Samples were run on 10% SDS-PAGE (20) and electroblotted onto polyvinylidene difluoride membrane (Immobilon P, Millipore, Bedford, MA) overnight in a electro-blotter (Pharmacia Biotech, Alameda, CA) at 1.0 A at 4°C in Towbin buffer *(26)* supplemented with 0.005% SDS as suggested by Aebersold et al. *(27)*. After transfer the membrane was blocked overnight with 10% (w/v) dry nonfat milk in PBS supplemented with 0.1% (v/v) Tween-20 (PBS-T). The membrane was washed twice with PBS-T and incubated with PAb421 hybridoma supernatant diluted 1:100 in PBS-T plus 1% (w/v) bovine serum albumin overnight. The membrane was washed two more times with PBS-T and incubated with protein A conjugated to horseradish peroxidase (Boehringer Mannheim) diluted 1:5000 in PBS-T for 1–5 h at room temperature. The membrane was washed four times in PBS-T supplemented with 0.1% (v/v) Triton X-100 and incubated with the ECL detection reagent (Amersham, Arlington Heights, IL) according to the manufacturer's instructions. The membrane was exposed to Kodak XAR-5 film for 3 s. **(B)** Silver stained gel of p53 eluted from the PAb421 column. Lane 1, high-mol-wt protein standards (1.0 μg) from Gibco-BRL, Gaithersburg, MD; lane 2, high-mol-wt protein standards (50 ng); lane 3, peptide eluted p53 (10 μL). The molecular weights of the protein standards are indicated on the left side of the gel in kilodaltons. Methods: Samples were run on 7.5% SDS-PAGE and stained with the Bio-Rad silver staining kit according to the manufacturer's instructions. The p53-bound column was stringently washed and treated with ATP to remove hsc70 prior to p53 elution.

after reapplying the lysate from the first pass. This indicated that residual p53 remaining in the lysate after the first pass could be captured by the same column. Lane 5 represents the p53 that remains bound to the column after the second peptide elution. This demonstrated that removal of p53 from the column with peptide was not 100% efficient (*see* Note 8). Lane 6 shows the level of p53 remaining in the lysate after the second pass through the column. There was very little p53 remaining in the lysate after the second pass through the column. This indicated that the majority of p53 could be removed from the initial lysate after two successive passes through the affinity column (compare lane 1 and lane 6). The purity of p53 was assessed by SDS-PAGE followed by silver staining (Fig. 1B). Lane 1 shows the silver staining pattern of 1.0 μg of mol-wt standards. Lane 2 shows the silver staining pattern of 0.05 μg of molecular weight standards. Lane 3 is the silver staining pattern of proteins eluted from the column with the epitope peptide. The minor polypeptides other than p53 may be copurifying proteins that form a tight complex with p53 in vivo.

4. Notes

1. One must be cautious when choosing a Mab for purification of proteins. As mentioned in the introduction a variety of mechanisms can prevent an antibody from binding to its target protein. Therefore it is important to ensure that the recombinant protein expresses the correct epitope prior to constructing the immunoaffinity column. To address this issue, metabolically radiolabel cells with ^{35}S-methionine and immunoprecipitate the recombinant protein with the antibody and protein A (or protein G) covalently bound to cross-linked agarose. The choice of protein A or protein G for immunoprecipitation depends on the IgG species and subclass (*see* ref. *21*; Tables 1 and 2 in Chapter 15). The recombinant protein is identified by its relative migration to protein standards on SDS-PAGE followed by autoradiography.

2. The epitope peptide was synthesized by the FMOC method on a semiautomated synthesizer *(22)*. The peptide cleaved from the solid phase support can be purified by HPLC reverse phase chromatography. However, we have observed that this purification step is not required.

3. PAb122 is another Mab with the same epitope on p53 as PAb421 and can be used as a substitute for PAb421 *(23)*. Hybridoma cells secreting PAb122 are available from the American Type Culture Collection.

4. Pipeting the coupled resin for SDS-PAGE analysis can be tricky. For quantitative analysis always start with a 50% (v/v) slurry. Cut 2–4 mm off the end of a 200-μL Pipetman yellow tip with a razor blade. Vortex the slurry so that the resin is homogeneously suspended and withdraw the required amount for analysis. Boil the resin in SDS-PAGE sample buffer *(20)* and use another clipped pipet tip to load all of the suspended resin plus sample buffer into one well of a Laemmli gel.

5. Three protease inhibitors (PMSF, E-64, and pepstatin) are used because they inhibit three major classes of mammalian proteases–serine, cysteine, and aspartic proteases, respectively *(24)*. *See* ref. *24* for storage conditions and shelf-life.

6. Although batch washing and elution is quicker than conventional column washing, manipulation of the resin outside of the column invariably leads to loss of some resin. We have recently altered our protocol so that these steps can be accomplished without removing the resin from the column. Preliminary evidence indicates that the purity and yield of the recombinant protein is as high as batch washing and elution. Remove the precolumn and wash the immunoaffinity column with four column volumes of each wash buffer at a flowrate of 2.4 mL/min. After the final wash step, drain the buffer such that the meniscus resides just above the wet column bed. Gently layer the epitope peptide solution onto the column bed with a Pasteur pipet. Connect the inlet and outlet lines to each other to make a closed loop and use the peristaltic pump to circulate the peptide solution through the column at a flowrate of 2.4 mL/min for 1.5 h. Recover the peptide-eluted protein by draining into a polypropylene tube.

7. Recombinant p53 expressed in mammalian cells sometimes forms a tight complex with the endogenous chaperone protein hsc70 *(25)*. To specifically remove hsc70, incubate the column-bound p53 in three column volumes of TBS and 1 mM ATP after the last wash step *(13)*. Hsc70 hydrolyzes the ATP and simultaneously dissociates from the p53. Gently rock the column in the TBS/ATP mixture for 1 h at room temperature and remove the free hsc70 by centrifugation or column elution. The p53 that remains bound to the column can be removed by epitope peptide elution.

8. We have observed some residual p53 remaining on the PAb421 column after peptide elution (*see* Fig. 1A, lane 5). It is possible that the tightly-bound p53 represents a distinct subpopulation of p53 with a higher affinity for the antibody than the major fraction of p53. Removal of the tightly-bound p53 by further peptide elution has failed. Attempts to remove the tightly-bound p53 with stripping buffers such as acid, ethylene glycol, or buffers containing a high salt concentration have either been ineffective or have resulted in damage to the column (assessed by the presence of antibody in the recovered stripping buffer).

Acknowledgments

This work was supported by an American Lung Association Research Grant (RG121N) and the Breast Cancer Research Program of the University of California, Grant Number IKB-0102.

References

1. Linzer, D. I. H. and Levine, A. J. (1979) Characterization of a 54K dalton cellular SV40 tumor antigen in SV40 transformed cells. *Cell* **17,** 43–52.
2. Lane, D. P. and Crawford, L. V. (1979) T antigen is bound to a host protein in SV40-transformed cells. *Nature* **278,** 261–263.
3. Harlow, E., Crawford, L. V., Pim, D. C., and Williamson, N. M. (1981) Monoclonal antibodies specific for simian virus 40 tumor antigens. *J. Virol.* **39,** 861–869.
4. Wade-Evans, A. and Jenkins, J. R. (1985) Precise epitope mapping of the murine transformation-associated protein, p53. *EMBO J.* **4,** 699–706.

5. Yewdell, J., Gannon, J. V., and Lane, D. P. (1986) Monoclonal antibody analysis of p53 expression in normal and transformed cells. *J. Virol.* **59,** 444–452.

6. Finlay, C. A., Hinds, P. W., Tan, T.-H., Eliyahu, D., Oren, M., and Levine, A. J. (1988) Activating mutations for transformation by p53 produce a gene product that forms an hsc70-p53 complex with an altered half-life. *Mol. Cell. Biol.* **8,** 531–539.

7. Milner, J. (1984) Different forms of p53 detected by monoclonal antibodies in non-dividing and dividing lymphocytes. *Nature* **310,** 143–145.

8. Ullrich, S. J., Mercer, W. E., and Appella, E. (1992) Human wild-type p53 adopts a unique conformational and phosphorylation state *in vivo* during growth arrest of glioblastoma cells. *Oncogene* **7,** 1635–1643.

9. Hupp, T. R. and Lane, D. P. (1994) Regulation of the cryptic sequence-specific DNA binding function of p53 by protein kinases. *Cold Spring Harb. Symp. Quant. Biol.* **59,** 195–206.

10. Takenaka, I., Morin, F., Seizinger, B. R., and Kley, N. (1995) Regulation of sequence-specific DNA binding function of p53 by protein kinase C and protein phosphatases. *J. Biol. Chem.* **270,** 5405–5411.

11. Arai, N., Nomura, D., Yokota, K., Wolf, D., Brill, E., Shohat, O., and Rotter, V. (1986) Immunologically distinct p53 molecules generated by alternative splicing. *Mol. Cell. Biol.* **6,** 3232–3239.

12. Kulesz-Martin, M. F., Lisafeld, B., Huang, H., Kisiel, N. D., and Lee, L. (1994) Endogenous p53 protein generated from wild-type alternatively spliced p53 RNA in mouse epidermal cells. *Mol. Cell. Biol.* **14,** 1698–1708.

13. Clarke, C. F., Cheng, K., Frey, A. B., Stein, R., Hinds, P. W., and Levine, A. J. (1988) Purification of complexes of nuclear oncogene p53 with rat and Escherichia coli heat shock proteins: In vitro dissociation of hsc70 and dnaK from murine p53 by ATP. *Mol. Cell. Biol.* **8,** 1206–1215.

14. Momand, J., Zambetti, G. P., Olson, D. C., George, D., and Levine, A. J. (1992) The *mdm-2* oncogene product forms a complex with the p53 protein and inhibits p53-mediated transactivation. *Cell* **69,** 1237–1245.

15. Wang, E. H., Friedman, P. N., and Prives, C. (1989) The murine p53 protein blocks replication of SV40 DNA in vitro by inhibiting the initiation functions of SV40 large T antigen. *Cell* **57,** 379–392.

16. Bargonetti, J., Friedman, P. N., Kern, S. E., Vogelstein B., and Prives, C. (1991) Wild-type but not mutant p53 immunopurified proteins bind to sequences adjacent to the SV40 origin of replication. *Cell* **65,** 1083–1091.

17. Kern, S. E., Kinzler, K. W., Bruskin, A., Jarosz, D., Friedman, P., Prives, C., and Vogelstein, B. (1991) Identification of p53 as a sequence-specific DNA-binding protein. *Science* **252,** 1708–1711.

18. Farmer, G., Bargonetti, J., Zhu, H., Friedman, P., Prywes, R., and Prives, C. (1992) Wild-type p53 activates transcription *in vitro. Nature* **358,** 83–86.

19. Bradford, M. M. (1976) A rapid and sensitive method for the quantitation of microgram quantities of protein utilizing the principle of protein-dye binding. *Anal. Biochem.* **72,** 248–254.

20. Laemmli, U. K. (1970) Cleavage of structural proteins during the assembly of the head of bacteriophage T4. *Nature* **227**, 680–685.
21. Harlow, E. and Lane, D. (1988) Antibodies, a laboratory manual. Cold Spring Harbor Laboratory Press, Cold Spring Harbor, NY.
22. Fields, G. B. and Noble, R. L. (1990) Solid phase peptide synthesis utilizing 9-fluorenylmethoxycarbonyl amino acids. *Int. J. Pept. Protein Res.* **35**, 161–214.
23. Gurney, E. G., Harrison, R. O., and Fenno, J. (1980) Monoclonal antibodies against Simian Virus 40 T antigens: evidence for distinct subclasses of Large T antigen and for similarities among nonviral T antigens. *J. Virol.* **34**, 752–763.
24. Beynon, R. J. and Salvesen, G. (1989) Commercially available protease inhibitors, in *Proteolytic Enzymes-A Practical Approach* (Beynon, R. J. and Bond, J. S., eds.), IRL Press, Oxford, pp. 241–249.
25. Hinds, P. W., Finlay, C. A., Frey, A. B., and Levine, A. J. (1987) Immunological evidence for the association of p53 with a heat shock protein, hsc70, in p53-plus-ras-transformed cell lines. *Mol. Cell. Biol.* **7**, 2863–2869.
26. Towbin, H., Staehelin, T. and Gordon, J. (1979) Electrophoretic transfer of proteins from polyacrylamide gels to nitrocellulose sheets: procedure and some applications. *Proc. Natl. Acad. Sci. USA* **76**, 4350–4353.
27. Aebersold, R. H., Leavitt, J., Saavedra, R. A., Hood, L. E., and Kent, S. B. H. (1987) Internal amino acid sequence analysis of proteins separated by one or two dimensional gel electrophoresis after in situ protease digestion on nitrocellulose. *Proc. Natl. Acad. Sci. USA* **84**, 6970–6974.

15

Yeast GAL4 Two-Hybrid System

A Genetic System to Identify Proteins That Interact with a Target Protein

Li Zhu

1. Introduction

The yeast two-hybrid system is a powerful in vivo method for identifying novel genes encoding proteins that interact with a protein of interest *(1,2)*. The system offers several innovations for identifying interacting proteins over conventional biochemical methods such as coimmunoprecipitation and affinity copurification. The two-hybrid system is the first genetic and molecular approach to detect interacting proteins, and the assay is performed in vivo rather than in vitro, which allows detection of interacting proteins in their native configurations. Consequently, the two-hybrid system has an unparalleled level of sensitivity to detect weak and transient protein interactions *(3)*. In addition, the molecular methodology used enables immediate access to the gene encoding the interacting protein of interest. The two-hybrid system can be used either to screen a library for a gene encoding a protein that interacts with a known target protein or to test known proteins for interaction. The significant impact of the two-hybrid system on this field is demonstrated by the many recent discoveries of interacting proteins using the system *(4–36)* and by the rapid proliferation of two-hybrid-based technologies *(37,38)*.

The yeast GAL4 two-hybrid systems (such as the MATCHMAKER Two-Hybrid System, Clontech, Palo Alto, CA) have been reported in the majority of successful two-hybrid studies. In addition, several modified two-hybrid systems for specific applications have been reported *(39,40)*. The underlying principles have also been used to identify proteins interacting with target DNA sequences (one-hybrid system; *41–48*), to identify a third protein that mediates

From: *Methods in Molecular Biology, vol. 63: Recombinant Protein Protocols: Detection and Isolation* Edited by: R. Tuan Humana Press Inc., Totowa, NJ

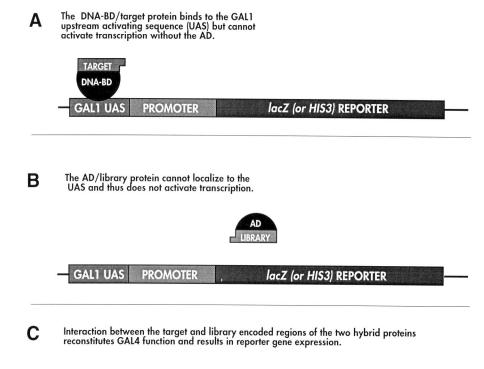

A The DNA-BD/target protein binds to the GAL1 upstream activating sequence (UAS) but cannot activate transcription without the AD.

B The AD/library protein cannot localize to the UAS and thus does not activate transcription.

C Interaction between the target and library encoded regions of the two hybrid proteins reconstitutes GAL4 function and results in reporter gene expression.

Fig. 1. Schematic diagram of the two-hybrid system.

the interaction of two proteins *(49,50)*, and to screen for drugs that block protein interaction *(51)*. More recently, Fields and Sternglanz proposed using two-hybrid principles to construct "protein linkage maps" that would provide a three-dimensional view of complex protein-protein interactions *(51)*. This chapter specifically describes using the yeast GAL4 two-hybrid system to screen a library for a gene encoding a protein that interacts with a known target protein.

The yeast two-hybrid system is based on the concept that many eukaryotic transcriptional activators are composed of structurally-separable and functionally independent domains (Fig. 1; *2,52*). Notably, a global transcriptional activator of the yeast galactose metabolic pathway, GAL4, consists of a DNA-binding domain (DNA-BD) and an activation domain (AD). The DNA-BD recognizes and binds to a sequence (UAS) in the upstream regions of GAL4-responsive genes, whereas the AD contacts other components of the machinery needed to initiate transcription. The presence of both domains, if physically separated, is not sufficient to activate the responsive genes *(53,54)*. In the two-hybrid system, the two GAL4 domains are separately fused to genes encoding proteins that interact with each other, and the recombinant hybrid proteins are expressed in yeast. Interaction of the two hybrid proteins brings the two GAL4 domains in close enough proximity to form a functional activator that activates transcription of reporter genes, making the protein interaction phenotypically detectable.

To use the two-hybrid assay, the gene encoding the target or "bait" protein must already be cloned. This gene is ligated into a DNA-BD vector to generate a fusion between the target protein and the DNA-BD. Likewise, a cDNA library (or when testing for an interaction between two known proteins, a second gene encoding a potentially interacting protein; *see* Note 1) is constructed and ligated into a special AD vector to generate fusions between the proteins encoded by the library cDNAs and the AD. A wide variety of cDNA libraries constructed in AD vectors are now commercially available. In the two-hybrid screening procedure, the two hybrid plasmids are cotransformed into a special yeast reporter strain, such as HF7c *(55)* used in the MATCHMAKER System.

The library screening procedure described here uses two different reporter genes, *HIS3* and *lacZ*, both of which are under the control of GAL4 responsive elements. Yeast transformed with both DNA-BD/target and AD/library plasmids are plated on a minimal medium lacking leucine (Leu) and tryptophan (Trp) to select for transformants that contain both plasmids. The medium also lacks histidine (His), to select for colonies expressing interacting hybrid proteins. If the DNA-BD/target protein interacts with an AD/library protein (both of which are targeted to the yeast nucleus), a functional GAL4 transcriptional activator is reconstituted and reporter gene expression is activated *(56)*. The *HIS3* nutritional reporter gene provides a simple yeast growth selection for cells expressing interacting proteins and is used for identifying positive clones in an initial library screening (Fig. 2). Primary transformants that express the *HIS3* gene (His[+]) are then tested for expression of the second reporter gene, *lacZ*, using a filter assay for β-galactosidase activity. This second assay reduces the background of false positives arising in the His selection. All double positive clones are further tested to eliminate any persisting false positives *(57*; Figs. 3 and 4).

2. Materials

2.1. Plasmids and Yeast Strains

1. The MATCHMAKER two-hybrid system is available from Clontech. This kit provides two yeast reporter strains (HF7c and SFY526), a standard GAL4-BD

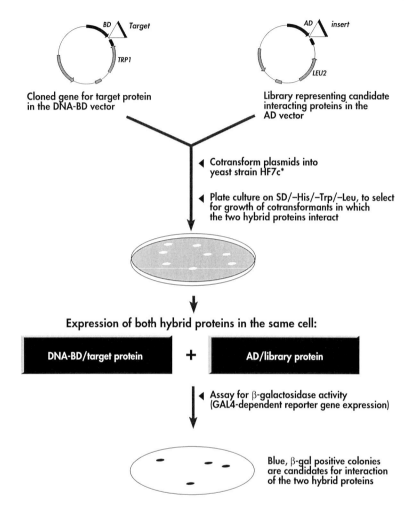

Fig. 2. Screening GAL4 AD fusion libraries for proteins that interact with a target protein.

vector (pGBT9), an AD vector (pGAD424), and several control plasmids (*see* Table 1). Figures 5 and 6 show vector maps for pGBT9 and pGAD424, respectively.

2. Yeast host strain HF7c has *lacZ* and *HIS3* reporter genes and *trp1* and *leu2* transformation marker genes that confer auxotrophic phenotypes to yeast (i.e., cannot grow in minimal –Trp/–Leu medium, unless functional *TRP1* and *LEU2* genes are introduced). The *trp1, leu2, his3, gal4,* and *gal80* mutations are all deletions. The complete yeast genotype is *ura3–52, his–200, lys2–801, ade2–101, trp1–*

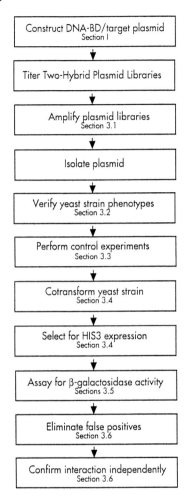

Fig. 3. Guide to the two-hybrid system protocols.

901, eu2–3, 112, gal4–542, gal80–538, LYS2 : : GAL1–HIS3, URA3 : : (GAL4 17-mers)ₛ–CYC1–lacZ (*see* Note 2; 55). *URA3* (but not *Lys2*) is functional in HF7c.

2.2. Yeast Growth and Maintenance

1. YPD medium: 20 g/L Difco peptone, 10 g/L yeast extract, 20 g/L agar (for plates only). Add H_2O to 950 mL. Adjust pH to 5.8, autoclave, and cool to ~55°. Add dextrose (glucose) to 2% (50 mL of a sterile, 40% stock solution per liter).

2. SD medium (*see* Note 3): 6.7 g/L Difco yeast nitrogen base without amino acids (Difco), 20 g/L agar (for solid medium only). Add H_2O to 850 mL/L, add 100 mL of the appropriate 10X dropout solution, adjust to pH 5.8, autoclave, and cool to ~55°C. Then add 50 mL of a sterile, 40% dextrose stock solution per liter of medium.

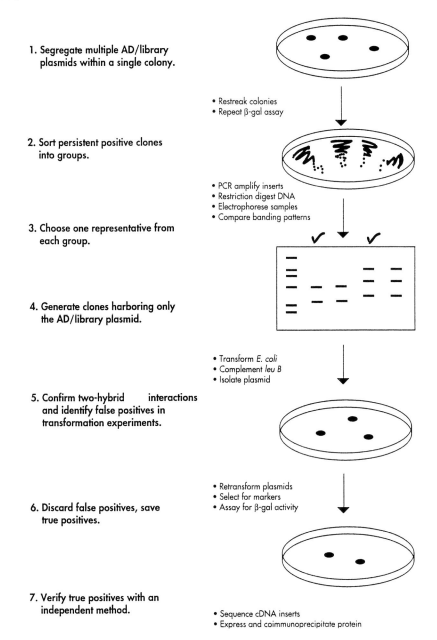

1. Segregate multiple AD/library plasmids within a single colony.

• Restreak colonies
• Repeat β-gal assay

2. Sort persistent positive clones into groups.

• PCR amplify inserts
• Restriction digest DNA
• Electrophorese samples
• Compare banding patterns

3. Choose one representative from each group.

4. Generate clones harboring only the AD/library plasmid.

• Transform *E. coli*
• Complement *leu B*
• Isolate plasmid

5. Confirm two-hybrid   interactions and identify false positives in transformation experiments.

• Retransform plasmids
• Select for markers
• Assay for β-gal activity

6. Discard false positives, save true positives.

7. Verify true positives with an independent method.

• Sequence cDNA inserts
• Express and coimmunoprecipitate protein

Fig. 4. Procedures for eliminating false positives.

3. 10X Dropout solution: 10X dropout solutions contain all but one or more of the components in Table 2; which components are omitted depends on the selection medium desired. To prepare SD/–Trp/–Leu, for example, use a 10X dropout

Table 1
Plasmids in the Two-Hybrid System

Vector	Description	Size, kb	Refs.
pGBT9	$GAL4_{(1–147)}$ DNA-BD,$TRP1$, amp^r	5.4	52
pGAD424	$GAL4_{(768–881)}$ AD, $LEU2$ amp^r	6.6	57
pVA3	GAL4 BD/mouse p53	6.4	10
pTD1	GAL4 AD /SV40 large T antigen	15	9
pLAM5$'^a$	GAL4 BD/human Lamin C	6.3	52
pCL1	wild-type, full-length GAL4	15	52

^{a}pLAM5' was modified by Clontech from the original plasmid pLm5' by subcloning a human Lamin C fragment from a HIS-marked plasmid to pGBT9.

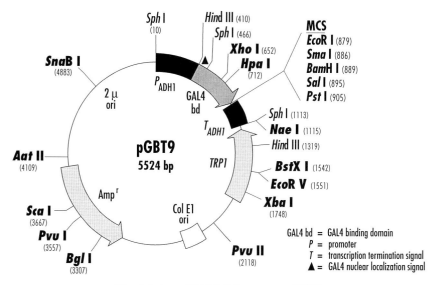

Fig. 5. Map of DNA-BD vector pGBT9.

solution lacking Trp and Leu. 10X dropout solutions can be autoclaved and stored at 4°C for up to 1 yr.
4. 1M 3-Aminotriazole (3-AT) (Sigma) filter-sterilized.
5. 40% Dextrose, autoclaved or filter-sterilized (avoid prolonged or repeated autoclaving).

2.3. Yeast Transformation

1. 10 mg/mL Herring testes carrier DNA: sheared and denatured (Clontech).
2. 50% PEG 4000: (polyethylene glycol, average 3350 kDa mol wt; Sigma). Filter sterilize or autoclave. Avoid repeated autoclaving.

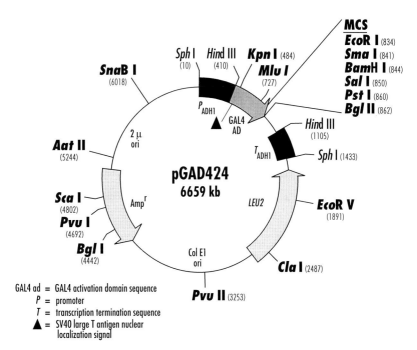

Fig. 6. Map of AD vector pGAD424.

3. 100% DMSO (Dimethyl sulfoxide; Sigma).
4. 10X TE buffer: 0.1M Tris-HCl, 10 mM EDTA, pH to 7.5, and autoclave.
5. 10X LiAc: 1M Lithium acetate (Sigma), pH to 7.5 with dilute acetic acid, and autoclave.
6. 1X PEG/LiAc/TE solution: (polyethylene glycol/lithium acetate). Prepare fresh from stock solutions just prior to use. Use 50% PEG 4000: 10X LiAc: 10X TE; 8:1:1, respectively.
7. 1X TE/LiAc solution: 40 % PEG 4000, 1X TE, and 1X LiAc. Prepare just before use from 10X stocks.

2.4. β-Galactosidase Assays

1. Z buffer: 6.1 g/L $Na_2HPO_4 \cdot 7H_2O$, 15.5 g/L $NaH_2PO_4 \cdot H_2O$, 0.75 g/L KCl, 0.246 g/L $MgSO_4 \cdot 7H_2O$. Adjust pH to 7.0 and autoclave.
2. X-gal stock solution: Dissolve 5-bromo-4-chloro-3-indolyl-β-D-galactopyranoside (X-GAL) in N,N-dimethylformamide (DMF) at 20 mg/mL. Store in the dark at –20°C.
3. Z buffer/X-gal solution: Prepare fresh daily as needed. 100 mL Z buffer, 1.67 mL X-gal stock solution.

Table 2
Component Stock Solutions for 10X Dropout Media

Component	Stock concentration, mg/L	Sigma cat. no.
1. L-Isoleucine	300	I-7383
2. L-Valine	1500	V-0500
3. L-Adenine hemisulfate salt	200	A-9126
4. L-Arginine HCl	200	A-5131
5. L-Histidine HCl monohydrate	200	H-9511
6. L-Leucine	1000	L-1512
7. L-Lysine HCl	300	L-1262
8. L-Methionine	200	M-9625
9. L-Phenylalanine	500	P-5030
10. L-Threonine	2000	T-8625
11. L-Tryptophan	200	T-0254
12. L-Tyrosine	300	T-3754
13. L-Uracil	200	U-0750

4. Z buffer with β-mercaptoethanol: Used only if the highest sensitivity level is desired. To 100 mL of Z buffer, add 0.27 mL of β-mercaptoethanol (Sigma).
5. Whatman #5 or VWR grade 413 paper filters: 75-mm diameter filters for use with 100-mm diameter Petri plates, 125-mm diameter filters for 150-mm plates. (Or, special order 85-mm and 135-mm filters directly from Whatman.)
6. Liquid nitrogen.
7. O-nitrophenyl β-D-galactopyranoside (ONPG) solution: Prepared fresh prior to use (Sigma) 4 mg/mL in Z buffer, mix well.
8. $1 M$ Na$_2$CO$_3$.

2.5. Plasmid Preparation from Yeast

1. Yeast lysis solution: 2% Triton X-100, 1% SDS, 100 mM NaCl, 10 mM Tris-HCl pH 8.0, 1.0 mM EDTA.
2. Phenol: chloroform: isoamyl alcohol (25:24:1): Prepare with neutralized pH 7.0 phenol (*see* ref. *58* for information on equilibrating phenol).
3. Acid-washed glass beads (average diameter, 425–600 µm; Sigma).
4. 95% and 70% EtOH.
5. $3 M$ NaOAc.

3. Method
3.1. cDNA Library Amplification

cDNA libraries suitable for two-hybrid screening are usually constructed in AD vectors, such as pGAD424, pGAD10 *(52)*, pGADGH, pGADGL *(8)*, and pACT *(7)*. The libraries are typically provided to the users as *Escherichia coli* transformants.

1. To obtain enough plasmid for yeast transformations, plate *E. coli* transformants on selective medium (e.g., LB+ amp agar) at a high density, 20,000 colonies/ 150-mm plate, so that the resulting colonies are nearly confluent. Plate enough transformants to obtain at least 2–3 times the number of independent clones in the library. Approximately 100–150 150-mm plates will be necessary (*see* Note 4).
2. Incubate plates at 37°C overnight.
3. Scrape the colonies into 1–2 L of LB+amp broth.
4. Incubate at 37°C for 2–4 h with shaking.
5. Perform a large-scale isolation of highly purified plasmid using a standard protocol *(58)*.

3.2. Yeast Maintenance

3.2.1. Yeast Phenotype Verification

1. Before starting any transformations, verify nutritional requirements by streaking 3–4 colonies of HF7c from the working stock onto four separate SD plates: SD/–Trp, SD/–Leu, SD/–His, and SD/–Ura.
2. Incubate plates at 30°C for 4–6 d for the phenotype to appear. Yeast strains grow slower on SD than on YPD medium.
3. Compare your results with the following: SD/–Trp, no growth; SD/–Leu, no growth; SD/–His, slow growth (*see* Note 5); and SD/–Ura, normal growth.
4. Use reddish colonies from the verified working stock plate to inoculate liquid cultures for preparing competent cells.

3.2.2. Yeast Strain Maintenance

1. To recover the strains, make a working stock plate of strain HF7c by streaking some frozen stock onto a YPD (or appropriate SD) agar plate. Incubate the plate at 30°C for 3-5 days to grow to 2 mm in diameter (*see* Note 6).
2. To prepare stock cultures of new yeast transformants, use a sterile inoculation loop to scrape a few isolated colonies from a plate (less than 3 wk old). Thoroughly resuspend the cells in 500 μL of H_2O or YPD medium. Transfer to 5–10 mL YPD and grow 14–18 h.
3. To prepare a long-term stock, add sterile glycerol to 50% volume of the saturated liquid culture. Tightly close the cap. Shake the vial before freezing at –70°C.
4. For short-term storage, yeast strains can be kept on culture plates. Seal with Parafilm and store plates at 4°C for up to 2 mo.

3.3. Control Experiments

Before committing to a full-scale library screening, perform control yeast transformations to verify two-hybrid system function. Also, be aware of

Table 3
β-gal Control Transformation Results

DNA BD	AD	SD medium	Expected colony color
pCL1	—	–Leu	blue
—	pGAD424	–Leu	white
pGBT9	—	–Trp	white
pVA3	—	–Trp	white
pVA3	pGAD424	–Leu, –Trp	white
—	pTD1	–Leu	white
pGBT9	pTD1	–Leu, –Trp	white
pVA3	pTD1	–Leu, –Trp	blue
pLAM5'	—	–Trp	white
pLAM5'	pTD1	–Leu, –Trp	white
pLAM5'	pCL1	–Leu, –Trp	blue

the occurrence of false positives, and the potential for false negative results (*see* Section 3.5.; Note 7) and design appropriate control experiments whenever possible. For example, bona fide interactions may be missed if the hybrid proteins are not stably expressed in yeast; are not localized to the yeast nucleus; are not expressed at a sufficient level; or if the GAL4 domains interfere with the ability of the test proteins to interact *(49)*. Another potential cause of false negatives, at least in the case of some mammalian proteins, is that yeast may not provide the proper posttranslational modifications required for native folding or interactions. To avoid unnecessary confusing results, the following controls should be conducted:

1. To verify two-hybrid system function, follow the small-scale transformation instructions in Section 3.3. and use control plasmid combinations recommended in Table 3.
2. To help eliminate false positives, demonstrate that the target protein cannot autonomously activate the reporter genes when fused to the DNA-BD, for example, owing to the presence of cryptic transcriptional activation sequences (*see* Note 8).
3. To use sequential transformation in a library screening (*see* Section 3.4.), you should demonstrate that your target protein is not toxic to yeast cells when expressed. Therefore, compare the growth curves of untransformed cells with cells transformed with your DNA-BD/target plasmid.
4. Perform a Western blot to verify that the target protein is expressed in yeast *(59)*. If the antibody against your bait protein in not available, you can use a GAL4 BD MAb (available from Clontech) to detect hybrid protein expression. It may be necessary to do immunoprecipitation using the same antibody in order to detect the hybrid protein by Western blot.

Table 4
Guide to Yeast Transformation Protocols

Scale	Suggested uses
Small	Perform control experiments. Verify that DNA-BD/target protein does not autonomously activate reporter genes. Look for toxicity effects of DNA-BD/ target protein. Transform with DNA-BD/target plasmid as first step of sequential transformation. Practice for larger-scale protocols.
Large	Only if no DNA-BD/target protein toxicity to yeast cells is found you may: Use for "sequential" method of library transformation (10–50 µg of plasmid). Practice for first attempt at library transformation.
Library	Used more often than large-scale transformation. Perform simultaneous cotransformation (100–500 µg of plasmid). Screen 1– 5 × 10^6 independent clones. Avoid possible toxicity of DNA-BD/target protein to yeast cells.

3.4. Two–Hybrid Library Screening

To perform a two-hybrid screening, you must introduce two fusion genes, either simultaneously or sequentially, into competent yeast reporter cells by transformation (*see* Table 4; Note 9). For example, you must transform competent HF7c cells with both the AD/library and the DNA-BD/target plasmids. Large-scale, sequential transformation (named LARGE below) is recommended if there is no selective disadvantage to cells expressing the DNA-BD/target protein (i.e., the protein is not toxic to the cell), because the transformation efficiency is higher than with simultaneous cotransformation. For large-scale sequential transformation, prepare competent HF7c that has been previously transformed with the DNA-BD/target plasmid, except use SD/–Trp medium for the overnight culture and YPD for the second culture. Then perform a large-scale transformation with the second plasmid using these cells.

Library-scale simultaneous cotransformation (*see* Table 5) is sometimes preferable to large-scale sequential transformation, even though the transformation efficiency is lower than with sequential transformation. Specifically, library-scale simultaneous cotransformation is preferable when expression of the DNA-BD/target protein is known to be toxic to the cell or to avoid potential toxicity problems. In either case, to screen 1 × 10^6 independent cDNA clones, 50 150-mm diameter plates are needed.

Table 5
Yeast Transformation Protocol at Three Different Scales

	Transformation scale		
	Small	Large	Library
1. Inoculate several 2–3-mm colonies into YPD.	0.5 mL	0.5 mL	0.5 mL
2. Vortex vigorously to disperse the clumps.			
3. Transfer this into a flask containing YPD: (*see* Note 10).	50 mL	50 mL	100 mL
4. Incubate at 30°C for 16–18 h with shaking at 250 rpm to stationary phase ($OD_{600} > 1.5$).			
5. Transfer enough overnight culture to produce an $OD_{600} = 0.2$–0.3 into YPD:	300 mL	300 mL	1 L
6. Incubate for 3 h at 30°C with shaking at 230 rpm.			
7. Centrifuge the cells at 1,000g for 5 min at room temperature (20–21°C).			
8. Discard the supernatant and vortex to resuspend the cell pellet in H_2O:	50 mL	50 mL	50 mL
9. Centrifuge the cells at 1,000g for 5 min at room temperature.			
10. Decant the supernatant.			
11. Resuspend the pellet in fresh 1X TE/LiAc:	1.5 mL	1.5 mL	8 mL
12. Prepare PEG/LiAc solution:	10 mL	10 mL	100 mL
13. Add each type of plasmid:	0.1 µg	10–50 µg	0.1–0.5 mg
with herring testes carrier DNA: to each tube and mix.	0.1 mg	2 mg	20 mg
14. Add yeast competent cells to each tube and mix well:	0.1 mL	1 mL	8 mL
15. Add sterile PEG/LiAc solution to each tube and vortex.	0.6 mL	6 mL	60 mL
16. Incubate at 30°C, 30 min, with shaking at 200 rpm.			
17. Add DMSO to 10% and mix gently by inversion.	70 µL	700 µL	7 mL
18. Heat shock for 15 min in a 42°C water bath. Swirl occasionally to mix (large- and library-scale only).			
19. Chill cells on ice.			
20. Pellet cells by centrifugation for:	5 s	5 min	5 min
(swinging bucket rotor best) at:	1000g	1000g	1000g
21. Remove the supernatant.			
22. Resuspend cells in 1X TE buffer:	0.1 mL	10 mL	10 mL

23. For library screening, spread 100 µL onto each 100-mm plate or 200 µL onto each 150-mm plate containing the appropriate SD medium (*see* Note 11).
24. Incubate plates, colony side down, at 30°C until colonies appear.
25. For library screening, after 2–3 d, some His[+] colonies will be visible, but incubate the plates for 8 d, to allow slower growing colonies (i.e., weak positives) a chance to appear (*see* Note 12).
26. Streak out His[+] colonies on fresh SD/– Trp/–Leu/– His master plates.
27. Perform β-galactosidase filter assay on the original plates containing freshly trans formed colonies and on fresh master plates containing restreaked His[+] colonies.
28. Seal master plates with Parafilm, and store at 4°C for up to 3–4 wk.
29. Calculate transformation efficiency to estimate the number of clones that you screened (*see* Note 11).

3.5. Confirmation of Positive Clones

3.5.1. Colony Lift β-galactosidase Filter Assay.

1. For blue/white colony screening, prepare Z buffer/X-gal solution (*see* Note 13).
2. Presoak one Whatman #5 or VWR grade 410 filter for each plate of transformants to be assayed in this solution by adding 1.75 mL of Z buffer/X-gal solution to a clean 100-mm plate. Then layer a 75-mm filter onto the liquid to soak it up (*see* Note 14).
3. Place a sterile, dry filter over the surface of the agar plate containing transformants.
4. Freeze/thaw permeabilize the cells by carefully lifting the filter off the agar plate with forceps and transferring it with colonies facing up into liquid nitrogen.
5. Completely submerge the filters for 10 s or until uniformly frozen.
6. Remove filter and thaw it at room temperature.
7. Carefully place the filter, colony side up, on a presoaked filter (step 2). Do not trap air bubbles under or between filters.
8. Incubate the filters at 30°C and check periodically for the appearance of blue colonies. Colonies producing β-galactosidase typically take 30 min to 8 h to turn blue. Incubation >8 h often gives false positives. However, certain strains, for example those transformed with pCL1, a wild-type GAL4 control, will turn blue within 20–30 min.
9. Pick the corresponding positive colonies from the original plates, and transfer them to fresh medium. If all of a colony was lifted onto the membrane, pick it from the filter, or incubate the plate for 1–2 d to regrow the colony.

3.5.2. Liquid Culture β-galactosidase Assay with ONPG as Substrate

The ONPG assay provided here is the standard β-galactosidase assay; however, the CPRG assay (*see* MATCHMAKER Library Protocol) may be preferable for detecting weak interactions, because they may not be quantifiable by ONPG *(18)*.

1. Inoculate a yeast HF7c colony into 5 mL of SD/–Leu/–Trp/–His liquid medium. Grow overnight at 30°C with shaking at 250 rpm (*see* Note 15).
2. Dissolve ONPG at 4 mg/mL in Z buffer with shaking for 1–2 h.
3. Inoculate 2 mL of overnight culture into 8 mL of YPD liquid medium.
4. Grow culture at 30°C for 3–5 h with shaking at 230–250 rpm, until the OD_{600} = 0.5–0.8.
5. Vortex for 0.5–1 min to disperse the culture. Record exact OD_{600}.
6. Put 1.5 mL of culture into each of three 1.5-mL tubes. Centrifuge at 14,000 rpm for 30 s.
7. Carefully remove supernatants. Wash and resuspend each pellet in 1.5 mL of Z buffer per tube 18.
8. Spin cells again, and resuspend in 300 µL of Z buffer, thereby concentrating cells 5X. Reread the OD_{600} to correct for sample variation. Try several dilutions of cells, if necessary, to remain within the linear range of the assay.

9. Add 100 µL of cell suspension into a fresh microcentrifuge tube.
10. Place tubes in liquid nitrogen until the cells are frozen.
11. Thaw in a 37°C water bath for 30 s–1 min.
12. Set up a blank tube with 100 µL of Z buffer.
13. Add 0.7 mL of Z buffer plus β-mercaptoethanol to reaction and blank tubes (do not add prior to freezing).
14. Start timer. Immediately add 0.16 mL of ONPG in Z buffer to reaction and blank tubes.
15. Incubate tubes at 30°C.
16. After yellow color develops, add 0.4 mL of $1 M$ Na_2CO_3 to the reaction and blank tubes. The time needed will vary: ~3–5 min for pCL1; ~1 h for pVA3/pTD1; and weaker interactions may take 24 h.
17. Record elapsed time in min.
18. Spin reaction tubes for 10 min at $10,000g$ to pellet cell debris.
19. Calibrate spectrophotometer against the blank at A_{420}.
20. Carefully remove supernatant and read OD_{420} of the samples relative to the blank (*see* Note 16).
21. Calculate β-galactosidase units, where 1 U β-galactosidase hydrolyzes 1 µmole of ONPG to *o*-nitrophenol and D-galactose per min *(60)*, as follows:

$$\beta\text{-galactosidase units} = 1000 \times OD_{420}/(t \times V \times OD_{600})$$

where t is the elapsed time (in min) of incubation, V is 0.1 mL X concentration factor, and OD_{600} is A_{600} of 1 mL of culture.

3.6. Elimination of False Positives

In the yeast two-hybrid screening, true positive clones exhibit reporter gene expression only when the AD/library plasmid is cotransformed with the DNA-BD/target plasmid. False positives are His+ or LacZ+ yeast transformant colonies that carry plasmids that do not encode hybrid proteins that directly interact. Such colonies arise in a two-hybrid system for several reasons (*see* ref. *57* for more information). Sometimes either the AD/library or DNA-BD/target plasmid alone can activate reporter gene transcription, or the activation comes from promoter regions other than the UAS_{GAL4}. Yeast strain HF7c employs *HIS3* and *LacZ* reporter genes with different promoters, which eliminates many false positives. However, you should test positives to determine that their activity is specific for your target protein hybrid (Fig. 4).

3.6.1. Sort Multiple Positive Clones into Groups

1. Restreak the initial positive clones (*see* Note 17).
2. Reassay completely isolated colonies to verify the β-galactosidase phenotype.
3. Isolate plasmids from a 5-mL culture of each positive clone as in Section 3.6.2., steps 2–10.
4. Dissolve plasmid in 10–20 µL of TE buffer.

5. PCR-amplify the cDNA inserts using Activation Domain Insert Screening Amplimer Set (Clontech).
6. Digest the PCR-amplified inserts with a frequent-cutting restriction enzyme, such as *Alu* I or *Hae* I.
7. Electrophorese a sample of the digest on an EtBr agarose gel.
8. Compare the restriction digest pattern of the clones.
9. Identify those positive clones that have similar-sized fragments, and sort into groups.
10. Choose one representative clone from each group for further analysis.

3.6.2. Segregate Plasmids by LEU2 Complementation in E. coli

Complementation of a bacterial *LEU2* gene deficiency can be used specifically to recover the AD/library plasmid without the DNA-BD/target plasmid. This procedure is necessary to test for false positives and to further characterize the AD/library plasmid of interest. The procedure involves isolating plasmids from yeast, transforming them into *E. coli*, and selecting for the AD/library plasmid using its *LEU2* to complement an *E. coli leuB* mutation in HB101. Plasmids isolated from yeast using the quick lysis procedure described here are contaminated with yeast genomic DNA; therefore, the plasmids are not suitable for restriction mapping or sequencing, and electroporation is recommended for a high transformation efficiency (*see* Note 18). (For protocols, *see* refs. *58* and *61*.)

1. Prepare electrocompetent *leuB E. coli* cells (i.e., HB101) *(61)*.
2. Culture 3–4 individual His$^+$, LacZ$^+$ transformant yeast colonies each in 5 mL of SD/– Leu liquid for only 1 d.
3. Fill a 1.5-mL tube with culture and spin at 10,000g for 5 s at room temperature to pellet the cells.
4. Decant the supernatant. Vortex to resuspend the pellet in residual liquid.
5. Add 0.2 mL of yeast lysis solution.
6. Add 0.2 mL of phenol:chloroform:isoamyl alcohol (25:24:1) and 0.3 g of acid-washed glass beads. Vortex vigorously for 2 min.
7. Spin at 10,000g for 5 min at room temperature.
8. Transfer the supernatant to a clean 1.5-mL tube.
9. Add 1/10 vol 3M NaOAc, pH 5.2, and 2.5 vol ethanol to precipitate the DNA.
10. Wash the pellet with 70% ethanol and dry under vacuum.
11. Resuspend the DNA pellet in 20 µL of TE buffer.
12. Add 1 µL of the plasmid to 40 µL of the *leuB E. coli* cells and perform the electroporation.
13. Plate cells on M9 medium containing 50 µg/mL ampicillin, 40 µg/mL proline, and 1 mM thiamine-HCl (*see* Note 19).
14. Isolate plasmids from Leu$^+$, Amp^r *leuB E. coli* transformants using a standard miniprep procedure *(58)*.
15. Verify plasmid genotype by restriction enzyme digestion.

**Table 6
Cotransformation Experiments to Eliminate False Positives**

No.	Purpose	Plasmid 1	Plasmid 2	SD medium	For true +
1	Confirm phenotype	None	AD/library	–Leu	White[a]
2	Confirm phenotype	DNA-BD/target	AD/library	–Trp/–Leu/–His	Blue[b]
3	Test for autonomous *lacZ* activation	DNA-BD/no insert	AD/library	–Trp/–Leu/–His	White[c]
4	Identify artifactual interactions	DNA-BD/control[e]	AD/library	–Trp/–Leu/–His	White[d]

[a]Reconfirm the LacZ⁻ Trp auxotrophic phenotypes.

[b]β-gal negatives could be due to the segregation of multiple AD/library plasmids present in one original colony: if one plasmid encodes an interacting hybrid protein, whereas the other does not, you will see the β-gal positive phenotype in the initial assay; however, subsequent plasmid segregation will result in both β-gal negative and positive phenotypes among the isolated colonies.

[c]False positives here may result from a chance interaction between the candidate library protein and the GAL4 DNA-BD, which activates *lacZ* expression.

[d]False positives here may be revealed when, for unknown reasons, the presence of a DNA-BD/control protein activates the reporter gene. This could happen if the AD/library plasmid encodes a protein that does not bind to the DNA-BD/target protein but interacts with promoter sequences flanking the GAL4 binding site, or to proteins bound to the flanking sequence.

[e]Use a GAL4 DNA-BD/any unrelated protein fusion plasmid.

3.6.3. Transformation Experiments to Eliminate False Positives

1. Using information provided in Table 6, conduct experiments 1–4, retransforming the plasmid(s) into yeast strain HF7c as shown.
2. Assay transformants for β-galactosidase activity.
3. Refer to Table 6 for the results expected when *lacZ* expression is truly dependent on interaction of the two hybrid proteins.
4. Discard false positives, save true positives.

3.6.4. Additional Tests to Confirm Genuine Protein–Protein Interactions

Some investigators may wish to conduct one or more of these two-hybrid tests to provide additional evidence for a genuine protein–protein interaction.

1. Move the library insert from the AD to the BD vector and vice versa, and then perform the two-hybrid assay again *(2,49)*.
2. Test the library and target inserts using a different system, such as the *lexA* two-hybrid system *(39,40)*.
3. Create a frameshift mutation just upstream of the library insert in the AD vector *(38)*. Using the unmutated AD/library vector in a positive control experiment, cotransform AD/mutated library and DNA-BD/target vectors into HF7c, and test for *lacZ* expression. The frameshift mutation should knock out the reporter gene expression.

3.6.5. Independent Methods to Verify True Positives

Although the yeast two-hybrid is an in vivo assay, the interactions detected could differ from those that occur under physiological conditions. Therefore, one should reconfirm the two-hybrid screening results with independent methods.

1. Sequence inserts in the AD/library plasmids using AD junction primers (Clontech).
2. Verify interactions identified by the two-hybrid approach by an independent methodology, for example, by overexpression and coimmunoprecipitation. You could transfer the gene into an overexpression vector, express the encoded protein at high levels, and purify it. Then fix the protein to a chromatographic column to bind and fractionate the target protein from a cellular extract. Elute bound proteins, and assay for the presence of the target protein by an immunoassay *(7, 62)*.
3. If possible, identify the functional relationship between the target protein and the interacting protein isolated from the yeast two-hybrid screening (such as described in refs. *16,34*, and *36*).

4. Notes

1. Procedures for using the two-hybrid system to test two known proteins for interaction, although similar to those described here, are not detailed in this chapter. When using two known proteins in the two-hybrid assay, the main difference is that nutritional selection (using the *HIS3* reporter gene) for interacting proteins is not necessary, because all cotransformants contain both potentially interacting proteins. Consequently, yeast strain SFY526 is often used because it has *lacZ* under the control of a stronger promoter.
2. The yeast *GAL1* promoter containing an UAS is fused to *lacZ* in SFY526 and to *HIS3* in HF7c. In addition, in HF7c, three copies of the *GAL4* 17-mer consensus sequence and the TATA region of the *CYC1* promoter are fused to *lacZ*. *GAL1* UAS and the *GAL4* 17-mer both respond to the GAL4 transcriptional activator. The native GAL1 is a stronger GAL4-responding promoter.
3. Synthetic dropout (SD) is a minimal medium used in yeast transformations to select for specific phenotypes. It includes a yeast nitrogen base, a carbon source, and a stock of "dropout" solution that contains essential nutrients, such as amino acids and nucleotides. One or more essential nutrients are often omitted to select for transformants carrying the corresponding nutritional gene.
4. Growing the transformants on solid instead of liquid medium minimizes uneven amplification of the individual clones.
5. HF7c will grow very slowly on SD/–His, owing to a low level, leaky expression of the *HIS3* reporter gene. If desired, this leaky expression can be effectively suppressed by 5 m*M* 3-aminotriazole.
6. Yeast colonies should appear slightly pink or red (from the *ade2–101* mutation) and grow to >2 mm in diameter. However, small white colonies will form at a rate of 1–2% owing to spontaneous mutations that eliminate mitochondrial function. Avoid these white colonies. For additional information on yeast genetics, ref. *63, Guide to Yeast Genetics and Molecular Biology* (Clontech), is recommended.

7. Failure to detect interaction between two proteins that normally interact in vivo will result in false negatives. If high-level expression of one or both of the hybrid proteins is toxic to the cell, transformants will not grow. Sometimes truncation of one of the proteins will alleviate the toxicity and still allow the interaction to occur. Alternatively, vectors providing a low level of expression may be used. It also may help to construct hybrids containing different domains of the target protein (*see* ref. *48*).

8. If autonomous activation by the DNA-BD/target construct is observed, 3-AT added to the growth medium may sufficiently suppress the background signal so that the target can be used. Alternatively, it may be possible to delete the activating regions from the target molecule before using it in a two-hybrid screening.

9. This LiAc method for preparing competent yeast cells was developed by Ito *(64)* and modified by Schiestl and Gietz *(65)*, Hill (66), and Gietz *(67)* to give a higher frequency of transformation (10^4–10^5 transformants per µg of single plasmid). Two plasmids can be introduced simultaneously into one strain but this reduces the efficiency to 10^3–10^4 transformants/µg of DNA. For best results, use competent cells for transformation immediately. However, if necessary, storage at room temperature for several hours should not significantly reduce competency.

10. Use SD/–Trp (instead of YPD) for overnight culture for second sequential transformation.

11. To estimate transformation efficiency in large- or library-scale experiments, spread 10, 1, and 0.1 µL of the transformation mixture plus 100 µL of H_2O on separate SD/–Trp/–Leu 100-mm plates. Calculate transformation efficiency as (#colonies/# µL spread) X (10,000 µL/# µg of DNA) = cfu/µg, and then calculate the amount of DNA needed to screen 1–5 × 10^6 clones.

12. Not all of the transformants surviving this selection will be "true" positives. During the initial library screening, beware of the following other types of colonies:
    a. Background growth: small colonies in which a low level of leaky *HIS3* expression in the HF7c host strain permits slow growth on SD/–His medium. The true His+ clones are robust, reddish-brown, and can grow to >2 mm in diameter. 5 mM 3-AT (Sigma) added to the selection medium is sufficient to suppress HF7c background growth.
    b. False positives may occur owing to the fortuitous cloning of *HIS3* transcriptional activators or other activators that appear to require the presence of both types of hybrid proteins, but are not dependent on interaction between the hybrid proteins *(57)*. This is why screening for the expression of the second reporter gene (*lacZ*) is necessary. (*See* Section 3.5. for more information on eliminating false positives.)

13. This assay can be done any time after colonies are visible, but best results will be obtained using freshly transformed colonies 1–2 mm in diameter. For best results, inoculate colonies onto filters placed on selection medium. Incubate plate for 1–2 days at 30°C, then lift out the filters and assay the colonies for β-galactosidase activity (Section 3.4.1., steps 4–6).

14. You can use nitrocellulose filters, but they often crack when frozen and thawed.

15. To reduce variability, assay five separate transformants, and with each, perform the assay in triplicate.
16. The cellular debris, if disturbed, will strongly interfere with the accuracy of this test. The OD should be between 0.02–1.0.
17. Several different AD/library plasmids may be present in each β-galactosidase positive colony; therefore, purify each colony by restreaking it on SD/–Trp/–Leu selection medium to segregate plasmids.
18. If you can obtain a transformation efficiency of at least 10^7 cfu/μg DNA, using standard chemical transformation procedures should be possible here. *See* Kaiser and Auer *(68)* for an alternative, rapid method for the transfer of a shuttle plasmid from yeast to *E. coli.*
19. For better recovery of transformants, add –Leu dropout solution supplement to 1X in the M9 medium (Section 2.2.).

Acknowledgments

I thank Stanley Fields, Paul Bartel, Gregory Hannon, Stephen Elledge, Rodney Rothstein, Roger Brent, and Kenneth Fong for invaluable instruction and encouragement. Ann Holtz, John Stile, John Ledesma, Paul Diehl, Anne Scholz, Kristen Mayo, and many other Clontech laboratory members participated in this work. I also acknowledge Sheila Colby for her excellent editorial assistance.

References

1. Fields S. and Song, O. (1989) A novel genetic system to detect protein-protein interactions. *Nature* **340,** 245–247.
2. Chien, C. T., Bartel, P. L., Sternglanz, R., and Fields, S. (1991) The two-hybrid system: a method to identify and clone genes for proteins that interact with a protein of interest. *Proc. Nat. Acad. Sci. USA* **88,** 9578–9582.
3. Guarente, L. (1993) Strategies for the identification of interacting proteins. *Proc. Natl. Acad. Sci. USA* **90,** 1639–1641.
4. Luban, J., Alin, K. B., Bossolt, K. L., Humaran, T., and Goff, S. P. (1992) Genetic assay for multimerization of retroviral *gag* polyproteins. *J. Virol.* **66,** 5157–5160.
5. Yang, X., Hubbard, E., and Carlson, M. (1992) A protein kinase substrate identified by the two-hybrid system. *Science* **257,** 680–682.
6. Chardin, P., Camonis, J. H., Gale, N. W., van Aelst, L., Schlessinger, J., Wigler, M. H., and Bar-Sagi, D. (1993) Human Sos1: a guanine nucleotide exchange factor for Ras that binds to GRB2. *Science* **260,** 1338–1343.
7. Durfee, T., Becherer, K., Chen, P. L., Yeh, S. H., Yang, Y., Kilburn, A. E., Lee, W. H., and Elledge, S. J. (1993) The retinoblastoma protein associates with the protein phosphatase type 1 catalytic subunit. *Genes Dev.* **7,** 555–569.
8. Hannon, G. J., Demetrick, D., and Beach, D. (1993) Isolation of the Rβ-related p130 through its interaction with CDK2 and cyclins. *Genes Dev.* **7,** 2378–2391.
9. Iwabuchi, K., Li, B., Bartel, P., and Fields, S. (1993) Use of the two-hybrid system to identify the domain of p53 involved in oligomerization. *Oncogene* **8,** 1693–1696.

10. Li, B. and Fields, S. (1993) Identification of mutations in p53 that affect its binding to SV40 T antigen by using the yeast two-hybrid system. *FASEB J.* **7,** 957–963.

11. Luban, J., Bossolt, K. L., Franke, E. K., Kalpana, G. V., and Goff, S. P. (1993) Human immunodeficiency virus type 1 Gag protein binds to cyclophillins A and B. *Cell* **73,** 1067–1078.

12. Standinger, J., Perry, M., Elledge, S. J., and Olson, E. N. (1993) Interactions among vertebrate helix-loop-helix proteins in yeast using the two-hybrid system. *J. Biol. Chem.* **268,** 4608–4611.

13. Vojtek, A., Hollenberg, S., and Cooper, J. (1993) Mammalian Ras interacts directly with the serine/threonine kinase Raf. *Cell* **74,** 205–214.

14. Guan, K., Jenkins, C. W., Li, Y., Nichols, M. A., Wu, X., O'Keefe, C. L., Matera, G. A., and Xiong, Y. (1994) Growth suppression by *p18,* a p16 [INK4/MTS1] -and p14[INK4B/MTS2]-related CDK6 inhibitor, correlates with wild-type pRb function. *Genes Dev.* **8,** 2939–2952.

15. Hannon, G. J., Casso, D., and Beach, D. (1994) KAP: a dual specificity phosphatase that interacts with cyclin-dependent kinases. *Proc. Natl. Acad. Sci. USA* **91,** 1731–1735.

16. Miyata, K. S., McCaw, S. E., Marcus, S. L., Rachubinski, R. A., and Capone, J. P. (1994) The peroxisome proliferator-activated receptor interacts with the retinoid X receptor in vivo. *Gene* **148,** 327–330.

17. Rothe, M., Wong, S. C., Henzel, W. J., and Goeddel, D. V. (1994) A novel family of putative signal transducers associated with the cytoplasmic domain of the 75 kDa tumor necrosis factor receptor. Cell **78,** 681–692.

18. Song, H. Y., Dunbar, J. D., and Donner D. B. (1994) Aggregation of the intracellular domain of the type 1 tumor necrosis factor receptor defined by the two-hybrid system. *J. Biol. Chem.* **269,** 22492–22495.

19. Boldin, M. P., Varfolomeev, E. E., Pancer, Z., Mett, I. L., Camonis, J. H., and Wallach, D. (1995) A novel protein that interacts with the death domain of Fas/APO1 contains a sequence motif related to the death domain. *J. Biol. Chem.* **270,** 7795–7798.

20. Cheng, G., Cleary, A. M., Ye, A-s., Hong, D. I., Lederman, S., and Baltimore, D. (1995) Involvement of CRAF1, a relative of TRAF, in CD40 signaling. *Science* **267,**1494–1498.

21. Coghlan, V. M., Perrino, B. A., Howard, M., Langeberg, L. K., Hicks, J. B., Gallatin, W. M., and Scott, J. D. (1995) Association of protein kinase A and protein phosphatase 2B with a common anchoring protein. *Science* **267,** 108–111.

22. Huang, Z. J., Curtin, K. D., and Rosbash, M. (1995) PER protein interactions and temperature compensation of a circadian clock in *Drosophila. Science* **267,**1169–1170.

23. Kitagawa, K., Masumoto, H., Ikeda, M., and Okazaki, T. (1995) Analysis of protein-DNA and protein-protein interactions of centromere protein B (CENP-B) and properties of the DNA-CENP-B complex in the cell cycle. *Mol. Cell Biol.* **15,**1602–1612.

24. Lee, J. W., Ryan, F., Swaffield, J. C., Johnson, S. A., and More, D. D. (1995) Interaction of thyroid-hormone receptor with a conserved transcriptional mediator. *Nature* **473**, 91-94.

25. Li, S., Janosch, P., Tanji, M., Rosenfeld, G. C., Waymire, J. C., Mischak, H., Kolch, W., and Sedivy, J. (1995) Regulation of Raf-1 kinase activity by the 14-3-3 family of proteins. *EMBO J.* **14**, 685-696.

26. MacDonald, P. N., Sherman, D. R., Dowd, D. R., Jefcoat, S. C., Jr., and Delise, R. K. (1995) The vitamin D receptor interacts with general transcription factor IIB. *J. Biol. Chem.* **270**, 4748–4752.

27. Matsuda, T., Saijo, M., Kuraoda, I., Kobayashi, T., Nakatsu, Y., Nagai, A., Enjoji, T., Masutani, C., Sugasawa, K., Hanaoka, F., Yasui, A., and Tanaka, K. (1995) DNA repair protein XPA binds replication protein A (RPA). *J. Biol. Chem.* **270**, 4152–4157.

28. Mosialos, G., Brikenbach, M., Yalamanchili, R., VanArsdale, T., Ware, C., and Kieff, E. (1995) The Epstein-Barr Virus transforming protein LMP1 engages signaling proteins for the tumor necrosis factor receptor family. *Cell* **80**, 389–399.

29. Sato, T., Irie, S., Kitada, S., and Reed, J. C. (1995) FAP-1: a protein tyrosine phosphatase that associates with Fas. *Science* **268**, 411 415.

30. Schreiber-Agus, N., Chin, L., Chen, K., Torres, R., Rao, G., Ginda, P., Skoultchi, A. I., and DePinho, R. A. (1995) An amino-terminal domain of Mxi1 mediates anti-Myc oncogenic activity and interacts with a homolog of the yeast transcriptional repressor SIN3. *Cell* **80**, 777–786.

31. Song, H. Y., Dunbar, J. D., Zhang, Y. X., Guo, D., and Donner, D. B. (1995) Identification o a protein with homology to hsp90 that binds the type 1 tumor necrosis factor receptor. *J. Biol Chem.* **270**, 3574–3581.

32. Staudinger, J., Zhou, J., Burgess, R., Elledge, S. J., and Olson, E. N. (1995) PICK1: a perinuclear binding protein and substrate for protein kinase C isolated by the yeast two-hybrid system. *J. Cell. Biol.* **128**, 263–271.

33. Strubin, M., Newell, J. W., and Matthias, P. (1995) OBF-1, a novel B cell-specific coactivator that stimulates immunoglobulin promoter activity through association with octamer-binding proteins. *Cell* **80**, 497–506.

34. Takayama, S., Sato, T., Krajewski, S., Kochel, K., Irie, S., Millan, J. A., and Reed, J. C. (1995) Cloning and functional analysis of Bag-1: a novel Bcl-2-binding protein with anti-cell death activity. *Cell* **80**, 279–284.

35. Uemura, H. and Jigami, Y. (1995) Mutations in GCR1, a transcriptional activator of *Saccharomyces cerevisiae* glycolytic genes, function as suppressors of *gcr2* mutations. *Genetics* **139**, 511–521.

36. Yang, E., Zha, J., Jockel, J., Boise, L. H., Thompson, C. B., and Korsmyer, S. J. (1995) Bad, a heterodimeric partner for Bcl-X_L and Bcl-2, displaces Bax and promotes cell death. *Cell* **80**, 285–291.

37. Harper, J. W., Adami, G. R., Wei, N., Keyomarsi, K., and Elledge, S. J. (1993) The p21 Cdk-interacting protein Cip1 is a potent inhibitor of G1 cyclin-dependent kinases. Cell **75**, 805–816.

38. Bendixen, C., Gangloff, S., and Rothstein, R. (1994) A yeast mating-selection scheme for detection of protein-protein interactions. *Nucleic Acids Res.* **22,** 1778–1779.

39. Zervos, A., Gyuris, J., and Brent, R. (1993) Mxi1, a protein that specifically interacts with *Max* to bind to *Myc-Max* recognition sites. *Cell* **72,** 223–232.

40. Vojtek, A., Hollenberg, S., and Cooper, J. (1993) Mammalian Ras interacts directly with the serine/threonine kinase Raf. *Cell* **74,** 205–214.

41. Alexandre, C., Grueneberg, D. A., and Gilman, M. Z. (1993) Studying heterologous transcription factors in yeast. *Methods* **5,**147–155.

42. Chan, J. Y., Han, X., and Kan, Y. W. (1993) Cloning of Nrf1, an NF-E2-related transcription factor, by genetic selection in yeast. *Proc. Natl. Acad. Sci. USA.* **90,**11371–11375.

43. Gstaiger, M., Knoepfel, L., Georgiev, O., Schaffner, W., and Hovens, C. M. (1995) A B-cell coactivator of octamer-binding transcription factors. *Nature* **373,** 360-362.

44. Li, J. J. and. Herskowitz, I. (1993) Isolation of *ORC6*, a component of the yeast origin of recognition complex by a one-hybrid system. *Science* **262,** 1870–1874.

45. Liu, J., Wilson, T. E., Milbbrndt, J., and Johnston, M. (1993). Identifying DNA binding sites and analyzing DNA-binding domains using a yeast selection system. *Methods.* **5,** 125-138.

46. Strubin, M., Newell, J. W., and Matthias, P. (1995) OBF-1, a novel B cell-specific coactivator that stimulates immunoglobulin promoter activity through association with octamer-binding proteins. *Cell* **80,** 497–506.

47. Wang, M. M. and Reed, R. R. (1993) Molecular cloning of the olfactory neuronal transcription factor Olf-1 by genetic selection in yeast. *Nature* (London) **364,** 121–126.

48. Wilson, T. E., Padgett, K. A., Johnston, M., and Milbrandt, J. (1993) A genetic method for defining DNA-binding domains: application to the nuclear receptor NGFI-B. *Proc. Natl. Acad. Sci. USA.* **90,** 9186–9190.

49. van Aelst, L., Barr, M., Marcus, S., Polverino, A., and Wigler, M. (1993) Complex formation between RAS and RAF and other protein kinases. *Proc. Natl. Acad. Sci. USA* **90,** 6213–6217.

50. Toyoshima, H., and Hunter, T. (1994) p27, a novel inhibitor of G1 cyclin-Cdk protein kinase activity, is related to p21. *Cell* **78,** 67–74.

51. Fields, S. and Sternglanz, R. (1994) The two-hybrid system: an assay for protein-protein interactions. *Trends in Gen.* **10,** 286–292.

52. Bartel, P. L., Chien, C.-T., Sternglanz, R., and Fields, S. (1993a) Using the two-hybrid system to detect protein-protein interactions, in *Cellular Interactions in Development: A Practical Approach* (D. A. Hartley, ed.), Oxford University Press, Oxford, pp. 153–179.

53. Ma, J. and Ptashne, M. (1987) A new class of yeast transcriptional activators. *Cell* **51,** 113–119.

54. Ma, J. and Ptashne, M. (1988) Converting a eukaryotic transcriptional inhibitor into an activator. *Cell* **55,** 443–446.

55. Feilotter, H. E., Hannon, G. J., Ruddel, C. J., and Beach, D. (1994) Construction of an improved host strain for two-hybrid screening. *Nucleic Acids Res.* **22,** 1502–1503.

56. Silver, P. A., Keegan, L. P., and Ptashne, M. (1984) Amino terminus of the yeast *GAL4* gene product is sufficient for nuclear localizaton. *Proc. Natl. Acad. Sci. USA* **81,** 5951–5955.

57. Bartel, P. L, Chien, C.-T., Sternglanz, R., and Fields, S. (1993b) Elimination of false positives that arise in using the two-hybrid system. *BioTechniques* **14,** 920–924.

58. Sambrook, J., Fritsch, E. F., and Maniatis, T. (1989) *Molecular Cloning: A Laboratory Manual* (Ford, N., Nolan, C., and Ferguson, M., eds.), Cold Spring Harbor Laboratory Press, Cold Spring Harbor, NY.

59. Printen, J. A. and Sprague, G. F. Jr. (1994) Protein interactions in the yeast pheromone response pathway: Ste5p interacts with all members of the MAP kinase cascade. *Genetics* **138,** 609–619.

60. Miller, J. H. (1972) *Experiments in Molecular Genetics,* Cold Spring Harbor Laboratory Press, Cold Spring Harbor, NY.

61. Dower, W. J., Miller, J. F., and Ragsdale, W. W. (1988) High efficiency transformation of *E. coli* by high voltage electroporation. *Nucleic Acids Res.* **16,** 6127–6145.

62. Zhang, X., Settleman, J., Kyriakis, J. M., Takeuchi-Suzuki, E., Elledge, S. J., Marshall, M. S., Bruder, J. T., Rapp, U. R., and Avruch, J. (1993) Normal and oncogenic p21^{ras} proteins bind to the amino-terminal regulatory domain of c-Raf-1. *Nature* **364,** 308–313.

63. Guthrie, C. and Fink, G. R. (1991) *Methods in Enzymology, vol. 194, Guide to Yeast Genetics and Molecular Biology,* Academic, San Diego, CA.

64. Ito, H., Fukada, Y., Murata, K., and Kimura, A. (1983) Transformation of intact yeast cells treated with alkali cations. *J. Bacteriol.* **153,** 163–168.

65. Schiestl, R. H. and Gietz, R. D. (1989) High efficiency transformation of intact cells using single stranded nucleic acids as a carrier. *Curr. Gen.* **16,** 339–346.

66. Hill, J., Donald, K. A., and Griffiths, D. E. (1991) DMSO-enhanced whole cell yeast transformation. *Nucleic Acids Res.* **19,** 5791.

67. Gietz, D., St. Jean, A., Woods, R. A., and Schiestl, R. H. (1992) Improved method for high efficiency transformation of intact yeast cells. *Nucleic Acids Res.* **20,** 1425.

68. Kaiser, P. and Auer, B. (1993) Rapid shuttle plasmid preparation from yeast cells by transfer to *E. coli. BioTechniques* **14,** 552.

16

Alternative Yeast Two-Hybrid Systems

The Interaction Trap and Interaction Mating

Erica A. Golemis and Vladimir Khazak

1. Introduction

Since the original demonstration that for proteins P1 and P2 known to interact, it was possible to detect interactions between a DNA-binding domain (DBD) fused P1 and activation-domain (AD) fused P2 by assaying transcriptional activation of one or more reporter genes containing cognate DNA binding site for the DBD (1), a number of groups have developed variants of this approach for purposes such as mapping interaction domains on proteins, screening cDNA libraries for novel interacting proteins, and screening novel proteins against predetermined sets of known proteins to identify which pairs interact. We here describe strategies for these purposes utilizing the interaction trap (2), a two-hybrid system variant developed in the laboratory of Roger Brent (Fig. 1).

In the interaction trap, the DBD function is provided by the bacterial protein LexA, and the AD function is provided by bacterial sequences encoding an amphipathic helix ("acid blob" B42). Two reporters are used to score activation: a LexA operator -LacZ plasmid, and a LexA-operator LEU2 gene present in single copy on the chromosome of an appropriate yeast strain. Plasmids required for the interaction trap are shown in Fig. 2, and summarized in the Section 2. The basic principles involved in the interaction trap are essentially the same as those in the two-hybrid system. Therefore, the question arises as to why to use one vs another system variant for studying protein-protein interactions. The interaction trap has several unique advantages. First, because it utilizes the bacterial LexA protein rather than the yeast GAL4 protein as DBD, it is not necessary to use gal4 yeast strains, which are frequently unhealthy. Second, in the case of library screens using yeast proteins, it is not necessary to

From: *Methods in Molecular Biology, vol. 63: Recombinant Protein Protocols: Detection and Isolation* Edited by: R. Tuan Humana Press Inc., Totowa, NJ

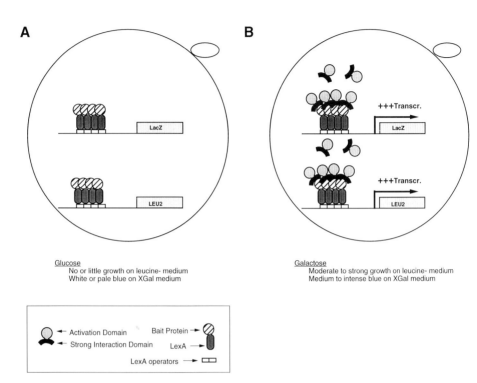

Fig. 1. The interaction trap. In the basic application, an EGY48 or EGY191 yeast cell contains two LexA-operator responsive reporters, the first a chromosomally integrated copy of the LEU2 gene (required for growth on media lacking leucine), and the second a plasmid bearing a GAL1 promoter-LacZ fusion gene (causing yeast to turn blue on media containing X-gal). These yeast additionally contain a plasmid constitutively expressing the DNA-binding domain of LexA-fused to the "bait" protein P1; and a plasmid which is induced by galactose to express an activation domain-fused cDNA library or specific protein P2. On glucose medium **(A)**, the DBD-fused protein is unable to activate expression of either of the two reporters, so yeast are unable to grow on medium lacking leucine, and are white on medium containing X-gal. On galactose medium **(B)**, the AD-fusion protein is induced: a positive interaction is shown in which the DBD and AD-fused proteins interact strongly , resulting in activation of the two reporters, thus causing growth on media lacking leucine, and blue color on media containing X-gal.

remove background, owing to recloning of GAL4. Third, the interaction trap utilizes a GAL1 inducible promoter to express the AD fusion, in contrast to the constitutive expression of the AD fusion in the two-hybrid system. The ability to conditionally express one of the fusions assists in eliminating background from library screens, and additionally is helpful in studying interactions in which one of two putative interacting proteins is toxic in yeast. Finally, all two-hybrid systems utilize chimeric fusion proteins to assay interactions. It occasionally happens that an interaction can be scored readily using the two-hybrid system and not at all with the interaction trap, or vice versa, presumably because of particular steric problems with either the GAL4 or the LexA DBD, or the GAL4 or the B42 AD. Having more than one option available increases the likelihood that a protein can be used effectively in a two-hybrid strategy.

The following describes three basic protocols for the interaction trap. The first describes how to detect protein-protein interactions between two predefined protein partners. The second describes how to rapidly screen a predefined panel of proteins against each other or against a novel protein to detect patterns of interaction. The third describes how to screen a library for novel interacting partners for a given protein.

2. Materials

Note: All the protocols utilize an overlapping set of reagents. Thus, all materials necessary for the three basic protocols are presented here. Plasmids, strains, and libraries are most readily obtained by writing to Roger Brent, whose address is included at the end. In addition, a number of variants of the LexA-fusion plasmids have been created to facilitate work with proteins that are toxic or require unobscured amino-terminal segments. These also are detailed in the Notes section at the end.

2.1. Plasmids (see Fig. 2)

1. pJK202 and peG202, *HIS3* plasmids for making LexA fusion protein with (JK202) or without (peG202) an incorporated nuclear localization sequence. Expression is from the constitutive ADH promoter.
2. pJG4-5, *TRP1* plasmid for making a nuclear localization sequence-activation domain-hemagglutinin epitope tag fusion to a unique protein or a cDNA library. Expression is from the GAL1 galactose-inducible promoter.
3. p1840, pJK103, or pSH18-34, *URA3* plasmids containing 1, 2, or 8 LexA operators upstream of the LacZ reporter gene, respectively.
4. pJK101, *URA3* UAS$_{GAL}$/LexA-operator/LacZ-reporter plasmid for repression assay.
5. pSH17-4, *HIS3* plasmid encoding LexA-GAL4, a positive control for activation.
6. pRFHM1, *HIS3* plasmid encoding LexA-bicoid, a negative control for activation and positive control for repression.

2.2. Strains

1. Yeast strain EGY48 *(MATα ura3 trp1 his3 3LexAop-leu2)*.
2. Yeast strain EGY191 *(MATα ura3 trp1 his3 1LexAop-leu2)*.

These strains are derivatives of the strain U457 (a gift of Rodney Rothstein) in which homologous recombination was used to replace the sequences

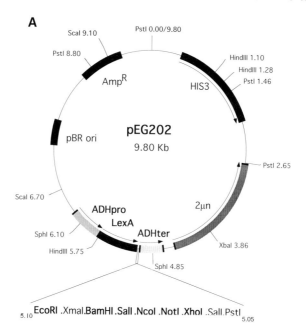

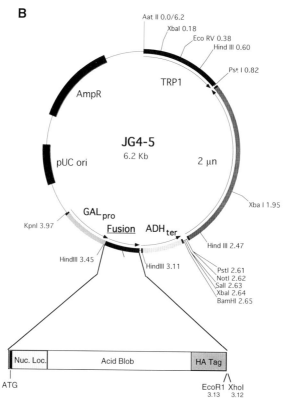

C

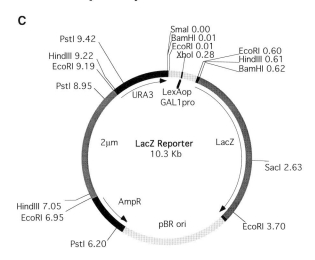

Fig. 2. Key Plasmids are shown. (**A**) pEG202 DBD-fusion plasmids. pEG202, a derivative of Lex202+PL, *(18)* uses the strong constitutive ADH promoter to express bait proteins as fusions to the DNA binding protein LexA. A number of restriction sites are available for insertion of coding sequences: those shown in bold type are unique. The reading frame for insertion is GAA TTC CCG GGG ATC CGT CGA CCA TGG CGG CCG CTC GAG TCG ACC TGC AGC. The sequence CGT CAG CAG AGC TTC ACC ATT G can be used to design a primer to confirm correct reading frame for LexA fusions. The plasmid contains the *HIS3* selectable marker and the 2-μm origin of replication to allow propagation in yeast, and the ampicillin-resistance gene and the pBR origin of replication to allow propagation in *E. coli*. Endpoints of the *HIS3* and 2-μm elements are drawn approximately with respect to the restriction data. pJK202 is identical, except that it incorporates nine amino acids corresponding to the SV40 T antigen nuclear localization motif between LexA and polylinker sequences. (**B**) JG4-5 AD-fusion plasmid. pJG4-5 *(2)* expresses cDNAs or other coding sequences inserted into the unique EcoRI and XhoI sites as translational fusion to a cassette consisting of the SV40 nuclear localization sequence (PPKKKRKVA), the acid blob B42 *(19)*, and the hemagglutinin (HA) epitope tag (YPYDVPDYA). Expression of sequences is under the control of the GAL1 galactose inducible promoter. The sequence CTG AGT GGA GAT GCC TCC can be used as a primer to identify inserts or confirm correct reading frame. The plasmid contains the *TRP1* selectable marker and the 2-μm origin to allow propagation in yeast, and the ampicillin resistance gene and the pUC origin to allow propagation in *E. coli*. (**C**) LacZ reporter plasmids standardly used are derivatives of LR1Δ1 (a deletion of RY121 described in ref. *20*). LexA operator sequences are cloned into a unique XhoI site upstream of a minimal (noninducible) GAL1 promoter fused to the LacZ gene. The plasmid contains the URA3 selectable marker and the 2-μm origin to allow propagation in yeast, and the ampicillin resistance gene and the pBR322 origin to allow propagation in *E. coli*. 1840 *(21)* contains a single LexA operator (binds two LexA monomers), JK103 contains an "overlapping" operator found upstream of the *colE1* gene *(3)* that binds four LexA monomers, and SH18-34 contains four colE1 operators and binds up to 16 LexA monomers.

upstream of the chromosomal LEU2 gene operators for LexA derived from the colE1 gene *(3)*. EGY48 contains three such operators, and can bind up to 12 LexA monomers: EGY191 contains one such operator, and can bind up to 4 LexA monomers.

3. Yeast strain RFY206 (*MATα his3 D200 leu2-3 lys2D201 ura3-52* trp1D: :hisG).
4. *E. coli* strain KC8 *(pyrF leuB600 trpC hisB463)*.

2.3. Reagents for LiOAc Transformation of Yeast

1. Sterile filtered [10 mM Tris-HCl pH 8.0, 1 mM EDTA, 0.1M lithium acetate].
2. Sterile filtered [10 mM Tris-HCl pH 8.0, 1 mM EDTA, 0.1M lithium acetate, 40% PEG4000].
3. Sterile DMSO.

2.4. Reagents for Minipreps from Yeast

1. STES lysis solution: 100 mM NaCl, 10 mM Tris-HCl, pH 8.0, 1 mM EDTA, 0.1% SDS.
2. Equilibrated phenol (pH ~7.0).
3. Chloroform.
4. 100% Ethanol.
5. 70% Ethanol.
6. Acid-washed sterile glass beads, 0.45-mm diameter.

2.5. Reagents for β-Galactosidase Assays

1. Lysis buffer: 0.1M Tris-HCl, pH 7.6 / 0.05%Triton-X-100.
2. ONPG (o-nitrophenyl β-D-galactopyranoside). Dissolve powder directly in Z buffer.
3. Z-buffer: 16.1 g $Na_2HPO_4 \cdot 7H_2O$, 5.5 g $NaH_2PO_4 \cdot H_2O$, 0.75 g KCl, 0.25 $MgSO_4 \cdot 7H_2O$, made up to 1 L, pH adjusted to 7.0.
4. 1M Na_2CO_3.

2.6. Plates for Growing Bacteria (100 mm)

1. LB supplemented with 50 μg/mL ampicillin.
2. KC8 plates for selecting library plasmids, made as follows:
   a. Autoclave 1 L of dH_2O containing 15 g agar, 1 g $(NH_4)_2SO_4$, 4.5 g KH_2PO_4, 10.5 g K_2HPO_4, and 0.5 g sodium citrate $\cdot 2H_2O$. Cool to 50°C.
   b. Add 1 mL sterile filtered 1M $MgSO_4 \cdot 7H_2O$, 10 mL sterile-filtered 20% glucose and 5 mL each of 40 μg/mL sterile-filtered stocks of L-histidine, L-leucine, and uracil. Pour.

2.7. General Directions for Defined Minimal Yeast Medium

All minimal yeast media, liquid, and plates, are based on the following three ingredients, which are sterilized by autoclaving for 15–20 min:

per L, 6.7 g Yeast Nitrogen base – amino acids (Difco, Detroit, MI)
20 g glucose or 20 g galactose + 10 g raffinose
2 g appropriate nutrient "dropout" mix *(see below)*

For plates, 20 g Difco bacto-agar (Difco) are also added.
A complete minimal nutrient mix includes the following:
2.5 g Adenine, 1.2 g L-arginine, 6.0 g L-aspartic acid, 6.0 g L-glutamic acid,
1.2 g L-histidine, 1.2 g L-isoleucine, 3.6 g L-leucine, L-lysine 1.8 g, 1.2 g L-methion-
ine, 3.0 g L-phenylalanine, 22.0 g L-serine, 12.0 g L-threonine, 2.4 g L-tryp-
tophan, 1.8 g L-tyrosine, 9.0 g L-valine, 1.2 g uracil.

Leaving out one or more nutrients selects for yeast able to grow in its absence,
i.e., containing a plasmid that covered the deficiency. Thus, "dropout medium"
lacking uracil (denoted ura- in the following recipes) would select for the presence
of plasmids with the URA3 marker, and so on. Note, the above quantities of
nutrients produce a quantity of dropout powder sufficient to make 40 L of medium:
it is advisable to scale down for most of the below dropout combinations. Note,
premade dropout mixes are available from some commercial suppliers.

2.8. Plates for Growing Yeast (100 mm)

1. Defined minimal dropout plates, with glucose as a carbon source: ura-; trp-; ura-
   his-; ura-his-trp-; ura-his-trp-leu-.
2. Defined minimal dropout plates, with galactose + raffinose as a carbon source:
   ura-his-trp-leu-.
3. YPD (rich medium): per L, 10 g yeast extract, 20 g peptone, 20 g glucose, 20 g
   Difco Bactoagar: autoclave for ~18 min, pour ~40 plates.

2.9. Liquid Medium for Growing Yeast

1. Defined minimal dropout media, with glucose as a carbon source: ura-his-; trp-.
2. YPD: per L, 10 g yeast extract, 20 g peptone, 20 g glucose: autoclave for ~15 min.

2.10. Plates
for Growing Yeast Library Transformations (240 × 240 mm)

Defined minimal dropout plates, with glucose as a carbon source:
ura-his-trp-. Each plate requires 200 mL medium.

2.11. X-Gal Plates

Start with the basic recipe for dropout plates, but mix basic ingredients in
900 mL of dH$_2$O rather than 1 L. Autoclave 20 min, and allow to cool to ~65°C;
then add 100 mL of sterile filtered 10X BU salts (1 L 10X stock = 70 g
Na$_2$HPO$_4$ · 7H2O, 30 g NaH$_2$PO$_4$, pH adjusted to 7.0, filter sterilized), and 2
mL of a 40 mg/mL solution of X-gal dissolved in dimethylformamide (DMF).
After pouring, plates should be stored at 4°C and protected from light.

X-Gal plates required for the experiments described below: ura-, made with glucose; ura-, made with 2% galactose + 1% raffinose.

2.12. Miscellaneous

1. Replica block and velvets or disposable plastic replica blocks.
2. Sterile microscope slide and/or
3. Sterile glass balls, 4 mm, #3000, Thomas Scientific.
4. Sterile glycerol solution for freezing transformants (65% sterile glycerol, 0.1M MgSO$_4$, 25 mM Tris-HCl, pH 8.0).

3. Methods

3.1. Assaying Interactions
Between Two Predetermined Protein Partners

1. Use standard methods to clone the gene encoding proteins to be tested for inter-action (P1 and P2) into either the pEG202 or the pJK202 vector (referred to as p-202 in the following instructions), and into the pJG4-5 vector. This will allow the testing of interactions in both orientations.
2. Transform the following combinations of plasmids into EGY48 yeast:
   a. p-202 -P1 + pJG4-5 -P2 + JK103.................plate to ura trp his .
   b. p-202 -P2 + pJG4-5 -P1 + JK103.................plate to ura-trp-his-.
   c. pRFHM1 + pJG4-5 -P2 + JK103.................plate to ura-trp-his-.
   d. pRFHM1 + pJG4-5 -P1 + JK103.................plate to ura-trp-his-.
   e. p-202 -P1 + pJG4-5 + JK103.................plate to ura-trp-his-.
   f. p-202 -P2 + pJG4-5 + JK103.................plate to ura-trp-his-.
   g. pRFHMI + pJG4-5 + JK103.................plate to ura-trp-his-.

   A standard, highly efficient means of performing such transformation involves the use of a lithium acetate solution to make yeast competent *(4–6)*. Briefly:
   i. Grow an overnight culture of EGY48 in ~5 mL YPD liquid medium at 30°C. Stocks of EGY48 and other yeast strains can be maintained on parafilm-wrapped YPD plates for several weeks at 4°C.
   ii. In the morning, dilute the culture into 50 mL fresh YPD liquid media to approximately 2×10^6 cell/mL (OD$_{600}$ ~0.10) culture. Grow until approx 1×10^7 cells/mL (OD$_{600}$ ~0.50).
   iii. Transfer culture to a 50 mL Falcon tube. Harvest by spinning at 1000–1500g in a low-speed centrifuge, at room temperature, for ~5 min. Resuspend in 30 mL sterile distilled water.
   iv. Repeat the spin at 1000–1500 g for 5 min. Resuspend in 0.3 mL TE/.1M LiAc pH 7.5.
   v. Take six 1.5-mL Eppendorfs. Into each, aliquot 1 µg of each plasmid DNA and 10 µg high quality sheared salmon sperm carrier DNA (see Schiestl and Gietz). For maximal efficiency, salmon sperm should be denatured by boiling for ~5 min immediately before addition to reactions. Total volume of DNA added per tube should be less than 20 µL, and preferably less than 10 micro-liters. Add 50 µL of the resuspended yeast solution.

   vi. Add 300 microliters of sterile TE/lithium acetate/40% PEG4000 solution, and mix thoroughly by inversion. Incubate at 30°C for at least 30 min: if desired, the incubation can then be placed at 4°C for storage overnight before proceeding.

   vii. Add DMSO to 10% (~40 µL) per tube: mix by inversion. Heat shock in 42°C heat block for 10 min, and plate ~200 µL of each tube's contents on a ura-his-trp- dropout plate, and incubate until colonies are > 1 mm in diameter (2–3 d).

3. For each transformation, restreak at least six colonies to a ura-his-trp- master plate (all six transformations can be pooled to a single plate with 42 streaks).

4. Restreak from the master plate to the following four tester plates: ura-his-trp-leu- with glucose; ura-his-trp-leu- with galactose; ura-his-trp- X-gal plate with glucose; ura-his-trp- X-gal plate with galactose; and incubate for 24–72 h at 30°C.

5. The AD-fusion is expressed under the control of the GAL1 promoter: Hence, it is made at high levels in yeast grown on medium containing galactose as a carbon source, but not on medium containing glucose as a carbon source. If there is a positive, specific interaction between the DBD and AD fused proteins, then colonies deriving from transformations a and b will grow better on leu- galactose plates, and will be darker blue on X-gal galactose plates than colonies derived from c,d,e,f, and g. In contrast, a,b,c,d,e,f, and g will grow comparably on leu- glucose and will turn comparably blue on X-gal glucose plates. Note, in very rare cases, if an interaction occurs with very high affinity, and P1 and P2 are very stable proteins in yeast, a weak enhancement of growth and X-gal activity will occur for a and b on glucose as well as galactose medium. This is because the very low levels of transcription from GAL1 promoter on glucose medium allow accumulation of sufficient levels of strongly interacting, stable AD-fusions for an interaction to be detected.

6. One potential problem to be alert for is that one or both of the LexA-fused proteins has significant intrinsic ability to activate transcription (in practice, this happens in 5–10% of cases). This would result in blue color on X-gal glucose plates and growth on leu- glucose plates by transformations a and e and/or b and f, which if substantial enough may block the ability to detect an interaction. In cases of weak to moderate activation by DBD fusions, it is frequently possible to suppress "background" by using less sensitive reporter plasmids and strains (EGY191 rather than EGY48, and 1840 rather than JK103). In cases of strong activation, it is sometimes necessary to truncate the DBD-fused protein to remove activating sequences.

7. It is possible to roughly quantitate rank order of interactions by performing a β-galactosidase assay *(7)*. This is done as follows:

   a. For each combination of plasmids to be assayed, pick at least three independent transformants and grow a 5-mL overnight culture in ura-his-trp- glucose medium at 30°C.

   b. The following morning, dilute the culture (which should be saturated, i.e., OD_{600} >2.0) into ura-his-trp- galactose medium to an OD_{600} of 0.1–0.15, and grow at 30°C until the culture has reached an OD_{600} of 0.45–0.65.

   c. Spin down 1 mL of cells from each culture in a microfuge at 13,000g for 2 min at room temperature. It is important to record for each culture the exact

OD_{600} of the culture at time of harvest *(see below)*. Note, it is generally wise to make duplicate samples of each culture as a guard against unforeseen accidents, as growing the cultures is the most labor-intensive part of the process.

d. There should be a visible pellet of yeast of 1–2 µL in volume: if not, spin another 2 min as above. Pour off supernatant. Add 100 µL of Tris/Triton lysis buffer to each microfuge tube. Vortex briefly, and place tubes directly onto dry ice to crack yeast. Samples may be frozen at –70°C overnight or longer at this stage, if desired.

e. Thaw samples at 30°C. For each sample, make up a solution of Z buffer containing ONPG at 1 mg/mL, and bring to 30°C as well. In addition, have ready 500 µL of $1M$ Na_2CO_3 for each sample.

f. Set a laboratory timer for ~30 min. Arrange all samples in a row, preferably in a 30°C water bath (although setting up the reactions at room temperature and moving them to a 30°C incubator is acceptable). Turning on the timer, start pipeting 750 µL of the Z buffer/ONPG solution into each sample in turn.

g. When a medium yellow color has appeared (which may be from 2 min to several hours, depending on the strength of the interaction), stop the reaction by adding 500 µL of $1M$ Na_2CO_3. Do not allow the reaction to become very dark yellow. As exactly as possible, record the length of the reaction. Note, when comparing many samples with different interaction strengths, it may be desirable to stop some after short periods, and allow others to go for longer periods of time.

h. Spin down the sample for 15 s at 13,000g at room temperature to precipitate cell debris.

i. Read the OD_{420} of all samples. If the OD_{420} is greater than 1.0, dilute the sample 1:10 to read.

j. Calculate β-galactosidase units using the formula

$$Units = (100/rxn\ time\ in\ minutes)/([OD_{420}]/[OD_{600}])$$

k. Average the values for independent colonies, and determine standard deviation. Variation should be less than 20% between independent colonies.

l. In general, it is advisable to repeat the entire process (starting with different independent colonies) on a separate day, to ensure maximal accuracy. Appropriate controls:

8. If an interaction is not detected, it is important to exclude the possibility that one or both of the fusion proteins is not expressed, or that the DBD-fused protein is unable to bind reporter operator sequences. Expression of the AD fusion can be determined by performing a Western blot of crude yeast lysate, using antibody to the hemagglutinin epitope tag. Operator binding of the DBD-fusion can be determined by performing a repression assay *(8)*.

9. To perform a repression assay, transform EGY48 yeast with the following combinations of plasmids:

p-202-P1 + JK101...plate to ura-his-.
p-202-P2 + JK101...plate to ura-his-.
pRFHM1 + JK101...plate to ura-his-.
JK101...plate to ura-.

For each transformation, restreak at least six colonies to a ura-his- or a ura- master plate.
10. Restreak from the master plates to a single ura- X-gal plate containing galactose/raffinose as a sugar source, and incubate 1–2 d at 30°C (It is not necessary to maintain selection for the HIS3 plasmid over this period, and because of variability of batches of X-gal plates, it is important to compare all transformations on the same plate). Inspect the plate. JK101 is similar to JK103, except that it contains the GAL1 upstream activating sequences (UAS_{GAL}) upstream of 2 LexA-operator sites. Yeast containing JK101 only will have significant β-galactosidase activity when grown on media in which galactose is the carbon source, which will result in dark blue colony color. LexA-fused proteins that are made, enter the nucleus and bind the LexA operator sequences, and will interfere with activation from the UAS_{GAL}, repressing β-galactosidase activity two- to 10-fold, resulting in reduced blue color (as with pRFHMI + JK101).
11. If a protein neither activates nor represses, the most likely reason is that it is not being expressed correctly. This can be determined using a Western blot of a crude lysate protein extract (as in ref. *9*) of EGY48 containing the plasmid, using anti-LexA antibodies as primary antisera. If this is the case, it may be possible to express truncated domains of the protein.

3.2. Interaction Mating

Interaction mating allows the testing of a large number of protein-protein interactions for predefined group of proteins *(10)*. In this method, a series of DBD-P(n) and AD-P(x) fusions are transformed independently into haploid yeast strains of opposite mating types, and cross-gridded together to select diploids. Combinations containing interacting fusion proteins can be rapidly detected by replica of diploids to X-gal and leu- selective plates (*see* Fig. 3).

1. Grow cultures of EGY48 and of RFY206 for transformation (*see* above for transformation protocols).
   Into EGY48 (MATα), transform the following combinations of plasmids

   > p-202-P1   + JK103................................plate to ura-his-.
   > p-202-P2   + JK103................................plate to ura-his-.
   > p-202-P(n) + JK103................................plate to ura-his-.
   > p-202   + JK103................................plate to ura-his-.

   Into RFY206 (MATa), transform the following plasmids:

   > pJG4-5-P1..plate to trp-.
   > pJG4-5-P2..plate to trp-.
   > pJG4-5-P(n)..plate to trp-.
   > pJG4-5...plate to trp-.

   Incubate transformation plates at 30°C for 2–3 d until colonies appear.

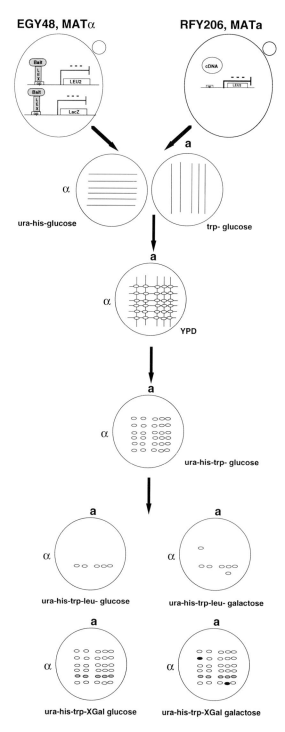

2. Make ura-his- and trp- master plates (as appropriate) containing at least two picked colonies for each transformation.
3. On a ura-his- plate, use a toothpick to make parallel streaks of EGY48 yeast containing p-202-P(n) and JK103. eight to ten separate colonies can conveniently be streaked on a single plate (*see* Fig. 3).

    On a trp- plate, use toothpicks to make parallel streaks of RFY206 containing a series of JG4-5-fused proteins.

    It is helpful to include the cross combination of p-202 and pJG4-5 on each plate, as negative control.
4. Replica plate these two sets of streaks onto YPD plates, where cells of different mating types can mate and form diploids at the intersections of the horizontal and vertical lines (Fig. 3). Note, it is important to use a separate velvet for each replica to avoid cross-contamination. Incubate plates overnight at 30°C
5. Using the YPD plates containing the mated yeast as master plates, replica plate to a ura-his-trp- glucose plate. Incubate plates overnight to two d at 30°C. This will allow outgrowth of diploid yeast containing all DBD- and AD-fused proteins, and reporters. Note, some researchers omit this step, and proceed directly from the YPD plate to selective media: In our hands, we have found that including this intermediate step to first select for diploids leads to clearer results.
6. Using the ura-his-trp- plate as a master, replica plate to the four following combinations of plates:
    a. ura-his-trp- glu XGAL.
    b. ura-his-trp- gal/raf XGAL.
    c. ura-his-trp-leu- glu.
    d. ura-his-trp-leu- gal/raf.
    Incubate plates 2–3 d at 30°C.
7. If two proteins interact specifically, then for plates containing galactose/raffinose (inducing the expression of the AD-fusion) that DBD/AD-fused combination of proteins will cause enhanced growth on ura-trp-his-leu- medium and will cause enhanced blue color on plates containing X-gal, relative to background (defined as the DBD-fusion coexpressed with JG4-5 vector). These same DBD/AD

Fig. 3. *(previous page)* Mating panel assay is illustrated. In separate transformations, MATa EGY48 yeast are transformed with a series of p-202 fusions and a LacZ reporter plasmid, whereas MATa RFY206 are transformed with a series of AD-fusions. Transformants are streaked as parallel lines on a ura-his- (EGY48) or trp- (RFY206) plate. A replica block is used to cross grid the EGY48 and RFY206 plates onto a YPD plate, where mating occurs. After 12 h, cross gridded colonies are optionally replica plated onto ura-his-trp- glucose to select for diploids. In a final replica plating, colonies are lifted from the ura-his-trp- glucose plate (or the YPD plate) to plates selective for LacZ and LEU2 expression (*see* text). Growth specific to galactose leu- medium and blue color specific to galactose X-gal medium indicate positive interactions. Note, any LexA-fusion with residual ability to activate transcription will also be positive on glucose media: one example is shown.

combinations will show no enhancement of growth or blue color on plates containing glucose as a sugar source.

8. Some AD-fused proteins enhance transcription relatively nonspecifically with a large number of other proteins (HSP70 is a good example, as is the fragment of bicoid encoded in RFHMI). These are likely to show a random pattern of interaction with multiple p-202 fused proteins in a mating grid, the significance of which is not known.

3.3. Performing a Library Screen

This protocol is initially similar to that for probing specific interactions. First, LexA is fused to the protein of interest, and a series of control experiments are performed to establish whether it is suitable as is, or must be further modified. These controls will establish whether the bait is made as a stable protein in yeast and is capable of entering the nucleus and binding LexA operator sites, and does not appreciably activate transcription of the LexA-operator-based reporter genes.

1. Insert the gene encoding the protein of interest (P1) into the polylinker of pEG202 or pJK202 to make an in-frame protein fusion.
2. Determine whether the LexA-fused probe protein activates transcription in yeast. Transform EGY48 using the following combinations of plasmids:

   p-202-P1 + JK103...plate to ura-his- at 30°C.
   pSH17-4 + JK103 (positive control).......................plate to ura-his-.
   pRFHM1 + JK103 (negative control).......................plate to ura-his-.

   Note: pEG202 itself is not a good negative control. The peptide encoded by the uninterrupted polylinker sequences is itself capable of very weakly activating transcription.

3. Streak a master plate as described above. After 24 h, restreak from the master plate to a ura-his- X-gal plate, and a ura-his-trp-leu- plate, both containing glucose as a carbon source. A good bait protein will be white on X-gal and have produced no visible colony growth on the leu- medium after 3 d. A bait protein that causes yeast to turn faintly blue and grow to a small degree on leu- plates after 3 d should be tested with the less sensitive reporters 1840 and EGY191. For baits that activate transcription strongly, it is frequently possible to generate a LexA-fusion that contains a truncation of the desired probe protein that does not activate.

4. For baits that do not activate, a repression assay should be performed as described above to ensure that the proteins are in fact binding operators on the reporters. If any dimerizing partners for the protein bait are known to exist, it may also be useful to synthesize them as AD-fusions, to test whether the bait can dimerize appropriately, although this is generally not necessary.

5. Once a bait has been made and tested, one can proceed to a library screen. A typical interactor hunt requires two, successive large platings of yeast containing

LexA-fused probes, reporters, and libraries. In the first, yeast are transformed with library, plated on dropout medium lacking uracil, histidine, and tryptophan, with glucose as a carbon source, and a frozen homogenized stock made. This stock is subsequently thawed and replated on medium lacking uracil, histidine, tryptophan, and leucine (the selection), with galactose as carbon source to induce the library. This two-step selection is helpful for two reasons. First, a number of interesting cDNA encoded proteins may be deleterious to the growth of yeast that bear them, and hence be competed out in an initial mass plating on selective medium. Second, it is possible that yeast bearing particular interacting proteins may not be able to express sufficient levels of these proteins after a simultaneous transformation and galactose induction to immediately support growth on media lacking leucine.

3.3.1. Library Transformation

1. Begin with an appropriate LexAop-LEU2 strain/LexAop-LacZ reporter combination expressing the LexA-fused bait. Grow up an overnight culture of ~20 mL in ura-his- glucose liquid medium.
2. In the morning, dilute the overnight culture into 300 mL ura-his-glu liquid media to approximately 2×10^6 cell/mL (OD_{600} ~0.10) culture. Grow until approx 1×10^7 cells/mL (OD_{600} ~0.50).
3. Harvest by spinning at 1000–1500g at room temperature in a low speed centrifuge. Resuspend in 30 mL sterile distilled water, transfer to 50 mL-Falcon tubes.
4. Spin at 1000–1500 g for 5 min, pour off supernatant, and resuspend cell pellet in 1.55 mL TE/.1M LiAc pH 7.5.
5. Take 30 1.5-mL-Eppendorfs. Into each, aliquot 1 µg of library DNA and 50 µg high-quality, sheared salmon sperm carrier DNA (average size range 1.5–7 kb, *see* Schiestl and Gietz). Total volume of DNA added should be less than 20 µL, and preferably less than 10 µL. Add 50 µL of the resuspended yeast solution.
6. *If the LexA-fused bait being used causes any background transcription with either reporter system*, take one additional 1.5 mL-Eppendorf. Add 1 µg of pure JG4-5 library vector and ~5 µg of carrier DNA, and add 50 µL of resuspended yeast solution.
7. Add 300 µL of sterile TE/L acetate/40% PEG4000 solution to all transformations, and mix thoroughly by inversion. Incubate at 30°C for 30 min.
8. Add DMSO to 10% (~40 µL) per tube. Mix by inversion. Heat shock in 42°C heat block for 10 min.
9. For 28 tubes containing library plasmid, plate complete contents of each tube on 24 × 24 cm ura-his-trp- glucose dropout plate (spread the complete contents of one tube per plate). This is most effectively accomplished by pipeting the contents of the tube onto the plate surface, dropping 5–10 sterile glass beads on the plate, and agitating vigorously for 1 minute.
10. For the remaining two tubes containing library plasmid, plate only 360 µL of the contents (9/10 of total) on a 24 × 24 cm plate, and use 40 µL (1/10 contents), and do a series of 1:10 dilutions in sterile, distilled water. Plate on normal sized ura-

his-trp- plates, and place at 30°C. This will give you some idea of the efficiency of the transformation, and allow you to accurately estimate the number of transformants obtained. Colonies should begin to appear in 2–3 d, and can be harvested when average colony size is 1–2 mm in diameter.

11. For the tube containing JG4-5 library vector, plate ~50 μL of heat shocked transformation to a ura-his-trp- glucose plate. This will yield sufficient colonies to pick controls for library screening (*see* step 18).

12. To harvest library plates, cool plates to 4°C for a few hours to harden agar. Wearing gloves, use a sterile glass microscope slide to gently scrape yeast off the plates. Alternatively, plate surface can be flooded with 1–1.5 mL of sterile TE, more glass beads added, and the plate agitated for several min. Pool the contents of the 30 plates into a sterile 50-mL Falcon tube.

13. Wash cells twice with 1X sterile TE or distilled water (pellet by spinning down for about 5 min at ~1000–1500g in a centrifuge). After the second wash, the pellet volume should be approx 25 mL of cells derived from ~1.5 million transformants. Resuspend by mixing thoroughly with an equal volume of glycerol freezing solution, and freeze in 1-mL aliquots at –70°C.

14. In general, 80–100% of yeast frozen as described above will remain viable for several months (at least). However, in order to accurately interpret the number of positives obtained in a screen as a percentage of total transformants plated, it is necessary to determine replating efficiency in parallel with plating on selective media.

15. Remove an aliquot of frozen transformed yeast, and dilute 1:10 with his-trp-ura-*CM* dropout gal/raff medium (2% galactose, 1% raffinose; raffinose aids in growth without altering induction of the *GAL1* promoter). Incubate, shaking, at 30°C for 4 h to induce the *GAL1* promoter on the library.

16. Read the OD_{600} of the culture. Assume 1 OD_{600} ~2×10^7 yeast cells. In general, library screening works best if each initial transformant obtained is represented by 3–10 cells plated on selective medium: further, it is easiest to visually scan for Leu+ colonies using cells plated at approx 10^6 colony forming units per 100 mm plate. Thus, in a hypothetical example, for a transformation in which 3×10^6 colonies had been obtained, you would plate ~2×10^7 cells, corresponding to 1 mL of a culture with an $OD_{600} = 1.0$, with 50 μL on each of a total of 20 selective plates. If any dilutions are necessary, they should be made using ura-his-trp-leu-galactose media, or sterile dH_2O.

17. In parallel, make serial dilutions of the culture (10^{-1}–10–4 is the most useful range) using his-trp-ura- CM dropout gal/raff medium, and plate ~20 μL of each dilution on his-trp-ura- CM dropout gal/raff plates to determine the number of colony-forming units per aliquot of transformed yeast.

18. If all is working well, new colonies will become visible daily over a period from 2–5 d. Carefully pick these colonies (noting on what day after plating they appeared) to a ura-his-trp glucose master plate. If a very large number of positives appear (>100), initially restrict subsequent analysis to those appearing earlier, as these are most likely to be biologically relevant. Where relevant, also pick

~10 colonies from the transformation plate containing JG4-5 vector only to this master plate.

19. Restreak or replica from the master plate to the following plates:
    a. ura-his-trp- glu X-gal.
    b. ura-his-trp- gal/raf X-gal.
    c. ura-his-trp-leu- glu.
    d. ura-his-trp-leu- gal.

    For LexA-fused baits that have no background ability to activate transcription of reporters, colonies which are blue on gal/raff but not on glu Xgal plates, and which grow on gal/raff but not glu leu- plates are considered positive. Note, for baits with intrinsic ability to activate transcription, it sometimes happens that equivalent blue color is seen with library plasmids on glucose and galactose X-gal plates. This is probably because the ADH promoter expressing the LexA-P1 protein works better on glucose media, causing greater levels of LexA-P1 to be expressed. To eliminate confusion, it is important to compare yeast containing library-encoded plasmids versus JG4-5 vector for color on X-gal plates and growth on leu- plates. If a library plasmid causes an enhancement over JG4-5 on galactose but not glucose medium, it may be a positive.

20. For positive colonies, isolate library plasmid DNA from yeast by a rapid miniprep protocol such as the following:
    a. Starting from the ura-his-trp- glucose master plate, pick colonies with the appropriate phenotype on selective plates into 5 mL of trp- glucose, and grow up overnight at 30°C. Omitting the his-ura- selection in this situation encourages loss of nonlibrary plasmids, and facilitates isolation of the desired library plasmid.
    b. Pellet 1 mL of the culture in a microfuge at 13,000g for 1 min at room temperature.
    c. Resuspend pellet in 200 µL of lysis solution STES, and add ~100 µL of 0.45-mm diameter sterile glass beads. Vortex vigorously for 1 min.
    d. Add 200 µL equilibrated phenol, and vortex vigorously for another minute.
    e. Spin the tube in a microfuge at 13,000g for 2 min at room temperature, and transfer aqueous phase to a new microfuge tube.
    f. Add 200 µL phenol and 100 µL chloroform, and vortex 30 s. Spin down as described in E., and transfer aqueous phase to a fresh tube.
    g. Add two volumes 100% ethanol (approx 400 µL), mix by inversion, and chill at –20°C for 20 min. Then spin down in a microfuge at 13,000 g for 15 min at 4°C.
    h. Pour off supernatant. Wash pellet with chilled 70% ethanol, and spin again at 13,000g, 2 min at 4°C. Pour off ethanol, and dry pellet briefly in speed vacuum. Resuspend in 5–10 µL TE (10 mM Tris-HCl pH 8.0, 1 mM EDTA).

21. Use the isolated DNA to transform by electroporation KC8 bacteria (*pyrF, leuB600, trpC-9830, hisB463,* constructed by K. Struhl), and plate on LB plates containing 50–100 mg/mL ampicillin.

22. The yeast TRP1 gene can successfully complement the bacterial *trpC* -9830 deficiency, allowing the library plasmid to be easily distinguished from the other two plasmids contained in the yeast. For each original yeast positive, pick ~10 colonies onto bacterial defined minimal bacterial plates (A medium) supplemented with uracil, histidine, and leucine but lacking tryptophan: those that grow will contain the library plasmid. Some investigators omit the initial plating on LB/ Amp media, plating directly onto trp- bacterial plates: In our own experience, we find the two-step procedure helpful, as the less stringent LB-Amp selection maximizes the likelihood of obtaining transformants.

23. Grow up KC8 containing library plasmid in LB-Amp, and do minipreps by standard methods. This will yield enough DNA for at least five transformations of yeast.

24. To eliminate false positives including mutations in the initial EGY yeast that favor growth on galactose media, library-encoded cDNAs that interact with the LexA DNA-binding domain, or library-encoded proteins that are "sticky" and interact with multiple biologically unrelated fusion domains, it is necessary to retransform putative positive interactors into yeast containing the original LexA-P1 and yeast containing LexA-fused to a nonspecific additional protein, and ascertain that transcriptional activation of the reporters is dependent on the library-encoded protein, and is specific to LexA-P1.

25. Use plasmids purified from KC8, as well as a JG4-5 vector control, to transform the following strains of yeast:
    a. EGY48 containing JK103 and LexA-P1.
    b. EGY48 containing JK103 and LexA-P2. (RFHM1, the control used above for the repression assay, is acceptable).
    c. EGY48 containing JK103 and pEG202.
    Plate transformants to ura-his-trp- medium. When colonies have grown up, create a ura-his-trp master plate so that for each library plasmid being tested, five to six independent colonies derived from each of A and B are streaked adjacently.

26. Restreak from this master plate to the same series of test plates used in the actual screen: that is,
    a. ura-his-trp- glu X-gal.
    b. ura-his-trp- gal/raf X-gal.
    c. ura-his-trp-leu- glu.
    d. ura-his-trp-leu- gal.

27. Yeast containing true "positive" cDNAs should generally be bluer than yeast containing JG4-5 on gal/raff but not on glu X-Gal plates, and grow on gal/raff leu- but not glu leu- plates *only* with the LexA-P1. Those cDNAs that are also positive with RFHM1 and pEG202 should be discarded. Note, RFHM-1 is a very rigorous control for specificity testing, because it has previously been demonstrated to be "sticky"—that is, to associate with a number of library-encoded proteins that are clearly nonphysiological interactors (R. Finley, unpublished results): thus, if a library protein is weakly positive with RFHMI but not with pEG202, it may be worthwhile to try more nonspecific baits before giving up on it. The three test plasmids outlined represent a minimal test series. If other LexA-

baits are available that are related to the bait used in the initial library screen, substantial amounts of information can be gathered by this approach. Note, very occasionally true positives will be strongly blue and grow well on leu- plates with galactose, but also weakly blue and grow on plates with glucose. Although the GAL1 promoter that expresses the library protein is repressed by glucose, a minimal amount of transcription does occur, which may in some cases lead to the production of sufficient library protein to produce a detectable interaction.

28. Those cDNAs that meet the criteria for specificity noted above are ready to be sequenced or otherwise characterized. Prior to sequencing, it is helpful to retransform from KC8 into another bacterial strain such as DH-5α or XL1-BLUE, which yields high quality DNA: preps from KC8 are generally poor for sequencing.

4. Notes

1. In hunts conducted to date, anywhere from none to practically all isolated plasmids passed the final specificity test. Note, some library-encoded proteins are known to be isolated repeatedly, using a series of unrelated baits, and to demonstrate at least some specificity. At least one of these, heat shock protein 70, might be explained by the fact that some LexA-fused baits are not normally folded. It is important to consider whether the protein isolated plausibly has physiological relevance to its bait.

2. Because the methodology of the screen involves plating multiple cells to ura-his-trp-leu-gal medium for each primary transformant obtained, multiple reisolates of true positive cDNAs are frequently obtained. If a large number of specific positives are obtained, it is generally a good idea to attempt to sort them into classes—for example, digesting minipreps of positives with EcoRI, XhoI, and HaeIII will generate a "fingerprint" of sufficient resolution to determine whether multiple reisolates of a small number of clones, or single isolates of many different clones have been obtained. The former situation is a good indication that the system is working well.

3. To maximize chances of an interactor hunt working well, there are a number of parameters to be taken into account. Bait proteins should be carefully tested before attempting a screen for low or no intrinsic ability to activate transcription. Baits must be expressed at reasonably high levels, and be able to enter the yeast nucleus and bind DNA (as confirmed by the repression assay). By extrapolation from these conditions, proteins that have extensive transmembrane domains or are normally excluded from the nucleus are not likely to be productively used in a library screen. Proteins that are moderate to strong activators will need to be truncated to remove activating domains before they can be used. Optimally, integrity and levels of bait proteins should be confirmed by Western analysis, using antibody either to LexA or the fused domain (*see* ref. *11* for a discussion of relevant variables for LexA-fused proteins).

4. It is extremely important to optimize transformation conditions before attempting a library screen. In contrast to *E. coli*, the maximum efficiency of transformation for *S. cerevisiae* is about $10^4–10^5/\mu g$ input DNA: small scale pilot

transformations should be performed to ensure this efficiency is attained, to avoid having to use prohibitive quantities of library DNA. As for any screen, it is a good idea to obtain or construct a library from a tissue source in which the bait protein is known to be a biologically relevant. Finally, Bendixen et al. *(12)* have recently described a mating protocol by which library is transformed once into a haploid strain of yeast, which can be frozen in aliquots and used subsequently in screens with multiple DBD-fused baits transformed into strains of the opposite mating type, by simply mating library against bait. Investigators intending to do multiple interactions screens may find it worthwhile to investigate this option.

5. In practice, very few proteins are isolated that are specific for LexA. However, it is generally informative to retest positive clones on more than one LexA-bait. Ideally, library-derived clones would be tested against: the LexA-fusion used for their isolation, several LexA-fusions to proteins that are clearly unrelated to the original fusion, and if possible, several LexA-fusions that there is reason to believe are related to the initial protein (for instance, if the initial probe was LexA-fos, a good related set would include LexA-jun and LexA-GCN4).

6. Depending on the protein used as a bait, anywhere from none to Lally hundreds of interactors will be obtained.

7. There are an increasing number of cases on record where for two proteins known by other means (e.g., coimmuneprecipitation) to interact, the interaction can only be detected with one specific component fused to the DBD and the other fused to the AD. Further, there are some cases where the interaction DBD-P1 x AD-P2 but not DBD-P2 x AD-P1 can be detected using the LexA-based interaction trap, whereas the combination DBD-P2 x AD-P1 but not DBD-P1 x AD-P2 can be detected in the GAL4-based two-hybrid system. Although there has been much hypothesis as to why this is, it remains an unsolved mystery. This is the reason we suggest testing interactions reciprocally, where possible. This phenomenon may also underly some failures to isolate interacting partners for proteins known by biochemical means to have them.

8. When examining interactions using the LEU2 and LacZ reporters in parallel, it will sometimes occur that a combination of proteins will strongly activate one of the reporters, but only very slightly activate the other. There are many possible explanations for this, including promoter preference, the fact that one reporter is integrated chromosomally and the other plasmid-borne, and so forth. In general, if a library plasmid activates both reporters to very disparate degrees, but specifically with one bait protein, it should not be arbitrarily discarded. Conversely, the fact that different reporters give different rank order of interaction affinity *(13)* should lead one to regard numbers from the two-hybrid system and interaction trap as only very crude quantitation of biochemical affinity between two proteins.

9. The fact that two proteins "interact" in the interaction trap is not conclusive proof that they associate directly or strongly in their native environment (e.g., mammalian cells). It should be noted that a number of proteins are conserved extremely well between higher and lower eukaryotes: Thus, an interaction between a DBD-P1

and an AD-P2 might also include as mediating factors yeast proteins P3, P4, and P5. Skepticism is helpful.

10. Although not detailed in this chapter, a number of other laboratories have developed two-hybrid system variants: some references describing these systems *(14–17)* are included below.

11. A number of additional LexA fusion plasmids have been constructed to permit use of two-hybrid methods with proteins that cannot be acceptably expressed from the basic system reagents described in this article. pGilda, made by David Shaywitz at MIT, puts the LexA-fusion cassette under control of the GAL1 promoter, and is useful for studying proteins toxic to yeast, as the LexA-fusion can then be induced only at the point of assaying activation of LacZ or LEU2 reporters. pEGE202I, made by Rich Buckholz of Glaxo, allows integration of the LexA-fusion plasmid, resulting in very low expression of the bait protein: this is useful for studying interactions that normally occur with very high affinity. pNLex, made by Manuel Sainz of University of Oregon and Ian York of the Dana-Farber Cancer Center, allows synthesis of a fusion in which P1 is located amino-terminally to LexA, useful in cases where domains thought to be important for protein function are located at the N-terminus. Further vectors are currently under construction by generous individuals using the system: all have been banked centrally and are available as described below.

12. Key plasmids, strains, and a number of cDNA libraries cloned into the JG4-5 vector (including HeLa cell, Drosophila embryonic, neural, and others) can be most readily obtained by contacting Roger Brent, Wellman 8, Department of Molecular Biology, Massachusetts General Hospital, 50 Blossom St., Boston, MA 02114, telephone (617) 726-5925, Fax (617) 726-6893, email Brent@frodo.mgh.harvard.edu. Reagents will also be on sale from Clontech in 1996–1997.

References

1. Fields, S. and Song, O. (1989) A novel genetic system to detect protein-protein interaction. *Nature* **340,** 245–246.
2. Gyuris, J., Golemis, E. A., Chertkov, H., and Brent, R. (1993) Cdi1, a human G1 and S phase protein phosphatase that associates with Cdk2. *Cell* **75,** 791–803.
3. Ebina, Y., Takahara, Y., Kishi, F., Nakazawa, A., and Brent, R. (1983) LexA is a repressor of the colicin E1 gene. *J. Biol. Chem.* **258,** 13,258–13,261.
4. Gietz, D., St. Jean, A., Woods, R. A., and Schiestl, R. H. (1992) Improved method for high efficiency transformation of intact yeast cells. *Nucleic Acids Res.* **20,** 1425.
5. Ito, H., Fukuda, Y., Murata, K., and Kimura, A. (1983) Transformation of intact yeast cells treated with alkali cations. *J. Bacter.* **153,** 163–168.
6. Schiestl, R. H. and Gietz, R. D. (1989) High efficiency transformation of intact yeast cells using single stranded nucleic acids as a carrier. *Curr. Genet.* **16,** 339–346.
7. Miller, J. (1972). *Experiments in Molecular Genetics*, Cold Spring Harbor Laboratory Press, Cold Spring Harbor, NY.
8. Brent, R. and Ptashne, R. (1984) A bacterial repressor protein or a yeast transcriptional terminator can block upstream activation of a yeast gene. *Nature* **312,** 612–615.

9. Samson, M.-L., Jackson-Grusby, L., and Brent, R. (1989) Gene activation and DNA binding by Drosophila *Ubx* and *abd-A* proteins. *Cell* **57,** 1045–1052.
10. Finley, R. and Brent, R. (1994) Interaction mating reveals binary and ternary connections between Drosophila cell cycle regulators. *Proc. Natl. Acad. Sci. USA* **91,** 12,980–12,984.
11. Golemis, E. A. and Brent, R. (1992) Fused protein domains inhibit DNA binding by LexA. *Mol. Cell. Biol.* **12,** 3006–3014.
12. Bendixen, C., Gangloff, S., and Rothstein, R. (1994) A yeast mating-selection scheme for detection of protein-protein interactions. *Nucleic Acids Res.* **22,** 1778–1779.
13. Estojak, J., Brent, R., and Golemis, E. (1995) Correlation of two-hybrid affinity data with *in vitro* measurements. *Mol. Cell. Biol.* **15,** 5820–5829.
14. Bartel, P. L., Chien, C.-T., Sternglanz, R., and Fields, S. (1994) Using the two-hybrid system to detect protein-protein interactions, in *Cellular Interactions in Development: A Practical Approach* (Hartley, D. A., ed.), Oxford University Press, Oxford, England, pp. 153–179.
15. Chien, C. T., Bartel, P. L., Sternglanz, R., and Fields, S. (1991) The two-hybrid system: a method to identify and clone genes for proteins that interact with a protein of interest. *Proc. Natl. Acad. Sci. USA* **88,** 9578–9582.
16. Dalton, S. and Treisman, R. (1992) Characterization of SAP-1, a protein recruited by serum response factor to the c-fos serum response element. *Cell* **68,** 597–612.
17. Durfee, T., Becherer, K., Chen, P. L., Yeh, S. H., Yang, Y., Kilburn, A. E., Lee, W. H., and Elledge, S. J. (1993) The retinoblastoma protein associates with the protein phosphatase type 1 catalytic subunit. *Genes Dev.* **7,** 555–569.
18. Ruden, D. M., Ma, J., Li, Y., Wood, K., and Ptashne, M. (1991) Generating yeast transcriptional activators containng no yeast protein sequences. *Nature* **350,** 250–252.
19. Ma, J. and Ptashne, M. (1987) A new class of yeast transcriptional activators. *Cell* **51,** 113–119.
20. West, R. W. J., Yocum, R. R., and Ptashne, M. (1984) *Saccharomyces cerevisiae* GAL1-GAL10 divergent promoter region: location and function of the upstream activator sequence UAS$_G$. *Mol. Cell. Biol.* **4,** 2467–2478.
21. Brent, R. and Ptashne, M. (1985) A eukaryotic transcriptional activator bearing the DNA specificity of a prokaryotic repressor. *Cell* **43,** 729–736.

17

Detection of Heterologous G$_s$-Coupled Receptor Activity in LLC-PK$_1$ Cells Based on Expression of Urokinase-Type Plasminogen Activator

Luigi Catanzariti and Brian A. Hemmings

1. Introduction

In renal epithelial cells LLC-PK$_1$ *(1)* expression of the urokinase-type plasminogen activator (uPA) gene is strongly induced by first messengers that stimulate intracellular cAMP synthesis *(2)*. The increase of cAMP leads to a 200-fold activation of uPA gene transcription *(4)*. In response to optimal concentrations of the peptide hormone calcitonin, i.e., for which the cells express a G$_s$-coupled receptor *(3,9)*, high levels of uPA protein are synthesized and secreted. The uPA protein is a serine/threonine protease that converts the zymogen plasminogen into plasmin, a protease with a wide substrate specificity *(6)*. The uPA activity is easily detected colorimetrically in the serum-free supernatant of induced cells *(5,7,10)*. Alternatively, it can be visualized *in situ* by overlaying colonies of stimulated cells with agar containing both plasminogen and casein (Fig. 1). The degradation of the milk-protein casein by plasmin leads to the formation of lytic zones that allow the identification of uPA secreting colonies

Recently, we showed that LLC-PK$_1$ cells and a derivative mutant (M18) can be used as an expression system for the study of heterologous G$_s$-coupled receptors *(10)*. The M18 cell line was derived from LLC-PK$_1$ following N-methyl-N'-nitro-N-nitrosoguanidine (MNNG) mutagenesis of wild-type cells *(5)*. The M18 cells express <3% of wild-type calcitonin and vasopressin receptor binding activity. Treatment with these peptide hormones does not activate adenyl cyclase *(5)*. We reasoned that transfected LLC-PK$_1$ cells expressing functional G$_s$-receptors would display high uPA activity in response

From: *Methods in Molecular Biology, vol. 63: Recombinant Protein Protocols: Detection and Isolation* Edited by: R. Tuan Humana Press Inc., Totowa, NJ

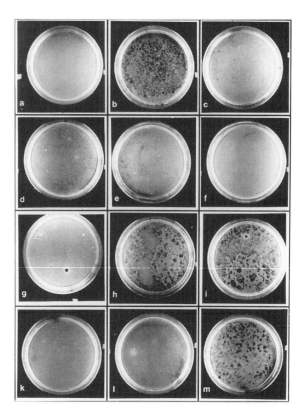

Fig. 1. uPA activity in the casein overlay: Cell lines were induced for 18 h with isoproterenol or calcitonin and subsequently overlaid with a plasminogen and casein containing agar as described in Sections 2. and 3. a,b,c: LLC-PK1 cells; d,e,f: M18 cells; g,h,i: LLC-PK1β3 cells; k,l,m: M18β2.1 cells. a,d,g,k: no induction; b,e,h,l: calcitonin (30 nM); c,f,i,m: isoproterenol ($10^{-5}M$). LLC-PK1 β3 was screened and isolated using the colorimetric assay. M18β2.1 was cloned using the casein overlay assay.

to appropriate agonists and therefore allow the rapid identification and isolation of cell-clones expressing heterologous receptors *(10)*.

Using this model-system, we functionally expressed a genomic clone of the mouse β$_2$AR in both wild-type LLC-PK$_1$ and mutant M18 cells *(10)*. The results showed that LLC-PK$_1$ cells apparently express all factors necessary for the transcription of the mouse β$_2$AR gene. The β$_2$ARs are also functionally expressed in the mutant cell line despite the lack of endogenous calcitonin and vasopressin receptor activity. Mechanisms for posttranslational modifications, targeting of plasma membrane-bound receptors ,and cAMP signaling are therefore normal in M18 cells *(9)*. This conclusion is also supported by the observation that expression of a cDNA clone encoding the porcine calcitonin receptor in the mutant cell line restored calcitonin binding and calcitonin-inducible uPA activity *(9)*. Either cell line, therefore, can be used as a recipient for heterologous receptors *(10)*.

LLC-PK$_1$ cells expressing heterologous G$_s$-receptors provide an in vitro system to test for agonist and/or receptor function. The enzymatic quantitation of uPA activity in the conditioned medium using a simple calorimetric assay represents a novel way for estimating agonist potencies nonradioactively. Since the conversion of plasminogen to plasmin by urokinase is an autocatalytic event (in which more plasmin is formed due to the conversion of plasminogen to plasmin by plasmin itself *(8)*, even small changes in the level of uPA-activity over basal are amplified. This chapter will focus on the usage of LLC-PK$_1$ and M18 cells as recipients for heterologous β_2AR and calcitonin receptors.

2. Materials

1. Casein, from bovine milk, technical grade (Sigma, St. Louis, MO).
2. Dulbecco's Modified Eagle's Medium (DMEM) GIBCO-BRL, Gaithersburg, MD).
3. Fetal calf serum (FCS), Flow Laboratories (UK).
4. Bovine serum albumin (BSA) (fatty-acid-free, fraction V; Sigma, St. Louis, MO.
5. Triton X-100, Sigma.
6. Porcine urokinase, Sigma.
7. Synthetic peptide S-2251 (D-Val-Leu-L-Lys-NH-NP) (Bachem Feinchemikalien AG, Bubendorf, Switzerland).
8. Hygromycin B (Boehringer Mannheim, Mannheim, Germany).
9. Phosphate-buffered saline without Mg^{2+}/Ca^{2+} (PBS$^-$).
10. Bio-Rad Gene Pulser (Hercules, CA).
11. Porcine plasminogen (Sigma).
12. Standard trypsin/EDTA tissue culture solution (0.25% trypsin/1 mM EDTA [GIBCO-BRL]).
13. Assay buffer: 90 mM Tris-HCl, pH 8.8, 0.41% Triton-X-100.
14. Microtiter plate reader (wavelength 495 nm).
15. G$_s$-receptor plasmid DNA.
16. Low melting agarose (Sigma).
17. (–) Isoproterenol (Sigma).
18. Salmon calcitonin (Sandoz AG, Basel, Switzerland).
19. Hygromycin B resistance plasmid pX343 *(7)*.

3. Method

3.1. Cell Culture

The LLC-PK$_1$ and derivative cell lines are cultured in Dulbecco's Modified Eagles Medium (DMEM) supplemented with 10% (v/v) fetal calf serum, 0.2 mg/mL streptomycin, and 50 U/mL penicillin, in a humidified incubator at

$37°C$ with a 95% air/5% CO_2 atmosphere. For electroporation, cells are grown to a density of 5×10^6 cells/15-cm dish.

3.2. Electroporation and Selection of Hygromycin B Resistant LLC-PK₁ Cells

The LLC-PK$_1$ cells can be routinely translected by electroporation. However, it is also possible to use standard calcium phosphate transfection techniques if an electroporator is not readily available (*see* Note 1).

1. Plate cells at a density of $1-2 \times 10^6$/15-cm plate in five plates and incubate for 18 h in DMEM/FCS.
2. To remove adherent cells, wash plates 3X with PBS$^-$, add 5 mL of trypsin/EDTA, and incubate at $37°C$ for 5–10 min. The cells will round up and detach from the surface of the plate. At this stage, they can also be removed from the dish by gently moving up and down the trypsin/EDTA solution using a 5-mL sterile pipet.
3. Pool the suspensions and transfer to a 50-mL Falcon tube.
4. Spin cells at $400g$ (room temperature, aspirate trypsin/EDTA solution, and wash three times with ice-cold PBS$^-$. Resuspend 2×10^7 cells in 950 µL of ice-cold PBS$^-$.
5. Add 20 µg of β_2AR DNA (1 µg/µL), and 5 µg of hygromycin B resistance plasmid pX343 (1 µg/µL). Mix gently and keep on ice for 15 min. Mix again and transfer to a previously cooled Gene-pulser cuvet. Electroporate the cell/DNA suspension using a Bio-Rad Gene Pulser (voltage: 0.26 V; capacitance: 960 µF). Keep cells on ice for 10 min.
6. Allow cells to recover for 5 min at room temperature, add them to 15 mL DMEM/FCS, resuspend gently, and plate the suspension (containing 2×10^7 cells) into a 15-cm dish. Incubate for 18 h at $37°C$.
7. Remove nonviable, floating cells by aspiration, wash 3X with PBS$^-$, add trypsin/EDTA, and count.
8. Plate cells in selective growth medium containing hygromycin B (1.2 µg/mL) at a density of 0.5×10^6 cells/10-cm dish. Select for 1–2 wk until appearance of drug-resistant colonies.

After appearance of resistant colonies, it is possible to screen the dishes directly for receptor activity using the agarose overlay assay (*see* Note 2).

3.3. Agarose Overlay for the Detection of uPA Activity

1. If cells have been transfected with a β_2AR-receptor clone, induce transfected colonies for 18 h with 10 µM of isoproterenol. Induce a control LLC-PK$_1$ (or M18) plate with 30 nM calcitonin. Induction is performed in DMEM containing 0.5% of BSA. Do not use DMEM containing FCS, since it will inhibit uPA activity! Use at least one plate containing normal LLC-PK$_1$ or M18 colonies as negative control (no inducer).
2. Heat a 2% agarose/H$_2$O (w/v) solution until homogenous melting. To keep it in a liquid state, put it into a water bath previously set to $40°C$.

3. Mix 920 µL of 2X DMEM with 440 µL of 8% casein/H$_2$O (w/v) in a 15-mL Falcon tube and keep tube in the water bath at 40°C until usage. One tube is needed for one overlay.

4. Take the 10-cm dish containing hygromycin B resistant colonies, remove supernatant through aspiration, and wash three times with prewarmed (37°C) PBS⁻.

5. Add 730 µL of the agarose solution to one 15-mL Falcon tube containing the previously mixed casein/DMEM solution. Add quickly 80 µL of plasminogen (40 µg/mL). Mix quickly and overlay the colonies rapidly and evenly. Allow the agar to solidify at room temperature (2–5 min).

6. Put plates into the incubator and check after 1 h for appearance of lytic zones. Use a negative/positive control plate to monitor appearance of lytic activity. If you do not see activity, put plates back into the incubator and keep on monitoring periodically (every 15–30 min).

7. To remove and expand colonies showing inducible uPA activity from the plate, the following procedure can be used (no need for cloning cylinders):

   a. Take up 10 µL of a trypsin/EDTA solution with a P10 or P20 Gilman pipet. Gently pierce the agar zone indicating uPA activity. Gently pipet 10–20 times up and down without releasing the entire volume of the trypsin/EDTA solution.

   b. Transfer the trypsinized cell suspension to a well (24-well plate) containing hygromycin B growth medium, check for the presence of cells and allow them to attach and expand for 1–2 wk. Repeat the procedure immediately if no cells can be detected.

3.4. Chromogenic Assay for Agonist-Inducible uPA Activity

This assay can be used to identify agonist-inducible uPA activity in wells containing transfected LLC-PK$_1$ or M18 cells that have been cloned through limiting dilutions, or expanded following cloning from the agar-overlay. In the case in which cell clones have been previously isolated using the agar-overlay technique, the induction of uPA activity can be assessed for a given clone by comparing induced versus noninduced cells. If limiting dilutions of cells have been plated into a 96-well dish directly following transfection, it is advisable to determine negative/positive cutoff by measuring uPA activities obtained from a series of noninduced clones.

1. Remove growth/selection medium by aspiration and wash cells three times with PBS- . Induce cells for 18 h with appropriate agonist in DMEM/BSA medium.

2. To test for uPA activity add on ice to *triplicate wells* (96-well plate):

   a. 50 µL ice-cold assay buffer (90 mM Tris-HCl, pH 8.8, .45% Triton-X-100).

   b. 10 µL supernatant from induced or control (noninduced) cells.

   c. 20 µL chromogenic substrate 2 mg/ml (in H$_2$O).

   d. 20 µL plasminogen (0.112 mg/mL (in assay buffer).

3. Take first reading (OD$_{495}$); then incubate plate at 37°C for 5–10 min and take second reading.

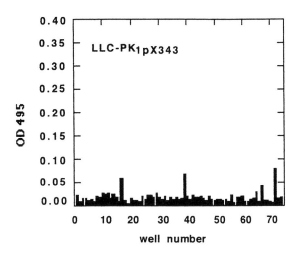

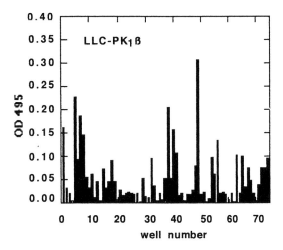

Fig. 2. uPA activities of LLC-PK1pX343 and LLC-PK1β cells plated under limiting dilution conditions: Pools of transfected cells were plated under limiting dilution conditions in 96-well plates and grown to confluence. Cells were induced for 18 h with 10–5M isoproterenol and uPA activity determined as described.

4. Read OD (495 nm) every 5 min for 30–45 minces and determine uPA activity (OD) as a function of time (*see* Note 3).
6. Analyze cell-lines with agonist-inducible uPA activity for receptor expression (mRNA, protein) and function (*see* Note 4).

4. Notes

1. The following procedure applies to both LLC-PK$_1$ and M18 cells. The M18 cells, however, are more sensitive to trypsin. For LLC-PK$_1$ cells, incubation with trypsin/EDTA at 37°C can be extended up to 10 min. Incubation for 5 min is normally sufficient to remove M18 cells from the plates.
2. Alternatively, following transfection (step 7), limiting dilutions of cells can be plated in 96-well plates in the presence of hygromycin B. Wells containing growing cells are assayed 1–2 wk later for inducible uPA activity using the calorimetric assay (Section 3.4.).
3. A typical OD$_{495}$ value for a 30 min read of uninduced (negative control cells) is in the range of 0.05–0.1. Induced cells will give values of >0.1 (depending on cell density and receptor expression levels, *see* Fig. 2).
4. If cell lines were cloned by limiting dilutions, sterile handling (induction, trypsinization, and so on) is important to assure that primary cultures of putative positives are not lost due to bacterial contamination.

References

1. Hull, R. N., Cherry, W. R., and Weaver, G. W. (1976) The origin and characteristics of a pig kidney cell strain, LLC-PK1. *In Vitro* **12**, 670–677.
2. Dayer, J.-M., Vassalli, J.-D., Bobbit, J. L., Hull, R. N., Reich, E., and Krane, S. M. (1981) Calcitonin stimulates plasminogen activator in porcine renal tubular cells. *J. Cell. Biol.* **91**, 195–200.
3. Goldring, S. R., Dayer, J.-M., Ausiello, D. A., and Krane, S. M. (1978) A cell strain cultured from porcine kidney increase cAMP content upon exposure to calcitonin or vasopressin. *Bochem. Biophys. Res. Commun.* **83**, 434–440.
4. Nagamine, Y., Sudol, M., and Reich, E. (1983) Hormonal regulation of plasminogen activator mRNA production in porcine kidney cells. *Cell* **32**, 1181–1190.
5. Jans, D. A., Resink, T. J., Wilson, L. E., Reich, E., and Hemmings, B. A. (1986) Isolation of a mutant LLC-PK1 cell line defective in hormonal responsiveness, A pleiotropic lesion in receptor function. *Eur. J. Biochem.* **160**, 407–412.
6. Granelli-Piperno, A. and Reich, E. (1978) A study of proteases and protease-inhibitor complexes in biological fluids. *J. Exp. Med.* **148**, 223–234.
7. Blochlinger, K., and Diggelmann, H. (1984) Hygromycin B phosphotransferase as a selectable marker for DNA transfer experiments with higher eucaryotic cells. *Mol. Cell. Biol.* **4**, 2929–2931.
8. Reich E., Rifkin, D. B., and Shaw, E. (1975) *Proteases and Biological Control.* Cold Spring Harbor Laboratory, Cold Spring Harbor, NY.
9. Chabre, O., Conklin, B. R., Lin, Y. H., Lodish, H. F., Wilson, E., Harlan, E. I., Catanzariti, L., Hemmings, B. A., and Bourne, H. R. (1992) A recombinant calcitonin receptor independently stimulates cAMP and Ca^{+2}/inositol phosphate signaling pathways. *Mol. Endocrinology* **4**, 551–556.
10. Catanzariti, L., Allen, J. M., and Hemmings, B. A. (1993) A novel expression system for G$_s$-coupled receptors. *Biotechniques* **15**, 474–479.

III

DETECTION OF RECOMBINANT GENE EXPRESSION *IN SITU*

18

Histochemical and Fluorochrome-Based Detection of β-Galactosidase

Ruth Sullivan and Cecilia W. Lo

1. Introduction

Bacterial β-galactosidase (β-gal the LacZ gene product) has been used extensively as a reporter gene in a wide variety of systems. It has been especially useful in transgenic and chimeric mouse studies. In such whole-animal applications, the experimental goals often include the evaluation of the spatial and tissue specific patterns of reporter gene expression *in situ*. In such instances, the LacZ reporter gene has the advantage over other reporters in that *in situ* histochemical staining for β-gal activity is relatively simple and sensitive. Additionally, fluorogenic substrates for β-gal are available that can be used to identify and isolate living cells expressing the β-gal reporter. Here, we discuss two protocols for localizing β-gal activity, one for staining fixed specimens, and the other for staining living cells. Each of these techniques has unique advantages and limitations, but together they offer a wide range of possibilities for taking advantage of β-gal as a reporter enzyme in a variety of applications.

1.1. Biochemical Properties of β-Gal

1.1.1. β-Galactosidase Protein Structure

The sequence, structure, and biochemical properties of the β-galactosidase enzyme are well characterized *(1,2)*. Recently the three dimensional structure of *E. coli* β-galactosidase has been solved by X-ray crystallography *(3)*. β-Galactosidase monomers are large (116 kDa) proteins that tetramerize to form the active site of the enzyme. Amino acids from regions throughout the β-gal monomer are involved in formation and/or stabilization of the active site. The domain consisting of the N terminal 39 amino acids (aa) (aa 3–41; the α pep-

From: *Methods in Molecular Biology, vol. 63: Recombinant Protein Protocols: Detection and Isolation* Edited by: R. Tuan Humana Press Inc., Totowa, NJ

tide) of the β-gal protein is of particular note because the β-gal polypeptide lacking this region is inactive. However, upon addition of the α peptide in trans enzyme activity is restored (i.e., α-complementation) *(4–6)*. The exact sequence of the first 26 amino acid residues of β-gal is not essential, however; these residues may be substituted with other sequences without affecting enzymatic activity *(5,7)*.

1.1.2. β-Galactosidase as a Reporter Enzyme in Eukaryotic Systems

Work in cultured eukaryotic cells has shown that β-galactosidase has no detectable deleterious effects on cell viability or growth *(8,9)*, a feature that is essential to any good reporter gene. Furthermore, the enzyme can be stably expressed in eukaryotic cells for many generations in the absence of selective pressure *(8)*. Despite stable β-gal expression, variable levels of β-gal activity have often been observed in clonal cell populations *(6,10,11)*. It has been suggested that the reason for this variability may be related to the metabolic state of the cells *(6)*. Finally, the bacterial β-gal enzyme is readily distinguishable from endogenous eukaryotic enzyme based on the difference in pH optimum of these two enzymes: The bacterial β-gal enzyme has a pH optimum at about pH 7–7.4, whereas the endogenous mammalian β-gal is a lysosomal enzyme with a pH optimum at 3–6 *(6)*. Thus, background staining is rarely a problem when eukaryotic cells are examined for the expression of bacterial β-gal activity. We have only observed one instance in which significant background β-gal activity is evident: the intestines of mouse embryos (beyond 14.5 d gestation) stain strongly for β-gal activity. However, this staining does not interfere with the evaluation of β-gal activity in other tissues.

1.1.3. β-Galactosidase Protein Fusions

Protein fusions can be made to the N terminus of β-galactosidase, often without disrupting β-gal activity *(7,10,12–17)*. The fact that the first 26 amino acids of the β-gal protein can be substituted without affecting enzyme activity facilitates the construction of such fusions *(7)*. Thus, the β-gal portion of the fusion protein can serve as a reporter for identifying cells expressing the fusion protein, or for tracking the intracellular localization and distribution of the fusion protein partner *(10,12–15)* (*see* Section 3.2.).

1.1.4. Perdurance of β-Gal

In using the β-gal enzyme as a reporter, it is important to take note of the fact that, at least in bacteria, β-gal is an extremely stable protein. In fact, it has been utilized to stabilize other proteins for purification in some bacterial expression systems *(2)*. Furthermore, it is likely that β-gal may be quite stable in eukaryotic systems as well; more stable, for example, than many endogenous gene products *(18)*. By analogy to the bacterial systems, one could imagine that the β-gal protein may help stabilize fusion proteins in eukaryotic cells. Given the perdurance of β-gal, the LacZ reporter must be used with some caution.

1.2. Applications Using *β-Gal* as a Reporter

1.2.1. Transcription Studies

The activity of potentially important *cis* acting DNA sequences can be examined by fusing them to the β-gal coding region and examining β-gal expression in tissue culture cells and/or transgenic mice *(2,9,18–21)*. Alternatively, homologous recombination strategies have been designed to insert β-gal coding sequences into the genomic DNA, disrupting the coding region of the gene of interest and placing β-gal under the regulation of the endogenous promoter and enhancer regions. As a result, not only can the loss of gene function phenotype be observed (due to disruption of the endogenous coding region by the β-gal sequences), but also those cells in which the disrupted DNA is normally expressed can be tracked and identified based on β-gal expression *(22)*.

1.2.2. Enhancer, Promoter, and Gene Trap Applications

In enhancer, promoter, and gene trap experiments, LacZ reporters that can be activated by flanking mouse genomic sequences are integrated into the genome of embryonic stem (ES) cells. The ES cell lines that exhibit interesting patterns of reporter gene expression either in culture (assessed by allowing cells to differentiate in culture) or in whole animals (examined by creating chimeras between ES cells and mouse embryos) are identified and used to generate chimeric animals. The transcriptional regulatory activity of the "trapped" DNA can then be examined by analyzing the expression of the reporter gene, and by characterizing any mutant phenotypes that arise from the insertion of the reporter into this region of the genome *(17)*. Similar experiments have been performed using transgenic mice in which LacZ transgenes with weak promoters have been used (either intentionally or fortuitously) to probe for regulatory sequences of developmentally important genes *(23,24)*.

1.2.3. Lineage Analysis

β-gal enzyme has been used as a cell marker in studies examining cell lineage in development. Lineage analyses fall into two main categories. In the first category, retroviruses containing the β-gal reporter are used to infect the cells of interest so that the progeny of these cells can be identified at later stages. In the second category, aggregation chimeras are generated between β-gal expressing ES cells or transgenic embryos and LacZ negative embryos in order to assess the contribution of cells of different genotypes to the developing embryo (for examples, *see* refs. *8,11,25–30*). The β-gal is a particularly useful marker for such lineage studies due to its ready visualization *in situ*. However, lineage studies suffer from the possibility that the β-gal marker gene could be shut off in some instances, making the determination of cell potency or lineage in development a conservative estimate *(31,32)*.

1.2.4. Fusion Proteins

The β-gal fusion proteins often retain reporter enzyme activity. Additionally, β-gal fusion proteins may retain the activity of the non β-gal fusion partner. In these cases, the fusion protein can be expressed in cells in order to examine gain of function of the fusion partner *(12)*. The β-gal portion of the fusion, if active, allows easy identification of expressing cells. Alternatively, β-gal fusion proteins made to polypeptides that normally oligomerize may retain the ability to associate with the wild type form of the fusion partner protein. In several instances, this association has been found to disrupt the wild-type protein function in a dominant negative fashion. In such cases, expression of the β-gal fusion protein in cells allows the examination of loss of function of the wild-type fusion partner protein *(10,13–16)*. The β-gal fusions have been used to examine the topology of membrane proteins by taking advantage of the fact that β-gal is only active if it is cytoplasmically localized *(2,10,15,16,33)*. Thus, if a β-gal fusion is made to an intracellular portion of a membrane protein, the reporter enzyme will be active, whereas if the fusion is made to an extracellular domain the enzyme will be inactive. Finally, β-gal fusions have been used to identify peptide sequences important in protein trafficking to various subcellular compartments *(2)*.

2. Materials

1. 40 mg/mL X-gal or Red-gal (Research Organics, Cleveland, OH) (US patent no. 5,358,854) (40X stock). Both X-gal and Red-gal are made up in DMSO and stored frozen at $-20°C$ in a light, tight container.
2. 0.5 Potassium ferricyanide (100X) made up in PBS and stored at room temperature in a light tight container. May be stored frozen at $-20°C$.
3. $0.5M$ Potassium ferrocyanide (100X) made up in PBS and stored at room temperature in a light, tight container. May be stored frozen at $-20°C$.
4. $1M$ MgCl$_2$ (500X) made up in PBS and stored at room temperature. May be stored frozen at $-20°C$.
5. 20% SDS (5000X) made up in water and stored at room temperature. May be stored frozen at $-20°C$.
6. 20% NP40 (1000X) diluted in PBS and stored at room temperature. May be stored frozen at $-20°C$.
7. PBS: Dulbecco's phosphate-buffered saline made up without Mg^{2+} or Ca^{2+} ions.
8. 2% Paraformaldehyde made up in PBS fresh for each use. The solution should be heated, but not allowed to boil, to allow the paraformaldehyde to dissolve. Be

sure to correct for any volume lost during heating by adding water. Cool the fix to room temperature prior to use. For fixing embryos, 0.02% glutaraldehyde may also be added although we do not find this necessary routinely.

9. Eosin Y stain: mix a solution of 176.3 mL 95% ethanol, 23.7 mL water, 1 mL glacial acetic acid. Use 150 mL of this solution to dilute 50 mL of commercial eosin stock for the staining solution.

10. FDG: 100 m*M* stock is made up by dissolving 5 mg FDG in 76 μL 1:1 water:DMSO. The stock is stored at –20°C. *See* Notes 1, 6, 7, and 8 regarding preparing FDG stocks.

11. FDG staining medium: 10 m*M* HEPES pH 7.2, 4% fetal calf serum, made up in PBS. Can be stored at –20°C.

3. Methods

In this chapter we consider methods for histochemically localizing β-gal activity *in situ*. *In situ* analysis is useful for evaluating the spatial and tissue specific patterns of β-gal expression. For this purpose, we describe standard histochemical methods for staining of fixed tissue samples (*see* Section 3.1.) (Fig. 1). These techniques are useful in the examination of β-gal expression in either whole mount specimens or in histological sections. Second, we will describe procedures for staining living cells (*see* Section 3.2.) (Fig. 2). These latter methods are useful if β-gal positive cells are to be cloned or cultured for further analysis.

3.1. Histochemical Detection of *β-Gal* in Fixed Specimens

In histochemical applications, X-gal (5-bromo-4-chloro-3-indolyl-β-D-galactopyranoside) is the most popular β-gal substrate. When cleaved by β-gal, it forms a blue precipitate that is robust to subsequent sectioning and histological examination. The precipitate is electron dense and thus the staining can be examined by electron microscopy *(34)*. Other chromogenic substrates have also become available and can be used in instances in which different colors are desirable. For example, the substrate red-gal (6-chloro-3-indolyl-β-D-galacto-pyranoside; Research Organics) forms a red precipitate when cleaved by β-gal. Below is a protocol for *in situ* X-gal staining that is appropriate for most applications, such as staining of adherent monolayers of tissue culture cells, or staining whole embryos or organs. This procedure produces strong staining with minimal background in most tissues with the exception of mouse embryo intestines, which exhibit considerable endogenous β-gal activity from d 14.5 (*see* Note 3). The same protocol is used for Red-gal staining by substituting Red-gal for X-gal. In the staining solution, X-gal or Red-gal acts as the indicator of enzyme activity. The products of X-gal cleavage are galactose and soluble indoxyl molecules. The indoxyl molecules are subsequently converted through oxidation to insoluble indigo to form a blue precipitate. Note that

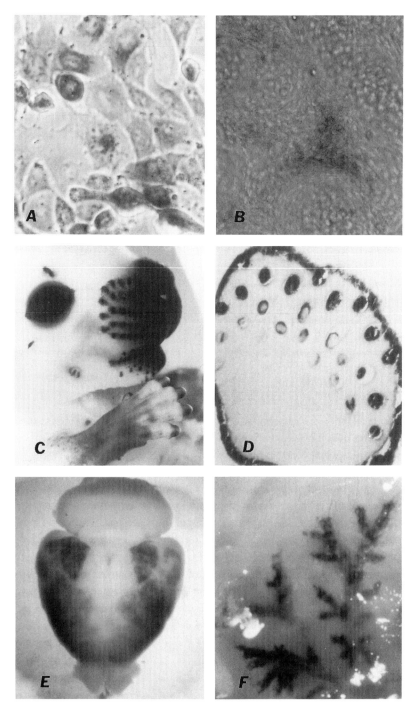

Fig. 1.

because the indoxyl molecules are soluble, they are subject to some diffusion, thereby resulting in a small degree of dispersion of the stain *(35)*. Sodium and magnesium ions act as activator and cofactor respectively, while the ferro/ferri cyanide mixture acts as an oxidation catalyst to help localize the stain and intensify the color of the reaction. They may further help buffer the pH *(6)*. The reaction is carried out in PBS, which provides the Na^+ ions and also helps buffer the staining solution to 7.0–7.4, the pH optimum for the bacterial enzyme (well above the pH optimum for the endogenous eukaryotic β-gal enzyme). This pH allows for optimal staining for bacterial β-gal and minimal background *(6,35)*. Care should be exercised in making up PBS stocks as we have noticed variability in staining associated with particular batches of PBS (*see* Note 2).

Prior to staining, we typically fix the cells, embryos, or tissues in 2% paraformaldehyde made up in PBS (made fresh each time) at room temperature. For fixation of embryos, glutaraldehyde can be added to the fix, although we do not routinely find that this is necessary.

Following X-gal staining, specimens may be paraffin embedded for subsequent sectioning. After sectioning, we typically stain the samples with eosin Y in order to best visualize the tissues while maintaining good resolution of the X-gal stain.

3.1.1. Fixation

1. Rinse samples in 1X PBS.
2. Place samples in 2% paraformaldehyde fixative at room temperature with shaking (care should be exercised when handling paraformaldehyde: *see* Note 1). For

---

Fig. 1. *(opposite page)* X-gal and Red-gal staining of fixed specimens: (**A**) A clonal population of β-gal expressing NIH3T3 cells was grown as a monolayer and stained for β-gal activity using X-gal. Variability in the intensity of X-gal staining may be related to the metabolic state of the cells (*see* Section 1.). (**B**) Limb buds were harvested from β-gal expressing transgenic embryos (*see* embryo in C) and grown in micromass culture. Subsequently the cells were fixed and stained with Red-gal. (**C**) 16.5 day embryos were collected from a transgenic mouse line expressing a β-gal reporter under the regulation of a connexin 43 (Cx43) gap junction promoter. Embryos were fixed and X-gal stained. Notice the blue staining indicating β-gal expression in the developing forelimb and whiskers (and thus Cx43 promoter activity in these areas). (**D**) Following X-gal staining, the embryo in C was paraffin-embedded, sectioned, and eosin Y-stained. Blue staining revealed β-gal expression in cells of the developing vibrissae. (**E,F**) Adult brain and lung were collected from a transgenic mouse expressing β-gal under the regulation of the Cx43 promoter. The tissues were fixed and X-gal stained. Notice the abundant blue staining of the cerebral hemispheres in E and of the bronchial tree of the lung in F.

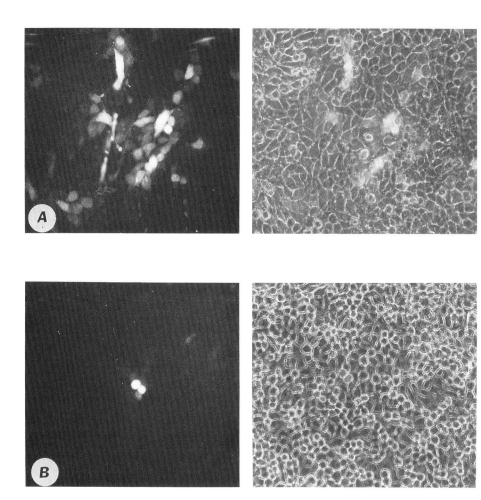

Fig. 2. FDG staining of living monolayers of NIH3T3 cells: NIH3T3 cells were transfected with a β-gal reporter plasmid **(A)** or a plasmid construct which provides for the expression of a β-gal/Cx43 fusion protein **(B)**. The latter protein retains β-gal activity but disrupts Cx43 function in a dominant negative fashion *(10)*. Transfection was performed by calcium phosphate precipitation, a method that typically results in a transfection efficiency of 1%. Thus, in both A and B, β-gal-expressing cells are found amongst a large background of β-gal-negative cells. Shown above are the transfected cells after incubation with the β-gal substrate FDG (according to the protocol described in Section 3.2.). The β-gal expressing cells are brightly fluorescent as a result of the release of fluorescein from the β-gal cleavage of FDG. Notice that in A, the brightly stained β-gal-expressing cells are surrounded by fluorescent neighbors with much lower fluorescence intensity. Fluorescence in these neighboring cells is due to gap

Table 1
Staining Solution

Component	Volume	Final concentration
40 mg/mL X-gal or Red-gal	1.25 mL	1 mg/mL
0.5M K ferricyanide	0.5 mL	5 mM
0.5M K ferrocyanide	0.5 mL	5 mM
1M MgCl$_2$	100 μL	2 mM
20%SDS	25 μL	0.01%
10% NP40	100 μL	0.02%
PBS	47.525 mL	
Final volume	50 mL	

monolayers of tissue culture cells, fix for 15–30 min. For small (<14.5 d) embryos, fix 30–45 min, and for larger embryos (>14.5 d), fix for 1 h. For proper fixative penetration in these larger embryos, use forceps to open the chest and abdominal cavities while embryos are in the fix. After 30 min, change the embryos to a new dish of fixative to help dilute out blood and other tissue debris that may inhibit X-gal staining. Continue fixation for a total of 1 h.

3. Rinse samples in PBS three times. For larger embryos, allow 20 min for each rinse (at room temperature with gentle shaking).
4. Following fixation and rinsing, embryo samples should be stained immediately. Tissue culture cells can be stored overnight in PBS at 4°C prior to staining.

3.1.2. Staining

1. Place samples in staining solution (*see* Table 1) at room temperature protected from light with shaking until blue staining is visualized. This can take from 1 h to overnight. For overnight staining, be sure the container is sealed to prevent evaporation. To increase the sensitivity of the assay, the samples can be incubated at 37°C. Care should be exercised when handling solutions containing DMSO, sodium ferro cyanide, and sodium ferri cyanide (*see* Note 1).
2. After staining, rinse the samples in PBS. Samples can be stored in PBS at 4°C if azide is included (azide is extremely toxic and should be handled with great care).
3. If samples are to be paraffin embedded, postfix them after the X-gal staining in 3.7% formaldehyde (formaldehyde stock diluted 1:10 in PBS) overnight at

---

junction mediated transfer of FDG derived fluorescein from the β-gal positive cells to adjacent nontransfected cells. In contrast, in the fusion protein-expressing cells in B, little or no transfer of fluorescein is observed from the brightly fluorescent β-gal-positive cells to their neighbors. The latter reflects the fact that gap junctional communication is inhibited by the β-gal/Cx43 fusion protein *(10)*.

room temperature with shaking. Following postfixation, rinse samples in PBS, then place in 50% ethanol for 1 h, followed by 70% ethanol. Samples can be stored in 70% ethanol at 4°C for prolonged periods prior to dehydration and paraffin embedding.

3.1.3. Embedding

1. Samples have been postfixed in formaldehyde (3.7% formaldehyde in PBS) and are in 70% ethanol at this point. Make sure they have been in 70% ethanol for at least 1 h at room temperature. We typically use 1-dram glass specimen vials for embedding. If embryo samples are <10.5 d, then they should be agar embedded (*see* Note 5).

    *Invert vial gently several times after each of the following steps to ensure good mixing and penetration. When changing solutions, remove as much of the old solution before adding the next solution (*see *Note 4).*
2. Change solution to 95% ethanol for 1 h at room temperature. If embryos are >14.5 d, then leave in 95% ethanol overnight.
3. Change to 100% ethanol at room temperature. Leave overnight.
4. Change to fresh 100% ethanol for 1 h at room temperature.
5. Under the hood, change to xylene for 1 h at room temperature. Repeat this two more times. However, if after the first 1 h the blue stain is rapidly eluting from the sample into the solvent, go immediately to step 6.
6. Change to 50% xylene/50% paraffin for 1 h at 56°C. Repeat this two more times. If embryos are >14.5 d, then leave in 50:50 xylene:paraffin overnight. For this step, fill tubes halfway with melted paraffin, then the rest of way with xylene. The paraffin will solidify: Place the tubes at 56°C and be sure to invert the tubes to mix after the paraffin has melted! Keep samples in heating blocks at 56°C, and never above 62°C.
7. Change to 100% paraffin for 1 h at 56°C.
8. Change to 100% paraffin for overnight at 56°C.
9. Change to fresh 100% paraffin at least two more times. If embryos are >14.5 d, then change the 100% paraffin every 2 or 3 h during the day and leave in 100% paraffin for one more overnight.
10. Embed: Prior to embedding, prepare paper labels identifying each sample and recording the orientation it will be embedded in. Fill embedding mold with paraffin. Transfer tissue into the molding block using prewarmed forceps or spatula if embryos are <10.5 d. Orient tissue, manipulating with the forceps. Insert label into molten paraffin. Let block cool overnight at room temperature.

3.1.4. Sectioning and Eosin Y Staining

1. Paraffin-embedded specimens are sectioned at 10 μm and mounted on ethanol-cleaned glass slides. Slides are air-dried overnight and then baked at 45–50°C overnight.
2. Following mounting, sections are stained with eosin Y:

Solution	Time
Histoclear	1 min
100% ethanol	1 min
95% ethanol	1 min
75% ethanol	1 min
diluted eosin Y	15 s
tap water	3 min
95% ethanol	1 min
100% ethanol	1 min

First stain one or two slides, mount them using water, and check the slides under the microscope to see if the desired staining has been achieved. Adjust the time of eosin staining to either increase or decrease staining intensity for the subsequent slides. If the slides are overstained, rinse in 100% ethanol until desired level of staining is achieved. After staining, allow the slides to air-dry for about 10 min prior to mounting.

3.2. Detection of β-Gal Activity in Living Cells

We will consider the fluorescent substrates that can be used to stain living β-gal expressing cells. FDG (fluorescein di-β-D-galactopyranoside) was the first of these various fluorescent substrates to be developed, and it has proven useful as a vital stain for the identification of β-gal positive cells for subsequent cloning. FDG is a very sensitive β-gal substrate due to the sensitivity of fluorescence detection (35; for further information, see catalog of Molecular Probes, Eugene, OR). The use of FDG to stain living cells has been important in various ES cell experiments, particularly in gene or enhancer trap studies (36). FDG analysis has also been used to examine the cellular identity of LacZ expressing cells in transgenic mice. Thus, tissues from transgenic mice can be dissociated into single cell suspensions, FDG stained, and subjected to fluorescence activated cell sorting (FACS) analysis allowing β-gal positive cells to be cultured in order to establish their identity (37).

FDG has numerous limitations that make it useful for only certain applications. The first problem is that FDG requires cell permeabilization for loading. Permeabilization is most often achieved by hypotonically shocking the cells (35) or using DMSO (38). However, either procedure may be poorly tolerated by the cells. The second problem is that fluorescein, the product of FDG cleavage by β-gal, has several properties that make FDG assay of β-gal activity problematic. Firstly, at room temperature, fluorescin rapidly leaks out of cells. This effect can be minimized by keeping the stained cells at 4°C (39). Secondly, fluorescein passes freely through gap junctions and thus can transfer between FDG+ and FDG– cells. Nevertheless, mixed populations of FDG+ and FDG–

cells have been successfully stained with FDG and subjected to fluorescence-activated cell sorting (FACS) allowing FDG+ cells to be selected and grown up *(36,39-41)*.

We have used FDG to stain monolayers of cells. In cells that are well coupled by gap junctions we found that the fluorescein product rapidly spread from the β-gal-positive cell to its neighbors through gap junctions. This property was in fact quite useful in rapidly assessing gap junctional communication in these cells. However, the method was difficult to standardize for quantitative studies and thus is not practical for the quantitative analysis of gap junctional communication.

Some of the drawbacks of FDG have been overcome by the development of modified FDG substrates. For example, the fatty acyl modified FDG's (Imagene green (C_{12} FDG), Molecular Probes) (US patent 5,208,148) do not require cell permeabilization for loading and the product of β-gal cleavage is a fluorescent lipophilic product that is efficiently retained within cells (probably due to incorporation of the fatty acyl tail within the cellular membranes). Imagene green (C_{12} FDG) has been used for whole-mount staining of zebra fish embryos for β-gal activity *(42)* as well as for staining of mammalian tissue culture cells for FACS analysis *(40,43)*. Indeed, C_{12} FDG has been suggested to be a more sensitive β-gal substrate for use in mammalian cells *(40)*, perhaps due to its easy loading and increased retention of the reaction product within cells. Various acyl chain lengths are available and the product is also available with a red instead of a FITC fluorophore (Imagene red, Molecular Probes) (US patent 5,242,805). Other FDG analogs have been developed in which the fluorescein portion has been modified with a chloro-methyl group so that it can react with cellular glutathione, thereby also allowing the product of β-gal cleavage to be efficiently retained within cells (DetectaGene [chloro-methyl-fluorescein di-β-D-galactopyranoside], Molecular Probes) (US patent 5,362,628). DetectaGene is also available with a blue product, 4-chloromethyl-7-hydroxycoumarin. However, DetectaGene, like FDG, still requires membrane permabilization for cell loading. Dichloro derivatives of FDG are also available and are phototoxic to cells. Thus, potentially they may also be useful in photoablation studies. Finally, carboxyfluorescein derivatives of FDG are available that result in the release of carboxyfluorescein instead of fluorescein. Carboxyfluorescein is better retained within the cells than fluorescein, although it can still pass through gap junctions. However, the carboxyfluorescein derivatives again require cell permeabilization for substrate loading. We outline a protocol for loading adherent cells in culture with FDG in Section 3.2.1. This procedure is useful for staining and subsequently cloning cells expressing the β-gal reporter gene.

3.2.1. FDG Staining Procedure

1. Wash 10-cm plate of adherent cells with PBS. If stained cells are to be subsequently cultured, all solutions should be kept sterile and sterile technique should be used.
2. Add 150 μL FDG staining medium/10-cm dish.
3. Add 150 μL FDG/H₂O (2 μL 100 m*M* FDG stock diluted in 100 μL sterile water).
4. Let sit 2 min at room temperature. The hypotonic shock loads cells with FDG during this incubation period. It is helpful to agitate the solution on the plate somewhat. (*See* Note 10 regarding alternate loading procedures if cells do not tolerate this hypotonic loading procedure.)
5. Remove staining solution, rinse plate with ice-cold PBS (*see* Note 9).
6. Add 2-mL ice-cold staining medium.
7. Put dish on ice covered with foil.
8. Observe under fluorescent microscope (*see* Note 11).
9. FDG-positive cells can be observed and, if they are present in isolated colonies, can be marked for subsequent cloning (*see* Note 15 regarding potential background staining in older cell cultures). The staining procedure can be modified for more quantitative applications (*see* Note 13) and/or for FACS analysis (Note 12) if desired.

4. Notes

1. Care should be exercised when handling paraformaldehyde, DMSO, potassium ferro and ferricyanide, and azide.
2. One problem we have noticed in X-gal staining is the accumulation of crystals in the staining solution. The presence of these crystals can greatly decrease the amount of staining obtained. As the presence of these crystals seems to be associated with particular batches of PBS, always make up the PBS in chromic acid clean bottles and adjust pH as needed.
3. Mouse embryo intestines have considerable β-gal activity from day 14.5. Although this staining makes analysis of β-gal expression in the intestines difficult, it does not interfere with analysis of expression in other tissues.
4. When changing solutions during embedding, be sure to drain all of the old solution from the vial before adding the next solution.
5. When embedding, if embryo samples are <10.5 d gestation, then they should be agar-embedded to facilitate handling during paraffin embedding. Do the agar embedding while the samples are still in 50% ethanol or allow the samples to equilibrate in 50% ethanol for at least 1 h prior to agar embedding. Use 0.75% low melting point agarose to minimize possible heat damage to embryo. Equilibrate melted agarose to 37°C in water bath. Place embryo in a Petri dish and remove any excess liquid with a pipet and/or paper towel. Place a drop of agarose over the embryo using a Pasteur pipet. Let cool. Cut out a block around embryo using a scalpel and place the block in 70% ethanol. This can be stored at –20°C or at 4°C until ready for embedding.

6. FDG is dissolved in a 1:1 mixture of DMSO:H$_2$O at a concentration of 100 mM or 200 mM (dissolve 5 mg of FDG in 76 μL of DMSO:H$_2$O for a 100 mM solution). The DMSO:water and FDG should be kept on ice as there is a strong exothermic reaction when DMSO and water are mixed. The limit of solubility of FDG is near 200 mM, so care must be taken to ensure that the powder goes completely into solution. If necessary, the solution can be placed at 37°C briefly to ensure that all of the FDG is dissolved.

7. We have noticed loss of activity of the FDG stock solution over time at −20°C. This can be minimized by avoiding multiple freeze-thaw cycles. The manufacturer recommends storage of the stock solution under Drierite.

8. FDG stocks can be contaminated with minor quantities of fluorescent monogalactoside and/or fluorescein, both of which will contribute to background fluorescence. FDG from Molecular Probes is quality controlled so that this is rarely a problem. However, if it is desirable to remove all background fluorescence, the stock can be photobleached by holding the Eppendorf containing the 2 mM working FDG solution directly in the path of the FACS 488 nm argon laser for 1 min or more. Move the tube slowly to avoid melting the plastic and hold the tube such that the laser strikes the solution at the meniscus to ensure optimal spread of the light throughout the solution. Laser goggles must be worn during this procedure. After photobleaching, the solution can be freeze/thawed without rebleaching *(35)*.

9. We have included the step of rinsing FDG stained cells with ice-cold PBS, without which we found excessive background fluorescence that complicated visualization of FDG positive cells.

10. Some cells and embryos, for example preimplantation embryos, do not tolerate the hypotonic shock used in the above protocol to load the FDG into the cells. Alternative procedures using DMSO to load the FDG into the cells may be helpful in such cases *(38)*.

11. Note that the short wave light source from the mercury light can be quite toxic to cells. It is often helpful to use cut off filters to eliminate light of short wavelength.

12. A similar staining procedure as outlined above can be used to stain cells for FACS analysis. We refer you to the following references for more specific details of FACS applications: *36,39–41*.

13. Some authors have used a competitive inhibitor of β-gal (phenylethyl β-D thiogalactoside) to stop the FDG reaction in conjunction with chloroquine, an inhibitor of mammalian (but not bacterial) β-gal, to inhibit endogenous β-gal activity. These modifications increase the accuracy of the FDG assay so that it can be used in quantitative applications *(14)*.

14. This FDG staining method can also be used to identify preimplantation embryos which express β-gal. However, the embryos do not withstand the hypotonic shock and die subsequently to the staining (*see* Note 10). Older embryos do not load well at all and there seems to be endogenous β-gal activity in damaged embryonic tissues that results in excessive and sometimes misleading background staining.

15. When using the fluorescent substrates with tissue culture cells, there is sometimes background fluorescence that becomes more pronounced in older confluent

cultures *(35)*. In our experience, background fluorescence was dim and could easily be distinguished from true staining. However, this problem may be cell-type specific.

5. Conclusions

In summary, bacterial β-gal has proven to be a very useful reporter gene in numerous applications. *In situ* staining for β-gal enzyme activity in fixed specimens is straightforward and has the advantage of being both reasonably sensitive and specific. The more recently developed vital fluorescent stains for β-gal activity are less widely useful at the present time both because of the limitations of the various stains and because of their high cost. However, the vital stains are quite useful in applications in which β-galactosidase-positive cells are to be cloned and/or characterized in culture. As these fluorescent dyes are modified and new dyes are developed, their utility will increase.

Acknowledgments

We would like to thank Dr. Emanuela Stringa for the photograph of the Red-Gal stained limb bud explants. This work was supported by grants from the NSF DCB6886 and NIH HD29573, and R. S. was supported in part by fellowships from the March of Dimes (#1023), and VMSTP Training Grant (GM07170).

References

1. Kalnins, A., Otto, K., Ruther, U., and Muller-Hill, B. (1983) Sequence of the lacZ gene of *Escherichia coli*. *EMBO J.* **2**, 593–597.
2. Silhavy, T. J. and Beckwith, J. R. (1985) Uses of lac Z fusions for the study of biological problems. *Microbiol. Rev.* **49**, 398–418.
3. Jacobson, R. H., Zhang, X.-J., DuBose, R. F., and Matthews, B. W. (1994) Three dimensional structure of β-galactosidase from *E. coli*. *Nature* **369**, 761–766.
4. Langley, K. E. and Zabin, I. (1976) β-galactosidase a complementation: properties of the complemented enzyme and mechanism of the complementation reaction. *Biochemistry* **15**, 4866–4875.
5. Zabin, I. (1982) β-galactosidase α-complementation. *Mol. Cell. Biochem.* **49**, 87–96.
6. MacGregor, G. R., Mogg, A. E., Burke, J. F., and Caskey, C. T. (1987) Histochemical staining of clonal mammalian cell lines expressing *E. coli* β-galactosidase indicates heterogeneous expression of the bacterial gene. *Somatic Cell Mol. Genet.* **13(3)**, 253–265.
7. Muller-Hill, B. and Kania, J. (1974) Lac repressor can be fused to β-galactosidase. *Nature* **249**, 561–563.
8. Sanes, J. R., Rubenstein, J. L. R., and Nicolas, J.-F. (1986) Use of a recombinant retrovirus to study post-implantation cell lineage in mouse embryos. *EMBO J.* **5**, 3133–3142.

9. Bonnerot, C., Rocancourt, D., Briand, P., Grimber, G., and Nicolas, J.-F. (1987) A β-galactosidase hybrid protein targeted to nuclei as a marker for developmental studies. *Proc. Natl. Acad. Sci. USA* **84,** 6793–6799.

10. Sullivan, R. and Lo, C. W. (1995) Expression of a Cx43/β-galactosidase fusion protein inhibits gap junctional communication in NIH3T3 cells. *J. Cell Biol.* **130,** 419–429.

11. Price, J., Turner, D., and Cepko, C. (1987) Lineage analysis in the vertebrate nervous system by retrovirus-mediated gene transfer. *Proc. Natl. Acad. Sci. USA* **84,** 156–160.

12. Hill, R. J. and Sternberg, P. W. (1992) The gene lin-3 encodes an inductive signal for vulval development in *C. elegans. Nature* **358,** 470–476.

13. Eyer, J. and Peterson, A. (1994) Neurofilament-deficient axons and perikaryal aggregates in viable trnasgenic mice expressing a neurofilament-β-galactosidase fusion protein. *Neuron* **12,** 389–405.

14. Govind, S., Whalen, A. M., and Steward, R. (1992) In vivo self-association of the *Drosophila* rel-protein dorsal. *Proc. Natl. Acad. Sci. USA* **89,** 7861–7865.

15. Yagi, T., Aizawa, S., Tokunaga, T., Shigetani, Y., Takeda, N., and Ikawa, Y. (1993) A role for Fyn tyrosine kinase in the suckling behavior of neonatal mice. *Nature* **366,** 742–745.

16. Umemori, H., Sato, S., Yagi, T., Aizawa, S., and Yamamoto, T. (1994) Initial events of myelination involve Fyn tyrosine kinase signaling. *Nature* **367,** 572–576.

17. Gossler, A., Joyner, A. L., Rossant, J., and Skarnes, W. C. (1989) Mouse embryonic stem cells and reporter constructs to detect developmentally regulated genes. *Science* **244,** 463–465.

18. Echelard, Y., Vassileva, G., and McMahon, A. P. (1994) Cis-acting regulatory sequences governing Wnt-1 expression in the developing mouse CNS. *Development* **120,** 2213–2224.

19. Zack, D. J., Bennett, J., Wang, Y., Davenport, C., Klaunberg, B., Gearhart, J., and Nathans, J. (1991) Unusual topography of bovine rhodopsin promoter-lacZ fusion gene expression in transgenic mouse retinas. *Neuron* **6,** 187–199.

20. Goring, D. R., Rossant, J., Clapoff, S., Breitman, M. L., and Tsui, L.-C. (1987) In situ detection of β-galactosidase in lenses of transgenic mice with a γ-crystallin/lacZ gene. *Science* **235,** 456–458.

21. Parr, B. A., Shea, M. J., Vassileva, G., and McMahon, A. P. (1993) Mouse Wnt genes exhibit descrete domains of expression in the early embryonic CNS and limb buds. *Development* **119,** 247–261.

22. LeMouellic, H., Lallemand, Y., and Brulet, P. (1990) Targeted replacement of the homeobox gene Hox-3. 1 by the *Escherichia coli* lacZ in mouse chimeric embryos. *Proc. Natl. Acad. Sci. USA* **87,** 4712–4716.

23. Allen, N. D., Cran, D. G., Barton, S. C., Hettle, S., Reik, W., and Surani, M. A. (1988) Transgenes as probes for active chomosomal domains in mouse development. *Nature* **333,** 852–860.

24. Kothary, R., Clapoff, S., Brown, A., Campbell, R., Peterson, A., and Rossant, J. (1988) A transgene containing lacZ inserted into the dystonia locus is expressed in neural tube. *Nature* **335,** 435–437.

25. Beddington, R. S., Morgernstern, J., Land, H., and Hogan, A. (1989) An in situ transgenic enzyme marker for the midgestation mouse embryo and the visualization of inner cell mass clones during early organogenesis. *Development* **106,** 37–46.

26. Beddington, R. S. P. (1994) Induction of a second neural axis by the mouse node. *Development* **120,** 613–620.

27. Epstein, M. L., Mikawa, T., Brown, A. M. C., and McFarlin, D. R. (1994) Mapping the origin of the avian enteric nervous system with a retroviral marker. *Developmental Dynamics* **201,** 236–244.

28. Tan, S.-S. and Breen, S. (1993) Radial mosaicism and tangential cell dispersion both contribute to mouse neocortical development. *Nature* **362,** 638–639.

29. Price, J. and Thurlow, L. (1988) Cell lineage in the rat cerebral cortex: a study using retroviral-mediated gene transfer. *Development* **104,** 473–482.

30. Luskin, M. B., Pearlman, A. L., and Sanes, J. R. (1988) Cell lineage in the cerebral cortex of the mouse studied in vivo and in vitro with a recombinant retrovirus. *Neuron* **1,** 635–647.

31. Kadokawa, Y., Suemori, H., and Nakatsuji, N. (1990) Cell lineage analysis of epithelia and blood vessels in chimeric mouse embryos by use of an embryonic stem cell line expressing the β-galactosidase gene. *Cell Differ. Devel.* **29,** 187–194.

32. Suemori, H., Kadodawa, Y., Goto, K., Araki, I., Kondoh, H., and Nakatsuji, N. (1990) A mouse embryonic stem cell line showing pluripotency of differentiation in early embryos and ubiquitous β-galactosidase expression. *Cell Differ. Devel.* **29,** 181–186.

33. Fire, A., Harrison, S. W., and Dixon, D. (1990) A modular set of lacZ fusion vectors for studying gene expression in *Caenorhabditis elegans. Gene* **93,** 189–198.

34. Bonnerot, C., Rocanacourt, D., Briand, P., Grimber, G., and Nicolas, J. F. (1987) A β-galactosidase hybrid protein targeted to nuclei as a marker for developmental studies. *Proc. Natl. Acad. Sci. USA* **84,** 6795–6799.

35. MacGregor, G., Nolan, G. P., Fiering, S., Roederer, M., and Herzenberg, L. A. (1991) Use of *E. coli* lacZ (β-galactosidase) as a reporter gene, in *Methods in Mol. Biol. vol. 7, Gene Transfer and Expression Protocols,* Humana, Clifton, NJ, pp. 217–235.

36. Reddy, S., Rayburn, H., VonMelchner, H., and Ruley, H. E. (1992) Fluorescnece-activated sorting of totipotent embryonic stem cells expressing developmentally regulated lacZ fusion genes. *Proc. Natl. Acad. Sci. USA* **89,** 6721–6725.

37. Hansbrough, J. R., Fine, S. M., and Gordon, J. I. (1993) A transgenic mouse model for studying the lineage relationships and differentiation program of type II pneumocytes at various stages of lung development. *J. Biol. Chem.* **268,** 9762–9770.

38. Nirenberg, S. and Cepko, C. (1993) Targeted ablation of diverse cell classes in the nervous system in vivo. *J. Neuro.* **13,** 3238–3251.

39. Nolan, B. P., Fiering, S., Nicolas, J.-F., and Herzenberg, L. A. (1988) Fluorescence-activated cell analysis and sorting of viable mammalian cells based on β-D-galactosidase activity after transduction of *Escherichia coli* lacZ. *Proc. Natl. Acad. Sci. USA* **85,** 2603–2607.

40. Plovins, A., Alvarez, A. M., Ibanez, M., Molina, M., and Nombela, C. *(1994)* Use of fluorescein-Di-β-D-galactopyranoside (FDG) and C_{14}-FDG as substrates for

β-galactosidase detection by flow cytometry in animal, bacterial and yeast cells. *Appl. Environ. Microbiol.*. **60,** 4638–4641.

41. Fiering, S. N., Roederer, M., Nolan, G. P., Micklem, D. R., Parks, D. R., and Herzenberg, L. A. (1991) Improved FACS-Gal: Flow cytometric analysis and sorting of viable eukaryotic cells expressing reporter gene constructs. *Cytometry* **12,** 291–301.

42. Westerfield, M., Wegner, J., Jegalian, B. G., DeRobertis, E. M., and Puschel, A. W. (1992) Specific activation of mammalian Hox promoters in mosaic transgenic zebrafish. *Genes Develop.* **6,** 591–598.

43. Peng, S., Somerfelt, M. A., Berta, G., Berry, A. K., Kirk, K. L., Hunter, E., and Sorscher, E. J. (1993) Rapid purification of recombinant baculovirus using fluorescence-activated cell sorting. *BioTechniques* **14,** 274–276.

19

Combined, Sequential *In Situ* Hybridization and Immunohistochemistry on the Same Tissue Section

Kenneth J. Shepley and Rocky S. Tuan

1. Introduction

The spatio-temporal study of gene expression in tissues is typically accomplished by using one of the following techniques: *in situ* hybridization (ISH), that localizes mRNA transcripts in particular cell populations using a labeled nucleic acid probe; and immunohistochemistry (IMH), that detects protein gene products via labeled primary or secondary antibodies. Often, ISH and IMH are performed separately on adjacent serial sections, allowing comparison of the cellular staining pattern for mRNA, and the presence or absence of the corresponding protein product. Combining these two procedures, by performing them on the same tissue section provides immediate, direct visualization of the expression pattern of a particular gene, avoiding the need to visually interpret and compare two separate sets of data.

Designing a simultaneous ISH/IMH methodology involves many alternatives. These include using cryopreserved or paraffin-embedded sections, DNA or riboprobes, fluorescent, radioactive, or colorimetric detection systems, and performing ISH or IMH first. Cryosections are generally regarded as having poor tissue and cellular morphology compared to formalin-fixed, paraffin-embedded sections. Riboprobes work at least as well as DNA probes, often with higher sensitivity, but may require subcloning of the sequence of interest into an appropriate vector. Fluorescent and radioactive tags do not permit simultaneous viewing of tissue morphology and signal. Additionally, radioactive detection is sometimes less desirable because of the need for autoradiographic processing and ecological concerns of radioactive disposal. Previously published procedures of combined ISH and IMH have described the perfor-

From: *Methods in Molecular Biology, vol. 63: Recombinant Protein Protocols:
Detection and Isolation* Edited by: R. Tuan Humana Press Inc., Totowa, NJ

mance of ISH first *(1–7)*, or IMH first *(8–14)*, using various combinations of all of the above protocols.

The main drawback of performing IMH first is the risk of mRNA degradation *(5)*, and steps to block ribonuclease activity are thus required. The primary risk in doing ISH first is the loss of antigenicity *(14)*. Procedures based on formalin-fixed, paraffin-embedded specimens, that promise better morphology than cryosections, for example, require proteinase digestion to remove interfering proteins and their crosslinking formed during fixation to allow the hybridization probe access to target mRNA *(2,3)*. Such a step significantly increases the risk of antigen degradation and loss of morphology.

Recently, new noncrosslinking fixatives have become available, including Histochoice (Amresco, Solon, OH) and Optiprope (Oncor, Gaithersburg, MD), that retain excellent morphology. These fixatives obviate the need for a proteinase digestion step for ISH, thereby lending themselves well to combined ISH and IMH on the same tissue section. This chapter describes a protocol based on paraffin-embedded specimens fixed in Histochoice, that involves a first step of ISH using nick-translated biotinylated DNA probes and streptavidin alkaline phosphatase NBT/BCIP detection (purple color). This is followed by IMH as detected with the AEC horseradish peroxidase system (red color). The water-insoluble ISH reaction product retains full signal strength throughout the IMH steps, and there is no loss of IMH signal due to proteinase digestion.

The procedure described here is applicable to various specimen types, with sensitivity comparable to separate ISH and IMH protocols performed on different serial sections *(15–19)*. To illustrate the technique, human sun-damaged skin was probed with a 3.2-kbp, full-length, biotinylated human elastin probe (cHDE1) *(20)*, and commercially available monoclonal anti-elastin antibodies (Sigma, St. Louis, MO). As a control, an adjacent serial section was probed with biotinylated pBluescript II (Stratagene, La Jolla, CA) and generic antimouse IgG antibodies.

2. Materials
2.1. Embedding and Sectioning

1. Histochoice (Amresco).
2. 50% Ethanol (EtOH).
3. 70% EtOH.
4. 95% EtOH.
5. 100% EtOH dehydrated, 200 proof (Pharmacia LKB, Piscataway, NJ).
6. Amyl acetate (Polysciences, Warrington, PA).
7. Paraplast X-tra (Fisher, Pittsburgh, PA).
8. Peel-Away molds (Polysciences, VWR).
9. Reichert-Jung 2050 Supercut Microtome.
10. Superfrost Plus slides (Fisher).

2.2. Nick Translation (see Notes 1 and 2)

1. Autoclaved, deionized, double distilled water (ddH$_2$O).
2. 1M Tris-HCl, pH 7.5.
3. 1M MgCl$_2$.
4. dATP, dTTP, dCTP, dGTP, 20 mM: Dilute 100 mM stock dNTPs, Ultrapure dNTP Set, 2'-deoxynucleoside-5'-triphosphate (Pharmacia, Uppsala, Sweden) 1:5 in 0.1M Tris pH 7.5.
5. Bovine serum albumin, fraction V (BSA) (Sigma) 10% in ddH$_2$O.
6. β-Mercaptoethanol (Sigma).
7. β-Mercaptoethanol working dilution: 4 µL β-Mercaptoethanol plus 67 µL ddH$_2$O.
8. Cloned DNA and DNA for quality control assessment: Double-stranded plasmid DNA containing various inserts.
9. Biotin-14-dATP (Gibco-BRL, Gaithersburgh, VA).
10. dATP-^{3}H (ICN, High Wycombe, UK).
11. DNA Polymerase I (Promega, Madison, WI).
12. DNase, RQ1 RNase-free (Promega).
13. DNase working dilution: 2 µL DNase plus 16 µL ddH$_2$O.
14. 0.5M Ethylene diamine tetraacetate tetrasodium salt (EDTA), pH 7.5.
15. Centricon-30 Microconcentrators (Amicon, Danvers, MA).
16. 0.1M EDTA.
17. Ecolume liquid scintillation fluid (ICN).

2.3. In Situ DNA-mRNA Hybridization (see Notes 1 and 2)

1. ddH$_2$O.
2. 1M KOH.
3. 100% EtOH.
4. Slide dipping chambers, 25 slide, 140-mL capacity (Tissue Tek, Baxter, Muskegon, MI).
5. Histo-Clear (National Diagnostics, Atlanta, GA).
6. Xylene (Fisher).
7. 95% EtOH.
8. 75% EtOH.
9. 50% EtOH.
10. 2X SSC: 0.3N NaCl, 30 mM sodium citrate, pH 7.0.
11. 0.2N HCl.
12. Deionized formamide (DI-formamide) (Fisher, *see* Note 3).
13. Mixed-bed ion exchange resin, AG 501-X8 (Bio-Rad).
14. 95% DI-formamide/0.1X SSC.
15. 0.1X SSC.
16. 0.1M EDTA.
17. 10X SSCP: 1.2N NaCl, 0.15M sodium citrate, 0.19M sodium phosphate, pH 6.0.
18. Dextran sulfate (Fisher) 50% in ddH$_2$O.
19. DNA probes: 125 µL/slide assembly; 100 ng nick translated, biotinylated DNA probe (approx 20 µL nick translation product from above) in 1.2X SSCP containing 7% dextran sulfate and 24% formamide.

20. Sigmacote (Sigma).
21. Probe-On slides (Fisher).
22. Carter's rubber cement.
23. Humid chamber (Medical Packaging Corporation, Camarillo, CA).
24. 12.5% DI-formamide/2X SSC.
25. 1X Phosphate-buffered saline (PBS).
26. 10% Triton X-100 in dd H_2O.
27. Blocking buffer: PBS containing 1.0% Triton X-100.
28. DETEK I-alk signal generating complex (ENZO Biochem, Syosset, NY).
29. Bovine serum albumin, fraction V (BSA) (Sigma) 10% in ddH_2O.
30. Streptavidin-alkaline phosphatase (SAP) solution: DETEK (1:100) in 1X PBS, 5 mM EDTA, 0.5% Triton X-100, and 0.1% BSA.
31. 1M Potassium phosphate, pH 6.5.
32. 4M NaCl.
33. Washing buffer: 10 mM potassium phosphate, 0.5M NaCl, 1.0 mM EDTA, pH 7.5, 0.5% Triton X-100, 0.1% BSA.
34. 1M Tris-HCl, pH 8.8.
35. 1M MgCl$_2$.
36. Substrate buffer: 42 mM Tris-HCl, pH 8.8, 0.1M NaCl, 0.1M MgCl$_2$.
37. Dimethylformamide (DMF, Sigma).
38. 5-Bromo-4-chloro-3-indolyl phosphate *p*-toluidine salt (BCIP) (Sigma).
39. Nitro blue tetrazolium (NBT) grade III (Sigma).
40. Substrate solution: 0.8 mM BCIP and 4.5 mM NBT in substrate buffer (*see* Note 4).

2.4. Immunohistochemistry

1. PBS.
2. Primary antibodies (selected by investigator).
3. Zymed Histostain SP Kit (Zymed Laboratories, San Francisco, CA).
4. Mouse IgG (Sigma).
5. Crystal/Mount (Biomeda, Foster City, CA).
6. Cytoseal 100 (Stephens Scientific, Riverdale, NJ).

3. Methods
3.1. Sample Fixation and Dehydration, Embedding, and Sectioning

1. Place samples in Histochoice fixative for 2–4 h, 5°C (*see* Note 5).
2. Transfer to 50% EtOH for 1 h, 5°C (*see* Note 6).
3. Transfer to 70% EtOH for 1 h, –20°C.
4. Transfer to 95% EtOH for 1 h, –20°C.
5. Transfer to 95% EtOH for 1 h, –20°C.
6. Transfer to 100% EtOH for 1 h, –20°C.
7. Store specimens in 100% EtOH, room temperature (*see* Note 7).
8. Transfer specimens to 100% amyl acetate for 1 h, room temperature.

9. Transfer to 100% amyl acetate for 1 h, 56°C (*see* Note 8).
10. Transfer to 50% amyl acetate/50% Paraplast X-tra for 1 h, 56°C.
11. Transfer to 50% amyl acetate/50% Paraplast X-tra for 1 h, 56°C.
12. Transfer to 100% Paraplast X-tra for 1 h, 56°C.
13. Transfer to 100% Paraplast X-tra overnight, 56°C.
14. Transfer to 100% Paraplast X-tra for 1-2 h, 56°C.
15. Position specimen in Peel-Away mold and allow to solidify overnight. Samples may be sectioned the following day.
16. Cut serial sections at 10 μm thickness (*see* Note 9).
17. Mount the sections by flotation in water at 42°C onto glass Superfrost Plus slides (*see* Note 10).
18. Bake mounted sections to dryness at 45°C for 1–2 d.
19. Store slides at room temperature for combined ISH/IMH.

3.2. Preparation of Biotinylated DNA Probes by Nick Translation

1. Thaw reagents and hold on ice (*see* Note 11).
2. Sample and quality control (QC) DNA reaction mixtures (*see* Note 12): To a sterile 1.5-mL Eppendorf tube held on ice, add:
   a. 4.0 μL $1M$ Tris-HCl, pH 7.5.
   b. 2.0 μL $1M$ $MgCl_2$.
   c. 2.0 μL   20 mM dCTP.
   d. 2.0 μL   20 mM dGTP.
   e. 2.0 μL   20 mM dATP. Add to QC sample only.
   f. 2.0 μL   20 mM dTTP
   g. 1.0 μL   10% BSA.
   h. 1.0 μL   β-Mercaptoethanol working dilution.
   i. Sample or QC DNA: volume to give 10 μg total (*see* Note 13).
   j. 25.0 μL   dATP-Biotin-14, omit in QC sample (*see* Note 14).
   k. 2.0 μL ^{3}H-dATP, (*see* Note 14).
   l. 3.5 μL   DNA Polymerase I.
   m. 3.0 μL   DNase working dilution.
   n. Sterile ddH$_2$O to bring total volume to 100 μL.
3. Vortex, pulse spin, and cool on ice. Avoid warming to room temperature.
4. Transfer to 16°C water bath, and incubate for 2 h.
5. Terminate the reaction with 5 μL 0.5M EDTA, vortex and hold on ice.
6. Purify product: Carefully transfer the reaction mixture to a Centricon-30 Microconcentrator; add 900 μL 0.1M EDTA; spin at 5000g. Final concentrated volume is usually 35–50 μL.
7. Measure final volume of retentate and transfer to a 1.5-mL Eppendorf tube.
8. Take aliquots for scintillation counting: 3 μL from the top retentate (labeled product) and 5 μL from the bottom filtrate (unincorporated label). Dilute each with 5 mL Ecolume (ICN) scintillation cocktail and count.
9. Calculate percent incorporation: Multiply the top and bottom counts by their respective dilution factors. Add the dilution corrected top and bottom counts to

derive total counts. Divide the corrected top counts by the total counts and multiply by 100 to obtain percent incorporation (*see* Note 15).
10. Dilute labeled product with 1.5 mL 0.1M EDTA for use in ISH. This yields a final probe of 200–500 bp in length, at a concentration of approx 5–7 ng/µL.

3.3. In Situ *DNA-mRNA Hybridization* (see *Notes 11,16, and 17*)

3.3.1. Day 1

1. Deparaffinize and rehydrate specimens: Histo-Clear, 20 min; xylene, 20 min; 100% EtOH, 20 min; 95% EtOH, 1 min; 70% EtOH, 1 min; 50% EtOH, 1 min; 2X SSC, 1 min.
2. Deproteinize and denature: 0.2N HCl, 20 min.
3. ddH$_2$O wash 3X (brief dips only).
4. Dehydrate: 70% EtOH, 2 min; 95% EtOH, 2 min.
5. Prehybridization: 95% formamide, 0.1X SSC, 15 min at 70°C.
6. 0.1X SSC, ice-cold: 2 min.
7. ddH$_2$O: 2 min.
8. Dehydrate: 70% EtOH, 2 min; 95% EtOH 1 min; 100% EtOH, 1 min.
9. Prepare DNA probes (*see* Note 18).
10. Denature probes by heating at 65°C for 5 min and immediately cool on ice.
11. Hybridize: Apply 120 µL DNA probe per slide; add a siliconized coverslip (*see* Note 19); seal with rubber cement using a 5-cc disposable plastic syringe; place in humid slide chamber (containing ddH$_2$O) for 36 h at 37°C.

3.3.2. Day 3

1. Very carefully disassemble slide assembly under ddH$_2$O.
2. Perform stringency washes to remove nonspecifically bound probe: 12.5% DI-formamide/2X SSC: 3X 3 min at 39°C; 2X SSC, 3 × 2 min at 39°C. The remaining hybridization detection steps are carried out at room temperature.
3. Blocking buffer: 2 × 4 min.
4. 1X PBS: 5 min.
5. SAP solution: add 125 µL per slide. Add siliconized coverslips or Probe-On slides without rubber cement; avoid air bubbles. Incubate for 1 h at room temperature in a humid slide chamber containing ddH$_2$O.
6. Remove coverslips.
7. Washing buffer: 5 × 5 min.
8. Substrate buffer: 2 × 2 min.
9. Substrate solution: Incubate at room temperature for up to 48 h, or less to obtain good signal to noise ratio.
10. Stop development by rinsing in 1X PBS.
11. Hold in 1X PBS for immunostaining.

3.4. Immunohistochemistry

1. React with primary antibodies, 30 min (*see* Note 20).
2. PBS rinses, 3 × 2 min.

3. Visualize immunolabeling by means of the Histostain-SP kit (Zymed Laboratories), which employs biotinylated secondary antibodies, a horseradish peroxidase-streptavidin conjugate and AEC substrate using the manufacturer's protocol.
4. Mount slides with Crystal/Mount (Biomeda) and coverslip with Cytoseal 100.

3.5. Microscopic Observation

Mounted specimens are best observed using either bright-field or Normarski differential interference contrast optics and photographed in color with appropriate correction filters (e.g., Kodak Wratten 80B filter). An example of the combined ISH/IMH gene expression protocol is shown in Fig. 1.

4. Notes

1. The following solutions are to be made fresh each run: β-mercaptoethanl working solution, DNase working dilution, 95% DI-formamide/0.1X SSC, DNA probe mixtures, 12.5% DI-formamide/2X SSC, blocking buffer, SAP solution, washing buffer, substrate buffer, and substrate solution.
2. All reagents are prepared in accordance with protocols in refs. 21 and 22 where applicable, and are molecular biology grade.
3. To make DI-formamide, mix 50 mL reagent grade formamide and 5 g of mixed-bed ion exchange resin. Stir for 30 min at room temperature. Filter twice though Whatman #1 filter paper using a Buchner funnel. Dispense into aliquots, and store at –20°C.
4. Weigh out 47 mg NBT, suspend in 0.43 mL DMF, and add 0.19 mL ddH$_2$O to fully dissolve. Weigh out 23 mg BCIP and dissolve in 2.3 mL DMF. Add both to 140 mL substrate buffer. Protect from light.
5. Dehydration performed at 5°C to –20°C vs room temperature prevented shrinkage and deformation of the specimens.
6. For harder tissue specimens, for example human skin, isopropanol may be substituted for EtOH to reduce the hardening effects of alcohol.
7. Specimens are stable for up to 2 yr in 100% EtOH.
8. For harder specimens, times may be reduced by half to minimize hardening.
9. For Histochoice fixed samples add 2 μm to normal sectioning thickness.
10. Superfrost Plus slides are electrostatically charged to adhere sections directly without having to use an adhesive.
11. Gloves should be worn throughout to avoid introduction of RNase from contact. All pipet tips, Eppendorf tubes, and so on, must be autoclaved prior to use.
12. To monitor performance of the enzymes, a clone that has been successfully nick translated previously should be run again as a quality control sample. The quality control sample is prepared by replacing the biotin label with unlabeled dATP.
13. Routinely, 10 μg of total DNA is processed at a time. As little as 0.5 μg may be labeled by scaling down the reaction volume.
14. Labeled nucleotide triphosphates: The choice of which nucleotides carry the tritium and biotin labels is not crucial. In the system described here, dATP carries both labels. The other dNTPs are present in excess and are not rate-limiting. The biotinylated dNTP is held close to the minimum concentration needed for 100% replacement in

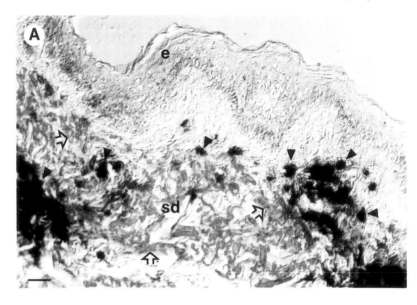

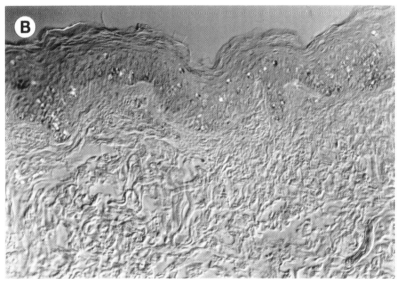

Fig. 1. Localization of elastin mRNA transcripts and protein using sequential ISH and IMH on a single section of human sun-damaged skin. **(A)** ISH was performed first using a human, biotinylated elastin probe showing intense purple staining (closed arrowheads) of fibroblast-like cells in the superficial dermis (sd). This was followed by IMH using monoclonal antielastin antibodies detecting a dense accumulation of large, clumped fibers staining red (open arrows) also in the superficial dermis. The epidermis (e) shows no signal. **(B)** Control section. Bar = 25 μm.

order to conserve the biotin label. A minimum concentration of 50 μ*M* must be present in the incubation mixture for 10 μg DNA. The final nucleotide concentrations are: ^{3}H-dATP, 0.8 μ*M*; biotin-14-dATP, 100 μ*M*; dCTP, dGTP, and dUTP, 400 μ*M* each.

15. Using this procedure, typical values for percent incorporation range from 15–25%.
16. All glassware and plasticware are cleaned by washing with soap and water, followed by rinses with 1*M* KOH, 100% EtOH and autoclaved, deionized distilled H_2O (ddH_2O).
17. Up to 25 slides can be processed together using the slide chamber system from Miles, Elkhart, IN.
18. Negative controls include biotinylated plasmids without inserts, and 0.1*M* EDTA in place of biotinylated plasmid.
19. Probe-On slides siliconized with Sigmacote may be used as coverslips to create a chamber, protecting sections mounted on standard slides.
20. Negative controls for IMH include the omission of primary antibodies, appropriate preimmune sera for polyclonal antibodies, or nonimmune mouse IgG for monoclonal primary antibodies.

Acknowledgments

The authors wish to thank Eric F. Bernstein (Thomas Jefferson University, Philadelphia, PA) for making available human sun-damaged skin samples and human elastin cDNA probes and antibodies. This work is supported in part by grants from the NIH (HD15822, HD29937, ES07005, and DE11327) and the Arcadia Foundation.

References

1. Kriegsmann, J., Keyszer, G., Geiler, T., Gay, R. E., and Gay, G. (1994) A new double labeling technique for combined *in situ* hybridization and immunohistochemical analysis. *Lab. Invest.* **71,** 911–917.
2. Hindrek, M., Lindh, A., and Sundler, F. (1993) Islet amyloid polypeptide gene expression in the endocrine pancreas of the rat: a combined *in situ* hybridization and immunocytochemical study. *Cell Tiss. Res.* **274,** 467–474.
3. Brousset, P., Butet, V., Chittal, S., Selves, J., and Delsol, G. (1992) Comparison of *in situ* hybridization using different nonisotopic probes for detection of Epstein-Barr virus in nasopharyngeal carcinoma and immunohistochemical correlation with anti-latent membrane protein antibody. *Lab. Invest.* **67,** 457–464.
4. Bugnon, C., Bahjaoui, M., and Fellmann, D. (1991) A simple method for coupling *in situ* hybridization and immunocytochemistry: application to the study of peptidergic neurons. *J. Histochem. Cytochem.* **39,** 859–862.
5. Chan-Palay, V., Yasargil, G., Hamid, Q., Polak, J. M., and Palay, S. L. (1987) Simultaneous demonstrations of neuropeptide Y gene expression and peptide storage in single neurons of the human brain. *Proc. Natl. Acad. Sci. USA* **85,** 3213–3215.
6. Hoefler, H., Childers, H., Montminy, M. R., Lechan, R. M., Goodman, R. H., and Wolf, H. J. (1986) *In situ* hybridization methods for the detection of somatostatin mRNA in tissue sections using antisense RNA probes. *Histochem. J.* **18,** 597–604.

7. Schalling, M., Hokfelt, T., Wallace, B., Goldstein, M., Filer, D., Yamin, C., and Schlesinger, D. H. (1986) Tyrosine 3-hydroxylase in rat brain and adrenal medulla: hybridization histochemistry and immunohistochemistry combined with retrograde tracing. *Proc. Natl. Acad. Sci. USA* **83,** 6208–6212.

8. Robben, H., Van Dekken, Poddighe, P. J. and Vooijs, G. P. (1994) Identification of aneuploid cells in cytological specimens by combined *in situ* hybridization and immunocytochemistry. *Cytopathology* **5,** 384–391.

9. Coouwenhoven, R. I., Luo, W., and Snead, M. L. (1990) Co-localization of EGF transcripts and peptides by combined immunohistochemistry and *in situ* hybridization. *J. Histochem. Cytochem.* **38,** 1853–1857.

10. Watts, A. G. and Swanson, L. W. (1989) Combination of *in situ* hybridization with immunohistochemistry and retrograde tract-tracing, in *Methods in Neurosciences*, vol. 1 (Conn, P. M., ed.), Academic, San Diego, CA, pp. 127–135.

11. Van Der Loos, C. M., Volkers, H. H., Rook, R., Van Den Berg, F. M., and Houthoff, H. (1989) Simutaneous application of *in situ* DNA hybridization and immunohistochemistry on one tissue section. *Histochem. J.* **21,** 279–284.

12. Shivers, B. D., Harlan, R. E., Pfaff, D. W., and Schachter, B. S. (1986) Combination of immunocytochemistry and *in situ* hybridization in the same tissue section of rat pituitary. *J. Histochem. Cytochem.* **34,** 39–43.

13. Brahic, M., Haase, A. T., and Cash, E. (1984) Simultaneous *in situ* detection of viral RNA and antigens. *Proc. Nat. Acad. Sci. USA* **81,** 5445–5448.

14. Gendelman, H. E., Moench, T. R., Narayan, O., Griffin, D. E., and Clements, J. E. (1985) A double labeling technique for performing immunocytochemistry and *in situ* hybridization in virus infected cell cultures and tissues. *J. Virol. Methods* **11,** 93–103.

15. Tuan, R. S., Shepley, K. J., Mulligan, M. M., Abraham, D., and Perler, F. B. (1991) Histochemical localization of gene expression in *Onchocerca volvulus*: *in situ* DNA histohybridization and immunocytochemistry. *Mol. Biochem. Parasitol.* **49,** 191–204.

16. McDonald, S. A. and Tuan, R. S. (1989) Expression of collagen type transcripts in chick embrionic bone detected by *in situ* cDNA-mRNA hybridization. *Develop. Biol.* **133,** 221–234.

17. Oshima, O., Leboy, P. S., McDonald, S. A., Tuan, R. S., and Shapiro, I. M. (1989) Developmental expression of genes in chick growth cartilage detected by *in situ* hybridization. *Calcif. Tiss. Int.* **45,** 182–192.

18. Tuan, R. S. (1993) *In situ* hybridization analysis of gene expression in skeletal tissues. *Bone* **14,** 309–314.

19. Oberlender, S. A. and Tuan, R. S. (1994) Spatiotemporal profile of N-cadherin expression in the developing limb mesenchyme. *Cell Adhes. Commun.* **2,** 521–537.

20. Bernstein, E. F., Chen Y. Q., Tamai, K., Shepley, K. J., Resnik, K. S., Zhang, H., Tuan, R., Mauviel, A., Uitto, J. (1994) Enhanced elastin and fibrillin gene expression in chronically photodamaged skin. *J. Invest. Dermatol.* **103,** 182–186.

21. Sambrook, J., Fritsch, E. F., and Maniatis, T. (1989) *Molecular Cloning: A Laboratory Manual*, 2nd ed., Cold Spring Harbor Laboratory, Cold Spring Harbor, NY.

22. Asubel, F. M., Brent, R., Kingston, R. E., Moore, D. D., Seidman, J. G., Smith, J. A., Struhl, K. (1995) *Current Protocols in Molecular Biology*, Wiley, Secaucus, NJ.

20

Whole Mount *In Situ* Hybridization to Embryos and Embryonic Tissues

Ronald A. Conlon

1. Introduction

In whole mount *in situ* hybridization, the distribution of RNA transcripts is visualized in the intact embryo. This allows for the rapid realization of transcript distribution throughout the embryo without the need for reconstruction from sections. In addition, embryos probed for different transcripts (or for the same transcript under different experimental conditions) can be readily compared. The procedure is limited to stages where the embryos are large enough to be conveniently processed, but not so large as to prevent penetration of the reagents (embryonic d 6–10 in the mouse). The following procedure for mouse embryos *(1)* is derived from whole mount *in situ* hybridization procedures for the embryos of other species *(2,3)*. This procedure also can be applied to isolated embryonic organs and tissues *(4)*.

2. Materials
2.1. Materials and Reagents Prepared Ahead of Time

1. An apparatus, such as a nutator (Adams), for gently rocking or rotating the embryos throughout the procedure.
2. PBS autoclaved with 0.1% v/v diethylpyrocarbonate (10X PBS stock, per liter: 80 g NaCl, 2 g KCl, 11.5 g Na_2HPO_4, 2 g KH_2PO_4).
3. PBT (PBS with 0.1% v/v Tween-20) autoclaved with 0.1% v/v diethylpyrocarbonate.
4. Methanol.
5. 30% Hydrogen peroxide.
6. 2 mL Plastic screw cap tubes.

From: *Methods in Molecular Biology, vol. 63: Recombinant Protein Protocols: Detection and Isolation* Edited by: R. Tuan Humana Press Inc., Totowa, NJ

7. Proteinase K.
8. Glutaraldehyde, electron microscopy grade.
9. Sodium borohydride.
10. Digoxigenin-labeled RNA probe, synthesized according to instructions of the digoxigenin-UTP supplier (Boehringer Mannheim, Mannheim, Germany). Store at –20°C in hybridization buffer.
11. Wash 1: 0.3M NaCl, 10 mM Pipes pH 6.8, 1 mM EDTA, 1% SDS.
12. Wash 1.5: 50 mM NaCl, 10 mM Pipes pH 6.8, 1 mM EDTA, 0.1% SDS.
13. RNase buffer: 0.5M NaCl, 10 mM Pipes pH 7.2, 0.1% Tween-20, autoclaved.
14. Wash 4: 500 mM NaCl, 10 mM Pipes pH 6.8, 1 mM EDTA, 0.1% Tween-20, autoclaved.
15. TBST: 137 mM NaCl, 25 mM Tris-HCl pH 7.6, 3 mM KCl, 0.1% Tween-20, autoclaved.
16. Levamisole (Sigma, St. Louis, MO).
17. Embryo powder (Homogenize gestational d 13 embryos in a minimum of PBS on ice. Add 4 vol of cold acetone and mix vigorously. Keep on ice for 30 min with occasional mixing. Collect the precipitate by centrifugation at 10,000g for 10 min at 4°C. Remove and discard the supernatant. Resuspend the pellet with cold acetone and mix vigorously. Keep on ice for 10 min. Spin at 10,000g for 10 min at 4°C. Decant the supernatant and transfer the pellet to a clean piece of filter paper, spread the precipitate, and allow to air dry. Store in an air-tight container at 4°C).
18. Antidigoxigenin antibody, alkaline phosphatase-conjugated (Boehringer Mannheim). Store at 4°C.
19. Normal goat serum. Store at 4°C.
20. BCIP: 50 mg/mL 5-bromo-4-chloro-3-indolyl phosphate, toluidine salt in 100% N,N-dimethylformamide, stored at –20°C in the dark.
21. NBT: 75 mg/mL nitroblue tetrazolium salt in 70% N,N-dimethylformamide, stored at –20°C in the dark.
22. PBTE: PBT plus 1 mM EDTA, autoclaved.
23. 50% Glycerol/50% PBTE.
24. 80% Glycerol/20% PBTE.
25. Sodium azide.

2.2. Reagents Prepared Fresh on Day of Use

2.2.1. Day 1

1. 20% Paraformaldehyde. (Heat 17 mL of water on a hot plate with stirring. Add 1 drop of 10N NaOH, then 4 g of electron microscopy-grade paraformaldehyde and stir until dissolved. Make up to 20 mL and filter.)
2. Fixative (4% paraformaldehyde in PBS made from 20% paraformaldehyde, 10X PBS, and diethylpyrocarbonate-treated water).
3. Hybridization buffer (50% formamide, 0.75M NaCl, 10 mM Pipes pH 6.8, 1 mM EDTA, 100 µg/mL tRNA, 0.05% heparin, 0.1% BSA, 1% SDS).

2.2.2. Day 2

1. Wash 2: 50% formamide, 300 mM NaCl, 10 mM Pipes pH 6.8, 1 mM EDTA, 1% SDS.
2. Wash 3: 50% formamide, 150 mM NaCl, 10 mM Pipes pH 6.8, 1 mM EDTA, 0.1% Tween-20.

2.2.3. Day 3

NTMT: 100 mM NaCl, 100 mM Tris-HCl pH 9.5, 50 mM MgCl$_2$, 0.1% Tween-20.

3. Method

Gentle agitation of the embryos through every step of the procedure (except steps 8 and 21) is critical for success. The instruction "rinse" is used to indicate that the solution should be gently added to the embryos, briefly and gently mixed, the tube set upright in a rack until the embryos settle and then the overlying solution carefully removed. The instruction "wash" is used to indicate the same, except that gentle mixing is performed for the indicated time and at the indicated temperature.

3.1. Embryo Preparation and Storage

1. Dissect gestational day 6–10 embryos free from extraembryonic tissues in cold PBS. A small puncture hole must be made in the anterior neural tube of day 9 and 10 embryos to prevent trapping of reagents inside the closed neural tube.
2. Fix for 2 h at 4°C in 10 mL of fresh cold fixative.
3. Rinse three times with cold PBT. Change directly into 100% methanol, invert the tube several times to mix. Store at −20°C, or proceed to step 4.
4. Treat with a 5:1 mixture of 100% methanol and 30% hydrogen peroxide for 2–3 h at room temperature. Rinse three times in methanol. Store at −20°C.

3.2. Hybridization

3.2.1. Day 1

5. Rehydrate through a methanol/PBT series (75, 50, and 25% v/v), 20 min per change at room temperature. Rinse through three changes of PBT.
6. Transfer the embryos to a 2 mL plastic screw-cap tube. Treat with 20 μg/mL proteinase K in PBT for 3 min at room temperature. Rinse 2X with PBT.
7. Fix the embryos in fresh 0.2% glutaraldehyde/4% paraformaldehyde in PBS at room temperature for 20 min.
8. Rinse through three changes of PBT. Treat with freshly prepared 0.1% sodium borohydride in PBT for 20 min in upright, uncapped tubes. **Do not cap the tubes—hydrogen gas is produced.** Rinse three times with PBT.
9. Rinse twice with hybridization buffer. Prehybridize for at least 1 h at 63°C.

10. Replace hybridization buffer and add the digoxigenin-labeled RNA probe to 0.5–2 μg/mL. Hybridize overnight at 63°C.

3.2.2. Day 2

11. Rinse once with Wash 1. Wash twice with Wash 1 for 30 min each at 63°C.
12. Wash twice with Wash 1.5 for 30 min each at 50°C.
13. Rinse once with RNase buffer. Treat with 100 μg/mL RNase A and 100 U/mL RNase T1 in RNase buffer for 60 min at 37°C. Rinse once with RNase buffer.
14. Wash with Wash 2 for 30 min at 50°C.
15. Wash with Wash 3 for 30 min at 50°C. (At this time, inactivate the goat serum at 68°C for 30 min, and inactivate the embryo powder by heating a few milligrams of powder in 1 mL of TBST to 68°C for 30 min.)
16. Rinse twice with Wash 4, then heat the embryos in wash 4 at 68°C for 20 min.
17. Incubate for at least 1 h at room temperature in TBST containing 0.5 mg/mL freshly added levamisole and 10% v/v heat-inactivated goat serum. (At this time, preabsorb the antibody by diluting to 1/5000 in cold TBST containing 0.5 mg/mL levamisole, 1% v/v heat-inactivated goat serum and the heat-inactivated embryo powder. Rock the tube for 30 min at 4°C. Centrifuge the mixture at 10,000*g* for 10 min at 4°C. The preabsorbed antibody is in the supernatant.)
18. Incubate with the preabsorbed antibody overnight at 4°C.

3.2.3. Day 3

19. Rinse three times with TBST containing freshly added 0.5 mg/mL levamisole, then wash five or six times, 1 h each, at room temperature in the same buffer.
20. Wash twice for 20 min each at room temperature with NTMT containing freshly added 0.5 mg/mL levamisole.
21. Incubate with color reagents (4.5 μL/mL NBT, 3.5 μL/mL BCIP and 0.5 mg/mL levamisole in NTMT). For most messages the color reaction needs to continue overnight at room temperature. Do not agitate the embryos during the overnight incubation. Protect from light.
22. Stop the color reaction with three rinses with PBTE. Clear the embryos by passing the embryos into 1:1 glycerol/PBTE for one hour, then into 4:1 glycerol/PBTE with 0.02% sodium azide. Store at 4°C.

4. Notes

1. Detailed descriptions of early mouse embryo morphology, and procedures for embryo dissection are contained in ref. *5*.
2. Because fibers and particles of dirt tend to accumulate and stick to the embryos, cleanliness throughout the procedure is important. Make sure solutions do not carry particulates or precipitates. If they do, remake or filter the solutions.
3. For optimal preservation of tissue structure and the target RNA, the embryos should be fixed as soon as possible after dissection. Work quickly, keep the embryos cold, and do not work with more embryos than can be quickly processed.

4. Fixed embryos can be stored in 100% methanol at $-20°C$ for several months.
5. The steps prior to hybridization are sensitive to contaminating RNases. Treat the PBS and PBT solutions with diethylpyrocarbonate to destroy RNases. Dedicate unopened bottles of paraformaldehyde, methanol, 30% hydrogen peroxide, proteinase K, glutaraldehyde, and sodium borohydride to RNA work and mark them clearly as such. Handle these reagents with appropriate precautions to prevent contamination with RNase (wear gloves, use disposable plasticware to handle, etc.). All plasticware and vessels used in the procedure prior to hybridization should also be free of RNase contamination.
6. The proper amount of protease digestion is critical for success. Too little digestion leads to poor penetration of reagents, causing low signal and high background, whereas too much digestion leads to loss of the target RNA and of tissue integrity manifested as low signal and fragmented embryos. If you have extra embryos, it is useful to test different protease conditions to determine the appropriate incubation period or protease concentration for your circumstances.
7. The treatment with sodium borohydride lowers background by reducing unreacted aldehydes to alcohols.
8. Measure the amount of digoxigenin-labeled RNA that was made by spotting some of the labeling reaction on a piece of membrane along with digoxigenin-DNA standards and detecting with the antibody. If the yield was not as high as expected (usually at least 1–20 μg of labeled RNA per reaction) it is unlikely that this probe will work in the procedure.
9. The embryos become transparent in the solutions containing formamide. Thus, greater care must be taken, especially with early embryos, so as not to lose or damage them.
10. Levamisole and the heat treatment of reagents and embryos at $68°C$ are used to inhibit endogenous alkaline phosphatases.
11. After stopping the staining reaction in step 22, the color of the signal may be intensified by dehydrating, then rehydrating the embryos, through a methanol series (30, 50, 70, and 100%). This converts the sometimes purplish reaction product to a deep blue that is more easily captured on color film. The embryos are quite permeable at this point, so the embryos can be taken through the solutions quickly.
12. After completion of the procedure, the embryos may be stored indefinitely in the glycerol plus azide solution at $4°C$ in a sealed container.
13. Photograph the whole mount embryos on a high quality stereo microscope, or on a compound microscope with low power objectives. For the best results, use a chamber or depression slide with a coverslip.
14. High background can result from insufficient protease digestion, failure to puncture the closed anterior neural tube, or poor probe preparation. Lack of signal can result from excessive protease digestion, contamination with RNase, loss of tissue integrity before fixation, or poor probe preparation.

References

1. Conlon, R. A. and Rossant, J. (1992) Exogenous retinoic acid rapidly induces anterior ectopic expression of murine *Hox-2* genes *in vivo*. *Development* **116,** 357–368.
2. Tautz, D. and Pfeifle, C. (1989) A non-radioactive *in situ* hybridization method for the localization of specific RNAs in *Drosophila* embryos reveals translational control of the segmentation gene hunchback. *Chromosoma* **98,** 81–85.
3. Hemmati-Brivanlou, A., Frank, D., Bolce, M. E., Brown, B. D., Sive, H. L., and Harland, R. M. (1990) Localization of specific mRNAs in *Xenopus* embryos by whole-mount *in situ* hybridization. *Development* **110,** 325–330.
4. Moens, C. B., Auerbach, A. B., Conlon, R. A., Joyner, A. L. and Rossant, J. (1992) A targeted mutation reveals a role for N-*myc* in branching morphogenesis in the embryonic mouse lung. *Genes Dev.* **6,** 691–704.
5. Hogan, B., Beddington, R., Costantini, F., and Lacy, E. (1994) *Manipulating the Mouse Embryo. A Laboratory Manual.* Second Ed. Cold Spring Harbor Laboratory, Cold Spring Harbor, NY.

21

Whole-Mount *In Situ* Hybridization for Developing Chick Embryos Using Digoxygenin-Labeled RNA Probes

Christopher W. Hsu and Rocky S. Tuan

1. Introduction

The design of recombinant gene constructs has proven to be a powerful technique in the field of developmental biology, and has allowed the investigator to analyze the regulation of expression and the function of specific genes during embryonic development. By mapping and correlating the spatiotemporal profile of gene expression in the developing embryo with events of morphogenesis and growth, useful information may be gathered on the functional role of specific genes.

In situ hybridization refers to the detection of specific gene transcripts at the site of their expression on the basis of hybridization of labeled DNA or RNA probes *(1,2)*. In particular, the method of whole-mount *in situ* hybridization, that detects gene expression in intact embryos or isolated organs and tissues, has recently received a great deal of attention *(3–5)*. This method allows the three-dimensional visualization of gene expression, and the integrated view permits the ready assessment of relative level of gene expression among different organ/tissue systems within the same embryo, as well as between different embryos. This chapter outlines the protocol developed for chicken embryos *(6–9)*, and should be applicable to similar size embryos of other species.

2. Materials

1. Buffers:
   a. PBS-T: Phosphate-buffered saline (PBS), pH 7.4, containing 0.3% Triton X-100 (Fisher, Pittsburgh, PA); composition per liter, 8 g NaCl, 0.2 g KCl, 1.15 g Na_2HPO_4, and 0.2 g KH_2PO_4.

From: *Methods in Molecular Biology, vol. 63: Recombinant Protein Protocols: Detection and Isolation* Edited by: R. Tuan Humana Press Inc., Totowa, NJ

b. TBS-T: Tris-buffered saline (TBS), pH 7.4, containing 0.3% Tween-20 (polyoxyethylene sorbitan monolaurate; Sigma, St. Louis, MO); TBS composition similar to PBS, except phosphates substituted with 20 mM Tris-HCl.

2. Paraformaldehyde fixative: 4% paraformaldehyde in 1X PBS. Add 4 g of paraformaldehyde (e.g., Fisher purified grade) to 80 mL of distilled water. Add 20 µL of a 10M NaOH solution (helps paraformaldehyde to dissolve). Stir and heat to 60°C until totally dissolved. Add 10 mL 10X PBS. Bring final volume to 100 mL with distilled water. Use only freshly prepared fixative. Storage at 4°C should be no longer than 1 d.

3. Dehydration: PBS-T and methanol series of compositions at 75:25, 50:50, 25:75 and 0:100, respectively.

4. Permeabilization solution: 150 mM NaCl, 1% Nonidet P-40 (Sigma, N-3516), 0.5% deoxycholate (deoxycholic acid, sodium salt, Sigma, D-6750), 0.1% sodium dodecyl sulfate (SDS; e.g., Ultrapure grade, Gibco-BRL, Gaithersburg, MD), 1 mM ethylene diamine tetraacetic acid (EDTA), 50 mM Tris-HCl, pH 8.0.

5. Digestion solution: Proteinase K (Boehringer-Mannheim, Mannheim, Germany) at a concentration of 10–50 µg/mL of PBS-T. Make fresh from 100X frozen stock.

6. Hybridization solution: 50% formamide (deionized biotechnology grade, e.g., Fisher), 5X SSC (1X SSC: 0.15M NaCl, 15 mM sodium citrate), pH 4.5, containing 1% SDS, 50 µg/mL yeast tRNA (biotechnology grade), and 50 µg/mL heparin, sodium salt (Grade II from porcine intestinal mucosa, Sigma).

7. Wash solution; same as hybridization buffer, but without tRNA and heparin.

8. Chick embryo powder *(10)*: Remove day 4 chick embryos from egg and drop into a ceramic mortar chilled with liquid nitrogen. Allow the embryos to freeze by adding liquid nitrogen, then grind with pestle into a fine powder. Let liquid nitrogen evaporate and lyophilize the embryo powder and store desiccated at –20°C.

9. Hydrogen peroxide (30%, e.g., Fisher Certified A.C.S. Grade). Store in refrigerator. Make up fresh 6% hydrogen peroxide in PBS-T.

10. Glutaraldehyde (8% or 25% electron microscopy grade, e.g., Ladd, Burlington, VT). Store stock in refrigerator. Prepare fresh 0.2% glutaraldehyde in 4% paraformaldehyde in PBS-T.

11. Ribonuclease A (RNase A) (molecular biology grade, e.g., Promega, Madison, WI; New England BioLabs, Beverly, MA).

12. Digoxigenin-labeled RNA probe for gene of interest per each specific application, prepared according to standard protocols (e.g., Genius Kit, Boehringer-Mannheim).

13. Alkaline phosphatase conjugated antidigoxigenin antibodies (Fab fragment, Boehringer-Mannheim).

14. Alkaline phosphatase reaction buffer: 100 mM Tris-HCl, pH 9.5, 100 mM NaCl, 50 mM MgCl$_2$

15. Endogenous alkaline phosphatase inhibitor: levamisol (Sigma).

16. Alkaline phosphatase substrate: 5-bromo-4-chloro-3-indolyl-phosphate (BCIP) (Boehringer-Mannheim), and 4-nitroblue tetrazolium chloride (NBT) (Boehringer-Mannheim). Stock concentrations (in reaction buffer): BCIP, 50 mg/mL; NBT, 100 mg/mL.

3. Methods

1. Embryo fixation, dehydration, and preparation: Harvest chicken embryos of appropriate ages from incubated eggs as follows. Crack open egg and drop content into a large Petri dish. Pick up embryos by laying a piece of filter paper (Whatman 3MM) ring, with the inner dimension trimmed to fit around the embryo, on top of the embryo. Trim the extraembryonic membranes along the outer edge of the filter paper ring, and lift the embryo from the underlying yolk. Fix the embryos on filter disks overnight at room temperature in paraformaldehyde fixative (*see* Note 1). Dehydrate embryos as follows: 10 min each in increasing methanol series (PBS-T:methanol 75:25, 50:50, 25:50, respectively), finally in 100% methanol (change 100% methanol at least twice) (*see* Note 2).

   Trim embryos in methanol, leaving about 1 mm of extraembryonic membrane around the embryo. Be sure to remove the vitelline membrane from the surface of the embryo (*see* Notes 3 and 4). Rehydrate embryo by transferring it into PBS-T/methanol series (25:50, 50:50, and 75:25, 10 min each) and finally into 100% PBS-T. Bleach embryo in 6% hydrogen peroxide in PBS-T for 40 min (*see* Note 5). Wash three times in PBS-T.

2. Permeabilization/digestion of embryo: This step is necessary to optimize access of labeled prove to all tissues and cells of the embryo.

   For younger, pre-limb bud stage embryos, a detergent-based permeabilization solution is used: incubate embryo in permeabilization solution twice (30 min each) at room temperature. For older, post-limb bud stage embryos, digest for 5–10 min at room temperature with digestion solution containing Proteinase K (*see* Note 6).

   After permeabilization/digestion, wash the embryo three times in PBS-T. Refix the embryo in the 0.2% glutaraldehyde/4% paraformaldeyde fixative for 25 min (*see* Note 7). Wash three times in PBS-T.

3. Hybridization steps: For prehybridization equilibration, rinse the embryo twice in hybridization buffer, then incubate for 45 min at 68°C. Replace buffer with hybridization buffer containing approximately 1 µg/mL of digoxygenin-labeled RNA probe and hybridize overnight at 68°C (*see* Notes 8 and 9).

   At the end of hybridization, sequentially wash embryo: two times with wash solution at 70°C; once in 1:1 mixture of wash solution and TE containing $0.5M$ NaCl; and once in TE containing $0.5M$ NaCl. After washing, excess unhybridized probe is removed by digestion with RNase A (50 µg/mL) in above solution for 30 min. Wash once in TE with $0.5M$ NaCl. Add 1 mL of 50% formamide, 2X SSC and let sit for 10 min (*see* Note 10), followed by incubation in fresh 50% formamide, 2X SSC for 60 min at 63°C. Perform final wash of embryos in TBS-T (three times, 5 min each).

4. Detection and development of hybridization signal: Incubate embryos overnight at room temperature with alkaline phosphatase-conjugated antidigoxigenin antibody adsorbed with embryo powder (*see* Note 11). Wash embryos with TBS-T (three times of 5 min each, and two times of 30 min each, *see* Note 12). Equilibrate embryos in alkaline phosphatase reaction buffer containing 2 mM levamisol

for 30 min (*see* Note 13). Add new reaction buffer containing BCIP (38 µL of stock per 10 mL of buffer; final concentration = 190 µg/mL) and NBT (50 µL of stock per 10 mL of buffer; final concentration = 500 µg/mL), and allow reaction to proceed in the dark. Inspect specimens using a stereo microscope at regular intervals after 15 min of reaction time to assess the progress of the reaction. Stop the reaction by washing the embryos in PBS (4 times, 5 min each). Clear the embryos by equilibration in 40% glycerol in PBS for 1 h, and store in 80% glycerol in PBS.

5. Examination and observation of hybridization: For young, pre-limb bud stage embryos, the specimens may be mounted on a glass slide in 80% glycerol and coverslipped. For low power examination, use a stereo microscope with incident illumination. For higher magnification, observe the embryos using an upright, analytical microscope with either bright-field or Nomarski differential interference optics (*see* Fig. 1A for whole-mount detection of *Pax-1* gene expression in chick embryo; refs. *8,9*). For older embryos, place embryos in the bottom of the well fully submerged in 80% glycerol and observe using a stereoscope. Hybridization signal appears as distinct purple deposits.

The whole-mount hybridized embryos may also be histologically processed for sectioning and further analysis of the cellular site of gene expression. For this purpose, first embed the embryos in agarose as follows for ease of handling. Melt low melt agarose (1% in PBS) at 45°C, and then place 0.5 mL in each well of a 24-well plate kept at 37°C. Place PBS-rinsed embryos, one per well, into the agarose (use an applicator to gently configure the embryos in the desired configuration). Allow the agarose to harden at room temperature for 15 min or more. Remove agarose plug from well and cut out embryos in minimal size agarose chunks. Process the agarose-embryo composites as for regular tissue specimens for histology with some minor modifications (*see* Note 14). Embed in paraffin and section at 8–10 µm sections, mount sections, and examine by standard microscopy (*see* Fig. 1B for localization of *Pax-1* gene expression).

4. Notes

1. Paraformaldehyde should be made fresh daily. Because whole-mount *in situ* hybridization requires the preservation of the whole, intact embryo without support, a crosslinking fixative is generally necessary. Other noncrosslinking fixatives, such as Histochoice (Amresco, Solon, OH), which perform well with tissue sections supported on a glass slide, usually do not provide sufficient tissue hardening for the whole-mount specimens. Do not use Histochoice, or your embryos will fall apart during hybridization.

2. Methanol appears to perform better than ethanol, yielding embryos that are less brittle. The embryos may be stored in the –20°C freezer for up to one month without compromising the hybridization signal. Parafilm seal the storage Petri dish and be sure that the methanol does not evaporate completely during storage; replenish if necessary.

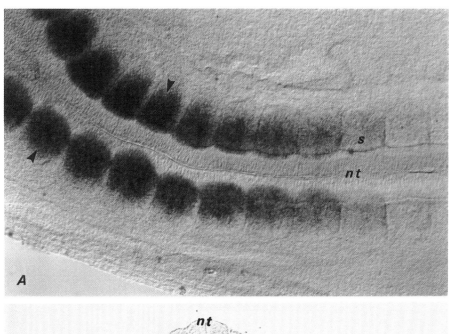

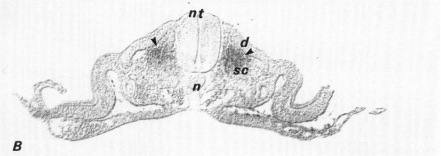

Fig. 1. Localization of chicken *Pax-1* gene expression in somite stage chick embryos using whole-mount *in situ* hybridization: **(A)** Whole-mount view of the caudal region of the embryo showing the somite-specific expression of chicken *Pax-1* gene expression (arrows), which increases progressively from the caudal somites to the rostral somites (right to left) as a function of somite growth and maturation *(7,8).* **(B)** View of transverse histological section derived from whole-mount hybridized embryos showing somitic sclerotomal localization of *Pax-1* gene expression (arrows). *s,* somite; *nt,* neural tube; *sc,* sclerotome; *d,* dermomyotome; *n,* notochord.

3. The vitelline membrane appears as a translucent film that can be detached from the embryo with gentle repeated sucking and releasing of the embryo with the wide end of a Pasteur pipet. Pull with fine forceps when the vitelline membrane starts to detach from the embryo.

4. After the embryos are trimmed, sort them according to age and/or experimental treatment and place in the wells of a 24-well culture dish (approx 3–4 embryos per well works well for pre-limb bud stage embryos; 1-2 embryos/well for older stages). All washes and rinses are 1 mL in volume unless stated otherwise, and should be done gently, i.e., slow and careful introduction and removal of solutions without harsh mechanical disturbance to the embryos in the well.

5. The hydrogen peroxide bleaching step removes contaminating blood-derived coloration that interferes with visualizing the hybridization signal. Generally, pre-limb bud stage embryos do not require bleaching.

6. The exact concentration of Proteinase K in the digestion solution needs to be established empirically. Under-digestion yields poor signal, whereas over-digestion results in disintegration of the specimen. Start with 10 µg/mL of Proteinase K and digestion time of 5 min.

7. The refixation step is crucial for the maintenance of morphology, particularly if the embryo is to be subsequently histologically sectioned after the *in situ* hybridization procedure. Glutaraldehyde is volatile and highly toxic; perform this step in the fume hood.

8. Antisense RNA probe is labeled with digoxygenin using standard protocols, such as those provided in the Genius Kit of Boehringer-Mannheim Biochemicals. Restriction linearize the insert-containing plasmid and use the appropriate RNA polymerase to synthesize the antisense RNA probe. For control, use labeled sense strand RNA. Probe size is generally not of great concern, except in the case of older embryos where permeability and/or background retention may become problematic. Mild alkaline hydrolysis may be used to reduce the size of the probe if necessary.

9. For all hybridization incubation steps, using the 24-well plate set-up, Parafilm seal the plate, place the plate inside a covered, small plastic box containing wet paper towels to maintain moisture, and float the box inside a water bath set to the appropriate temperature. Other similar set-ups may also be used as long as sufficient humidity is maintained to prevent evaporation of the hybridization solution.

10. This short equilibration step allows the embryo to reach the same density of the formamide solution to minimize "floating."

11. The adsorption step reduces nonspecific binding of antibodies to the embryo. Prepare 1 mg/mL chick embryo powder in TBS-T, vortex briefly, and centrifuge for 2 min to remove the debris. Discard pellet, and add antidigoxygenin antibodies to supernatant for adsorption purpose. The antibodies-embryo powder extract mixture is used for localization of hybridization signal.

12. Phosphate buffers are inhibitory to alkaline phosphatase activity. A Tris-buffered saline is therefore used here for alkaline phosphatase enzyme histochemistry.

13. Endogenous alkaline phosphatase activity should be inhibited with levamisol to reduce background noise. This is particularly important for chicken embryos, since the extraembryonic yolk sac membrane exhibits high alkaline phosphatase activity *(11)*. In general, background staining appears as diffuse, brownish deposits or coloration.

14. Standard histological processing protocol may be used *(12,13)*. However, use xylene instead of amyl acetate as an organic impregnating solvent, since amyl acetate tends to dissolve some of the alkaline phosphatase reaction product. In general, expose the specimen, both whole as well as in sections, to minimal time in the organic solvents to minimize loss of signal. Alternatively, the embryos may be processed as follows for cryosectioning (Note: The histology is generally less well maintained in the cryosectioned specimen). Equilibrate embryos with 10–20% sucrose and freeze-embed in Tissue-Tek O.C.T. Compound (Miles, Eikhart, IN). Section embryos with a cryostat microtome at 10–15 μm thickness, mount in PBS-glycerol or other aqueous mounting agents and observe by means of Nomarski differential interference contrast microscopy.

Acknowledgment

The authors thank Claudia Iannotti, Brian Mariani, and George Barnes for their participation in development of the protocol. This work is supported in part by grants from the NIH (HD15822, HD29937, ES07005, and DE11327), and the Orthopaedic Research and Education Foundation.

References

1. Wilkinson, D. G. (1992) In Situ *Hybridization: A Practical Approach.* IRL, Oxford.
2. Chesselet, M.-F. (1990) In Situ *Hybridization Histochemistry.* CRC, Boca Raton, FL.
3. Conlon, R. A. and Rossant, J. (1992) Exogenous retinoic acid rapidly induces anterior ectopic expression of murine *Hox-2* gene *in vivo. Development* **116,** 357–368.
4. Shimamura, K., Hirano, S., McMahon, A. P., and Takeichi, M. (1994) *Wnt-1* dependent regulation of local E-cadherin and αN-catenin expression in the embryonic mouse brain. *Development* **120,** 2225–2234.
5. Shepherd, I., Luo, Y., Raper, J. A., and Chang, S. (1996) The distribution of collapsin-1 mRNA in the developing chick nervous system. *Develop. Biol.* **173,** 185–199.
6. Coutinho, L., Morris, J., and Ivarie, R. (1992) Whole mount *in situ* detection of low abundance transcripts of the myogenic factor *qmf1* and myosin heavy chain protein in quail embryos. *Biotechniques* **13,** 722–724.
7. Spann, P., Ginsburg, M., Rangini, Z., Fainsod, A., Eyal-Giladi, H., and Gruenbaum, Y. (1994) The spatial and temporal dynamics of SAX1 (CHox3) homeobox gene expression in the chick's spinal chord. *Development* **120,** 1817–1828.
8. Smith, C. A. and Tuan, R. S. (1996) Functional involvement of *Pax-1* in somite development: Somite dysmorphogenesis in chick embryos treated with *Pax-1* paired-box antisense oligonucleotide. *Teratology* **52,** 333–345.
9. Barnes G. L., Hsu, C. W., Mariani, B. D. and Tuan, R. S. (1996) Chicken *Pax-1* gene: structure and expression during embryonic somite development. *Differentiation* **61,** 13–23.

10. Riddle, R., Johnson, R., Laufer, E., and Tabin, C. (1993). *Sonic hedgehog* mediates the polarizing activity of the ZPA. *Cell* **75,** 1401–1416.

11. Love, J., and Tuan, R. S. (1993) Pair-rule gene expression in the developing avian embryo. *Differentiation* **54,** 73–83.

12. Humason, G. L. (1967) *Animal Tissue Techniques,* W. H. Freeman, San Francisco.

13. Kiernan, J. (1981) *Histological and Histochemical Methods: Theory and Practice.* Pergamon, Elmsford, NY.

22

Methods for Double Detection of Gene Expression

Combined In Situ *Hybridization and Immunocytochemistry or Histochemistry*

Ronald A. Conlon

1. Introduction

The distribution of two different molecules can be analyzed within the same embryo using the procedures described below. The protocol for combined whole mount *in situ* hybridization and immunocytochemistry allows for simultaneous detection of mRNA and protein. The protocol for combined whole mount *in situ* hybridization and β-galactosidase staining allows for simultaneous detection of mRNA and transgene-directed β-galactosidase expression. Simultaneous detection allows for the most direct comparison of expression patterns. These procedures are derived from protocols used in *Drosophila (1)* and mice *(2,3)*.

2. Materials

The required materials include those listed in Chapter 20 for whole mount *in situ* hybridization, plus the following reagents.

2.1. Combined Protein and RNA Detection

1. Primary antibody against the protein of interest.
2. An appropriate secondary antibody conjugated to horseradish peroxidase.
3. Blocking reagent for nucleic acid hybridization (Boehringer-Mannheim).
4. DAB (30 mg/mL 3, 3'-diaminobenzidine tetrahydrochloride in 10 mM Tris-HCl, pH 7.6: Stored at –20°C in the dark in single use aliquots). Handle with extreme caution: DAB is carcinogenic.

2.2. Combined β-Galactosidase and RNA Detection

1. 100 mM EGTA pH 7.3 treated with 0.1% diethylpyrocarbonate and autoclaved.
2. 1M MgCl$_2$ treated with 0.1% diethylpyrocarbonate and autoclaved.

From: *Methods in Molecular Biology, vol. 63: Recombinant Protein Protocols: Detection and Isolation* Edited by: R. Tuan Humana Press Inc., Totowa, NJ

3. Fixative G. Prepare fresh 0.2% glutaraldehyde, 2 mM MgCl$_2$, 6 mM EGTA pH 7.3 in PBS.
4. Wash G: PBT containing 2 mM MgCl$_2$, treated with 0.1% v/v diethylpyro-carbonate and autoclaved.
5. X-gal: 25 mg/mL 5-bromo-4-chloro-3-indolyl β-D-galactopyranoside in *N,N*-dimethylformamide stored at –20°C.
6. Potassium ferricyanide (K$_3$Fe[CN]$_6$).
7. Potassium ferrocyanide (K$_4$Fe[CN]$_6$ · 3H$_2$O).

3. Methods
3.1. Combined Protein and RNA Detection

Follow the protocol for whole mount *in situ* hybridization in Chapter 20 to step 14 in Section 3.2.2., then substitute the following for step 15 and all subsequent steps. Note that the antidigoxigenin and secondary antibodies are preabsorbed against the embryo powder, whereas the primary antibody against the protein of interest is not.

3.1.1. Day 2

15. Wash with Wash 3 for 30 min at 50°C. (At this time, inactivate the embryo powder by heating a few milligrams of powder in 1 ml of TBST to 68°C for 30 min. Prepare a 1% w/v solution of blocking reagent [Boehringer Mannheim] in TBST. The reagent must be stirred and heated for some time to make a milky solution. Once in solution, cool to 4°C.)
16. Rinse twice with Wash 4, then heat in wash 4 at 68°C for 20 min.
17. Incubate for at least 1 h at room temperature in TBST containing 0.5 mg/mL freshly added levamisole and 1% blocking reagent. (At this time, preabsorb the antidigoxigenin antibody by diluting to 1/5000 in cold TBST containing 0.5 mg/mL levamisole, 1% blocking reagent, and the heat-inactivated embryo powder. Rock the tube for 30 min at 4°C. Centrifuge the mixture at 10,000*g* for 10 min at 4°C. Remove the supernatant containing the preabsorbed antidigoxigenin antibody, and add the primary antibody against the protein of interest to the appropriate dilution.)
18. Incubate the embryos with the antibodies overnight at 4°C.

3.1.2. Day 3

19. Rinse three times with TBST containing freshly added 0.5 mg/mL levamisole, then wash five or six times, 1 h each, at room temperature in the same buffer. (Once again, inactivate some embryo powder and prepare a 1% solution of blocking reagent in TBST)
20. Incubate for at least 1 h at room temperature in TBST containing 0.5 mg/mL freshly added levamisole and 1% blocking reagent. (At this time, preabsorb the secondary antibody by diluting to the appropriate concentration in cold TBST containing 0.5 mg/mL levamisole, 1% blocking reagent and the heat-inactivated embryo powder. Rock the tube for 30 min at 4°C. Centrifuge the mixture at 10,000*g* for 10 min at 4°C. The preabsorbed secondary antibody is in the supernatant.)

21. Incubate with the secondary antibody overnight at 4°C.
22. Rinse three times with TBST containing freshly added 0.5 mg/mL levamisole, then wash five or six times, 1 h each, at room temperature in the same buffer.
23. Incubate for 20 min in the dark with TBST containing freshly added 0.5 mg/mL levamisole and 0.3 mg/mL DAB.
24. Develop the peroxidase reaction by adding hydrogen peroxide to 0.03%. The reaction typically generates signal for the first 10 or 15 min, and then background staining begins to become evident. Stop the reaction by rinsing with TBST containing 0.5 mg/mL levamisole.
25. Wash twice for 20 min each at room temperature with NTMT containing freshly added 0.5 mg/mL levamisole.
26. Incubate with the alkaline phosphatase color reagents (4.5 μL/mL NBT, 3.5 μL/mL BCIP and 0.5 mg/mL levamisole in NTMT). For most messages the color reaction needs to continue overnight at room temperature. Do not agitate the embryos during the overnight color reaction. Protect from light.
27. Stop the color reaction with three rinses with PBTE. Clear the embryos by passing the embryos into 1:1 glycerol/PBTE for 1 h, then into 4:1 glycerol/PBTE with 0.02% sodium azide. The peroxidase reaction product fades with exposure to light. Store at 4°C in the dark.

3.2. Combined β-Galactosidase and RNA Detection

The procedure for β-galactosidase staining decreases the sensitivity of the *in situ* hybridization procedure somewhat, so this combined procedure works best for prevalent target mRNAs.

3.2.1. Embryo Preparation, β-Galactosidase Staining and Storage

1. Dissect gestational day 6–10 embryos free from extraembryonic tissues in cold PBS. A small puncture hole must be made in the anterior neural tube of day 9 and 10 embryos.
2. Fix in 10 mL of fresh cold Fixative G for 10 min on ice.
3. Rinse three times with wash G. Wash with Wash G for 60 min at 4°C.
4. Transfer to a 2-mL plastic screw-cap tube. Incubate in freshly made staining solution (1 mg/mL X-gal, 2 mg/mL potassium ferrocyanide, 1.6 mg/mL potassium ferricyanide in Wash G) at 37°C until desired staining intensity is achieved. The incubation period can vary from minutes to hours depending on the level of expression of β-galactosidase. Use the minimum incubation period possible.
5. Rinse twice with PBT. Fix for 2 h at 4°C in fresh fixative (4% paraformaldehyde in PBS, Chapter 20).
6. Rinse three times with cold PBT. Change directly into 100% methanol, invert the tube several times to mix. Store at –20°C, or proceed to step 7.
7. Treat with a 5:1 mixture of 100% methanol and 30% hydrogen peroxide for 2–3 h at room temperature. Rinse three times in methanol. Store at –20°C.
8. The detection of RNA by *in situ* hybridization may be resumed at step 5 of Section 3.2.1. of Chapter 20.

4. Notes

1. The peroxidase reaction products may be intensified by addition of metal salts to the reaction. If this is desired, make a 0.3% w/v stock solution of $NiCl_2$ or $CoCl_2$. Add to the DAB staining solution for a final concentration of 0.03%, filter, and use immediately.
2. The accumulated background from two combined procedures may obscure signal somewhat. Better visualization may be possible with a stronger clearing agent, for example 1:2 benzyl alcohol/benzyl benzoate (BABB). In glass or polypropylene tubes, dehydrate the embryos quickly through an alcohol series to 100% ethanol. Transfer to 1:1 100% ethanol/BABB until the embryos sink, then into BABB. The BABB dissolves polystyrene, so the embryos must be observed in glass dishes. The BABB also slowly dissolves the colored reaction products of alkaline phosphatase and β-galactosidase, so the embryos cannot be kept in this clearing agent for very long. Reverse the solvent series to return the embryos to an aqueous storage solution.
3. The combined procedures give their best results when the probed expression patterns are largely nonoverlapping, since it is difficult to distinguish double-labeled cells.

References

1. Cubas, P., de Celis, J.-F., Campuzano, S., and Modolell, J. (1991) Proneural clusters of achaete-scute expression and the generation of sensory organs in the *Drosophila* imaginal wing disc. *Genes Dev.* **5,** 996–1008.
2. Davis, C. A., Holmyard, D. P., Millen, K. J., and Joyner, A. L. (1991) Examining pattern formation in mouse, chicken and frog embryos with an *En*-specific antiserum. *Development* **111,** 287–298.
3. Conlon, R. A. and Rossant, J. (1992) Exogenous retinoic acid rapidly induces anterior ectopic expression of murine *Hox-2* genes in vivo. *Development* **116,** 357–368.

23

In Situ PCR

Current Protocol

Omar Bagasra, Muhammad Mukhtar, Farida Shaheen, and Roger J. Pomerantz

1. Introduction

Since we first reported our findings regarding the *in situ* amplification of HIV-1 *gag* gene in a HIV-1 infected cell line in 1990 *(1)*, there has been an explosion of research in the area of *in situ* PCR. There are over 200 publications describing various forms of *in situ* gene amplifications (selected bibliography refs. *1–105)*, identifying various infectious agents *(1–17,20,23,32,44,47–78,81–88,96)*, tumor marker genes *(23,78–79,92)*, cytokines, growth factors and their receptors *(37,39,40,65,77)*, and other genetic elements of interest, in peer reviewed journals *(18,23,84,94)*. The polymerase chain reaction (PCR) method for amplification of defined gene sequences has proved to be a valuable tool not only for basic researchers but also for clinical scientists *(2,4,5,20,23,79–82)*. Using even a minute amount of DNA or RNA and choosing a thermostable enzyme from a large variety of sources, one can amplify the amount of the gene of interest, which can be analyzed and/or sequenced. Thus, genes or portions of gene sequences present only in a small sample of cells or small fraction of mixed cellular populations can be examined. However, one of the major drawbacks of standard PCR technique is that the procedure does not allow the association of amplified signals of a specific gene segment with the histological cell type(s). For example, it would be advantageous to determine what types of cells in the peripheral blood circulation carry HIV-1 provirus at various stages of HIV-1 infection *(2–6,62,78)* and what percent of HIV-1-infected cells actually are expressing viral RNA *(2–16,35,42)*. Similar approaches have been used to detect the presence of other gene sequences in tissue materials and pathological specimens *(1–105)*.

From: *Methods in Molecular Biology, vol. 63: Recombinant Protein Protocols: Detection and Isolation* Edited by: R. Tuan Humana Press Inc., Totowa, NJ

The ability to identify individual cells, expressing or carrying specific genes of interest in a tissue section, under the microscope, provides a great advantage in determining various aspects of normal, as opposed to pathological, conditions. For example, this technique could be used in determination of tumor burden, before and after chemotherapy, in lymphomas or leukemias, in which specific aberrant gene translocations are associated with certain types of malignancy *(23,44,79,80)*. In the case of HIV-1 infection or other viral infections, one can determine the effects of therapy or putative antiviral vaccination by evaluating the number of cells still infected with viral agent, postvaccination. Similarly, one can potentially determine the preneoplastic lesions by examining p53 mutations associated with certain tumors, oncogenes, or other aberrant gene sequences which are known to be associated with certain types of tumors *(2,23,44,79,80)*. In the area of diagnostic pathology determination of origin of metastatic tumors are a perplexing problem. By utilizing the proper primers for genes which are expressed by certain histological cell types, one can potentially determine the origin of metastatic tumors by performing reverse transcriptase-initiated *in situ* PCR *(80)*.

Our laboratories are using *in situ* PCR (ISPCR) techniques since 1988 and we have developed a simple, sensitive ISPCR which has proved reproducible in multiple double-blinded studies *(1–16,40,52,53,79–82,86,96)*. One can use this method for amplification of both DNA or RNA gene sequences. By use of multiple labeled probes, one can detect various signals in a single cell. In addition, under special circumstances, one can perform immunohistochemistry, RNA and DNA amplification at a single cell level (the so-called "triple-labeling").

To date, we have successfully amplified and detected HIV-1, SIV, HPV, HBV, CMV, EBV, HHV-6, HSV, LGV, p53 and its mutations, mRNA for surfactant Protein A , estrogen receptors, and inducable nitrous oxide synthesis (iNOS)- gene sequences associated with multiple sclerosis, by DNA and/or RNA (RT-ISPCR), in various tissues, including: PBMCs, lymph nodes, spleen, brain, skin, breast, lungs, cytological specimens, tumors, cultured cells, and numerous other formalin-fixed, paraffin-embedded tissues. In this chapter, we have provided a detailed account of the ISPCR procedure, which can be used for routine research investigations. In addition, we have also provided a detailed procedure for special applications of ISPCR. For example, its use in cytogenetics, to localize a single gene in the chromosomal bands; its use in dual and triple labeling of cells, where more than one signal can be detected at the single cell level; and its use in combination with EM, immunohistochemistry, and in other special situations *(15)*.

2. Materials

1. Slides: Heavy, teflon-coated glass slides with three 10-, 12-, or 14-mm diameter wells for cell suspensions, or single oval wells for tissue sections, are available

from Cel-line Associates of New Field, NJ, or Erie Scientific of Portsmouth, NH. These specific slide designs are particularly useful, for the teflon coating serves to form distinct wells, each of which serves as a small reaction "chamber" when the coverslip is attached. Furthermore, the teflon coating helps to keep the nail polish from entering the reaction chamber, and multiple wells allow for both a positive and negative control on the same slide.

2. Coplin jars and glass staining dishes: Suitable vessels for washing, fixing, and staining 4–20 glass slides simultaneously are available from several vendors, including Fisher Scientific, Pittsburgh, PA, and VWR Scientific (Bridgeport, NJ).

3. 2% Paraformaldehyde:
   a. Take 12 g paraformaldehyde (Merck, Darmstadt, Germany, ultra pure Art No. 4005) and add to 600 mL 1X PBS.
   b. Heat at 65°C for 10 min.
   c. When the solution starts to clear, add 4 drops $10N$ NaOH and stir.
   d. Adjust to neutral pH and cool to room temperature.
   e. Filter through Nalgene 0.45-μ filtration unit.

4. 10X PBS stock solution pH 7.2–7.4: Dissolve 20.5 g NaH_2PO_4 H_2O and 179.9 g Na_2HPO_4 $7H_2O$ (or 95.5 g Na_2 HPO_4) in about 4 L of double-distilled water. Adjust to the required pH (7.2–7.4). Add 701.3 g NaCl and make up to a total volume of 8 L.

5. 1X PBS: Dilute the stock 10 X PBS at 1:10 ratio (i.e., 100 mL 10X PBS and 900 mL of water for 1 L). Final concentration of buffer should be $0.01M$ phosphate and $0.15M$ NaCl.

6. 0.3% Hydrogen peroxide (H_2O_2) in PBS: Dilute stock 30% hydrogen peroxide (H_2O_2) at a 1:100 ratio in 1X PBS for a final concentration of 0.3% H_2O_2.

7. Proteinase K: Dissolve powder from Sigma in water to obtain 1 mg/mL concentration. Aliquot and store at –20°C. Working solution: Dilute 1 mL of stock (1 mg/mL) into 150 mL of 1X PBS.

8. 20X SSC: Dissolve 175.3 g of NaCl and 88.2 g of sodium citrate in 800 mL of water. Adjust the pH to 7.0 with a few drops of $10N$ solution of NaOH. Adjust the volume to 1 L with water. Sterilize by autoclaving.

9. 2X SSC: Dilute 20X SSC. 100 mL of 20X SSC and 900 mL of water.

10. Solutions for amplifying chromosomal bands:
    a. RPMI Medium 1640: Per 100 mL, supplement with 15 mL fetal bovine serum, $1.5M$ HEPES buffer (IM), 0.1 mL gentamycin (0.1 mL heparin is optional).
    b. Velban: Reconstitute vial with 10 mL sterile H_2O From this solution, dilute 0.1 mL into 50 mL distilled H_2O. Store in refrigerator.
    c. EGTA hypotonic solution: Dissolve 0.2 g EGTA powder, 3.0 g KCL, 4.8 g HEPES buffer into 1000 mL of distilled H_2O. Adjust pH to 7.4. Store in refrigerator, prior to use prewarm to 37°C.
    d. PHA-C (phytohemagglutinin): Reconstitute with 5 mL sterile H_2O. Aliquot into five 1-mL insulin syringes. Freeze four for later use, leave one in refrigerator.

11. Streptavidin peroxidase: Dissolve powder from Sigma in PBS to make a stock of 1 mg/mL. Just before use, dilute stock solution in sterile PBS at a 1:30 ratio.

12. Color solution: Dissolve one AEC (3-amino-9-ethyl-carbazole from Sigma) tab let in 2.5 mL of N.N. Dimethyl formamide. Store at 4°C in the dark. Working solution:
    a. 50 mM Acetate buffer, pH (5.0); 5 mL.
    b. AEC solution; 250 μL.
    c. 30% H_2O_2; 25 μL.
    Make fresh before each use, keeping solution in the dark.
13. 50 mM Acetate buffer pH 5.0: Add 74 mL of 0.2N acetic acid (11.55 mL glacia acid/L) and 176 mL of 0.2M sodium acetate (27.2 g sodium acetate trihydrate ir 1 L) to 1 L of deionized water and mix.
14. In situ hybridization buffer (for 5 mL): Formamide, 2.5 mL; salmon sperm DNA (ssDNA) (10 mg/mL), 500 μL; 20X SSC, 500 μL; 50X Denhardt's solution, 1 mL 10% SDS, 50 μL; water, 450 μL.
    Note: Heat denature ssDNA at 94°C for 10 min before adding to the solution.
15. Preparation of glass slides: Before one can perform *in situ* reactions by this protocol, the proper sort of glass slide with teflon appliqué must be obtained (various sources are described in Section 3.). Then, the glass surface must be treated with the proper sort of silicon compound. Both of these factors are very important, and the following are the reasons why.

    First, one should always use *glass* slides that are partially covered by a teflon coating. Not only does the glass withstand the stress of repeated heat-denaturation, but it also presents the right chemical surface—silicon oxide—that is needed for proper silanization.

    Furthermore, slides with special teflon coatings that form individual wells are useful because vapor-tight reaction chambers can be formed on the surface of the slides when cover slips are adhered with coatings of nail polish around the periphery. These reaction chambers are necessary because within them, proper tonicity and ion concentrations can be maintained in aqueous solutions during thermal cycling—conditions that are vital for proper DNA amplification. The teflon coating serves a dual purpose in this regard. First, the teflon helps keep the two glass surfaces slightly separated, allowing for reaction chambers about 20 μm in height to form between the slides and cover slips. Secondly, the teflon border helps keep the nail polish from entering the reaction chambers when the polish is being applied. This is important, for any leakage of nail polish into a reaction chamber can compromise the results in that chamber. Even if one is using an advanced thermal cycler with humidification, use of the teflon-coated slides is still recommended, as the hydrophobicity of the teflon combined with the pressure applied by a coverslip helps spread small volumes of reaction cocktail over the entire sample region, without forcing much fluid out the periphery. In order to prepare glass slides properly, follow this procedure:
    a. Prepare the following 2% AES solution just prior to use: 3-aminopropyltriethoxysilane (AES: Sigma A-3648), 5 mL; and acetone, 250 mL.
    b. Put solution into a Coplin jar or glass staining dish and dip glass slides in 2% AES for 60 s (*see* Section 3. for sources of both Coplin jars and the proper glass slides).

c. Dip slides five times into a different vessel filled with 1000 mL of distilled water.

d. Repeat step 3 three or four times, changing the water each time.

e. Air dry in laminar-flow hood from a few hours to overnight, then store slides in sealed container at room temperature. Try to use slides within 15 d of silanization; 250 mL of AES solution is sufficient to treat 200 glass slides.

16. Preparation of tissue:

a. Paraffin-fixed tissue: Routinely-fixed paraffin tissue sections can be amplified quite successfully. This permits the evaluation of individual cells in the tissue for the presence of a specific RNA or DNA sequence. For this purpose, tissue sections are placed on specially-designed slides that have *single* wells— which are described further in Section 2., step 15. In our laboratory, we routinely use placental tissues, CNS tissues, cardiac tissues, and so on, which are sliced to a 3–5 μm thickness. Other laboratories prefer to use sections up to 10 μm thickness, but in our experience, amplification is often less successful with the thicker sections, and multiple cell layers can often lead to difficult interpretation due to superposition of cells. However, if one is using tissues that contain particularly large cells—such as ovarian follicles—then thicker sections may be appropriate.

   i. Place tissue section upon the glass surface of the slide.

   ii. Incubate the slides in an oven at 60–80°C (depending on type of paraffin used to embed the tissue) for 1 h, to melt the paraffin.

   iii. Dip the slides in EM grade xylene solution for 5 min, then in EM grade 100% ethanol for 5 min (EM grade reagents are benzene-free). Repeat these washes two or three times, in order to rid the tissue of paraformaldehyde completely.

   iv. Dry the slides in an oven at 80°C for 1 h.

b. Cell suspensions: To use peripheral blood leukocytes, first isolate cells on a Ficoll-Hypaque density gradient. Tissue-culture cells or other single-cell suspensions can also be used. Prepare all cell suspensions with the following procedure:

   i. Wash cells with 1X PBS twice.

   ii. Resuspend cells in PBS at 2×10^6 cells/mL.

   iii. Add 10 μL of cell suspension to each well of the slide using a P_{20} micropipet.

   iv. Air dry slide in a laminar-flow hood.

c. Frozen sections: It is possible to use frozen sections for *in situ* amplification; however, the morphology of the tissue following the amplification process is generally not as good as with paraffin sections. It seems that the cryogenic freezing of the tissue, combined with the lack of paraffin substrate during slicing, compromises the integrity of the tissue. Usually, thicker slices must be made, and the tissue "chatters" in the microtome. As any clinical pathologist will relate, definitive diagnoses are made from paraffin sections, and this rule-of-thumb seems to extend to the amplification procedure as well.

   It is very important to use tissues that were frozen in liquid nitrogen or were placed on dry ice immediately after they were harvested before autolysis began to take place. If tissues were frozen slowly by placing them in –70°C, then eventually ice crystals will form inside the tissues creating a gap which will distort the morphology.

To use frozen sections, use as thin a slice as possible (down to 3–6 μm), apply to slide, dehydrate for 10 min in 100% methanol (exception to methanol is when surface antigens are lipoprotein and will denature in methanol, then use 2% paraformaldehyde or other reagent), and air dry in a laminar-flow hood. Then, proceed with heat treatment described (*see* Section 3.1.1.).

3. Methods: *In Situ* Amplification—DNA and RNA Targets
3.1. Basic Preparation, All Protocols

For all sample types, the following steps comprise the basic preparatory work which must be done before any amplification-hybridization procedure.

3.1.1. Heat Treatment

Place the slides with adhered tissue on a heat block at 105°C for 5–30 s, to stabilize the cells or tissue on the glass surface of the slide.

This step is absolutely critical, and one may need to experiment with different periods in order to optimize the heat treatment for specific tissues. Our laboratory routinely uses 90 s for DNA target sequences, and 5–10 s for RNA sequences. The shorter incubation is recommended for RNA targets, because certain mRNAs may be unstable at high temperature.

3.1.2. Fixation and Washes

1. Place the slides in a solution of 4% paraformaldehyde in PBS (pH 7.4) for 4 h at room temperature. Use of the recommended Coplin jars or staining dishes facilitates these steps.
2. Wash the slides once with 3X PBS for 10 min, agitating periodically with an up and down motion.
3. Wash the slides with 1X PBS for 10 min, agitating periodically with an up and down motion. Repeat once with fresh 1X PBS.
4. At this point, slides with adhered tissue can be stored at –80°C until use. Before storage, dehydrate with 100% ethanol.

If biotinylated probes or peroxidase-based color development are to be used, the samples should further be treated with a 0.3% solution of hydrogen peroxide in PBS, in order to inactivate any endogenous peroxidase activity. Once again, incubate the slides overnight—either at 37°C or at room temperature. Then, wash the slides once with PBS.

If other probes are to be used, proceed directly to the following proteinase K digestion, which is perhaps the most critical step in the protocol.

3.1.3. Proteinase K Treatment (The most rate limiting step)

1. Treat samples with 6 µg/mL proteinase K in PBS for 5–60 min at room temperature or at 55°C. To make a proper solution, dilute 1.0 mL of proteinase K at (1 mg/mL) in 150 mL of 1X PBS.
2. After 5 min, look at the cells under the microscope at 400X. If the majority of the cells of interest exhibit uniform, small, round "salt-and-pepper" dots on the cytoplasmic membrane, then stop the treatment immediately with step 3. Otherwise, continue treatment for another 5 min and re-examine.
3. After proper digestion, heat slides on a block at 95°C for 2 min to inactive the proteinase K.
4. Rinse slides in 1X PBS for 10 s.
5. Rinse slides in distilled water for 10 s.
6. Air dry.

3.2. Comments on Optimizing Digestion

The time and temperature of incubation should be optimized carefully for each cell line or tissue-section type. Too little digestion, and the cytoplasmic and nuclear membranes will not be sufficiently permeable to primers and enzyme, and amplification will be inconsistent. Too much digestion, and the membranes will either deteriorate during repeated denaturation or worse, signals will leak out. In the first case cells will not contain the signal—high background will result. In the last case, many cells will show pericytoplasmic staining, representing the leaked signals going into the cells not containing the signals. Attention to detail here can often mean the difference between success and failure, and this procedure should be practiced rigorously with extra sections before continuing on to the amplification steps.

In our laboratory, proper digestion parameters vary considerably with tissue type. Typically, lymphocytes will require 5–10 min at 25°C or room temperature, CNS tissue will require 12–18 min at room temperature, and paraffin-fixed tissue will require 15–30 min at room temperature. However, these periods can vary widely and the appearance of the "peppery dots" is the important factor.

The critical importance of these dots should not be underestimated, since an extra 2–3 min of treatment after the appearance of dots will result in leakage of signals.

An alternate to observation of "dots" method is to select a constant time and treat slides in varying amount of proteinase K. For example treat slides for 15 min in 1–6 µg/mL of proteinase K.

3.3. RT Variation: In Situ *RNA Amplification*

One has two choices in order to detect an RNA signal. The first and more elegant method is to simply use primer pairs that flank the junctions of spliced

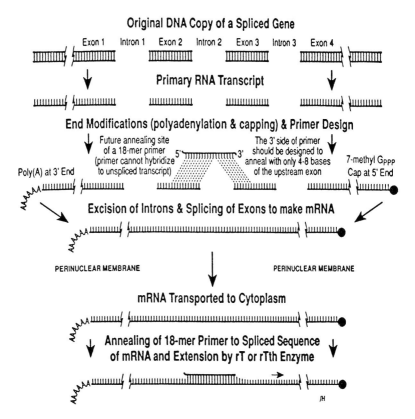

Fig. 1. Most eukaryotic genes are split into segments, as there are numerous introns in the DNA that are excised during the synthesis of mRNA in the nucleus. This characteristic can be exploited in the design of primers in order to amplify mRNA signals without interference from the DNA. Simply design primers so that their sequences flank spliced regions where two exons are fused, such that the homologous annealing sites exist only in mRNA, not in DNA. This allows the elimination of a DNase treatment during slide preparation, as well as simultaneous amplification of both RNA and DNA signals.

sequences of mRNA, as these particular sequences will be found *only* in mRNA (*see* Fig. 1). Thus, by using these RNA-specific primers, one can skip the following DNase step and proceed directly to reverse transcription.

The second approach is to treat the cells or tissue with a DNase solution subsequent to the proteinase K digestion. This step destroys all of the endogenous DNA in the cells so that only RNA survives to provide signals for amplification.

Note: All reagents for RT *in situ* amplification should be prepared with RNase-free water (i.e., DEPC-treated water). In addition, the silanized glass slides and all glassware should be RNase-free, which we ensure by baking the glassware overnight in an oven before use in the RT-amplification procedure.

3.3.1. DNase Treatment

Prepare a RNase-free, DNase solution: 40 mM Tris-HCl, pH 7.4, 6 mM MgCl$_2$, 2 mM CaCl$_2$, 1 U/μL final volume of DNase. (Use RNase-free DNase, such as 10 U/μL RQ1 DNase, cat. no. 776785 from Boehringer.)

1. Add 10 μL of solution to each well.
2. Incubate the slides overnight at 37°C in a humidified chamber. If one is using liver tissue, this incubation should be extended an additional 18-24 h.
3. After incubation, rinse the slides with a similar solution that was prepared *without* the DNase enzyme.
4. Wash the slides 2X with DEPC-treated water.

Note: Some cells are particularly rich in ribonuclease, in this circumstance, add the following ribonuclease inhibitor to the DNase solution: 1000 U/mL placental ribonuclease inhibitor (e.g., RNasin) plus 1 mM DTT.

3.3.2. Reverse Transcriptase Reaction

Next, one wishes to make DNA copies of the targeted RNA sequence so that the signal can be amplified. The following are typical cocktails for the reverse-transciptase reaction.

1. If using AMVRT or MMLVRT enzyme:
   a. 10X Reaction buffer *(see below)*; 2.0 μL.
   b. 10 mM dATP; 2.0 μL.
   c. 10 mM dCTP; 2.0 μL.
   d. 10 mM dGTP; 2.0 μL.
   e. 10 mM dTTP; 2.0 μL.
   f. RNasin at 40 U/μL (Promega)*; 0.5 μL.
   g. 20 μM downstream primer; 1.0 μL.
   h. AMVRT 20 U/μL; 0.5 μL.
   i. DEPC-water; 8.0 μL.
   (10X Reaction buffer: 100 mM Tris-HCl, pH 8.3, 500 mM KCl, 15 mM MgCl$_2$.)
2. If using Superscript II enzyme from BRL:
   a. 5X Reaction buffer (as supplied with enzyme); 4.0 μL.
   b. 10 mM dATP; 2.0 μL.
   c. 10 mM dCTP; 2.0 μL.
   d. 10 mM dGTP; 2.0 μL.
   e. 10 mM dTTP; 2.0 μL.
   f. RNasin at 40 U/μL (Promega)*; 0.5 μL.
   g. 20 μM downstream primer; 1.0 μL.
   h. Superscript II, 200 U/μL; 0.5 μL.

   i. 0.1M dTT; 1.2 μL.
   j. DEPC-water; 4.8 μL.
3.
   a. Add 10 μL of cocktail to each well. Carefully place the coverslip on top of the slide.
   b. Incubate at 42°C or 37°C for 1 h in a humidified atmosphere.
   c. Incubate slides at 92°C for 2 min.
   d. Remove coverslip and wash twice with distilled water. Proceed with the amplification procedure, which is the same for both DNA- and RNA-based protocols.
   *RNAsin inhibits ribosomal RNases—use for optimal yields.

3.4. Comments on RT Enzymes

Avian myoblastosis virus reverse transcriptase (AMVRT) and Moloney murine leukemia virus reverse transcriptase (MMLVRT) give comparable results in our laboratory. Other RT enzymes will probably work also. However, it is important to read the manufacturer's descriptions of the RT enzyme and to make certain that the proper buffer is used.

An alternative RT enzyme that lacks RNase activity is available. Called "Superscript II," it is available from BRL Lifesciences (Gaithersburg, MD), and it is suitable for reverse transcription of long mRNAs. It is also suitable for routine RT amplification, and in our laboratory it has proven to be more efficient than the two enzymes described above.

3.4.1. Comments on Primers and Target Sequences

In our laboratory, we simply use antisense downstream primers for our gene-of-interest, as we already know the sequence of most genes we study. However, one can alternatively use oligo (dT) primers to first convert all mRNA populations into cDNA, and then perform the *in situ* amplification for a specific cDNA. This technique may be useful when one is performing amplification of several different gene transcripts at the same time in a single cell. For example, if one is attempting to detect various cytokine expressions, one can use an oligo (dT) primer to reverse transcribe all of the mRNA copies in a cell or tissue section. Then, one can amplify more than one type of cytokine and detect the various types with different probes which develop into different colors (*see* Section 3.9.2.).

In all RT reactions, it is advantageous to reverse-transcribe only relatively small fragments of mRNA (<1500 bp). Larger fragments may not completely reverse-transcribe due to the presence of secondary structures. Furthermore, the RT enzymes—AMVRT and MMLVRT, at least—are not very efficient in transcribing large mRNA fragments. However, this size restriction does not apply to amplification reactions that are exclusively DNA, for the polymerase enzyme copies nucleotides better. In *in situ* DNA reactions, we routinely amplify genes up to 300 bp.

The following are several additional points one should keep in mind:

1. The length for both sense and antisense primers should be 14–22 bp.
2. At the 3' ends, primers should contain a GC-type base pairs (e.g., GG, CC, GC, or CG) to facilitate complementary strand formation.
3. The preferred GC content of the primers is from 45–55%.
4. Try to design primers so they do not form intra- or interstrand base pairs. Furthermore, the 3'-ends should not be complementary to each other, or they will form primer dimers.
5. One can design an RT-primer so that it does not contain secondary structures.

3.4.2. Comments on Annealing Temperatures

Annealing temperatures for reverse transcription and for DNA amplification can be chosen according to the following formula:

$$T_m \text{ of the primers} = 81.5°C + 16.6 (\text{Log } M) + 0.41 (G + C\%) - 500/n$$

In this equation, n = length of primers; and M = molarity of the salt in the buffer, usually $0.047M$ for DNA reactions and $0.070M$ for RT reactions (*see below*).

If using AMVRT, the value will be lower according to the following formula:

$$T_m \text{ of the primers} = 62.3°C + 0.41 (G + C\%) - 500/n$$

Usually, primer annealing is optimal at 2°C above its T_m. However, this formula provides only an approximate temperature for annealing, since base-stacking, near-neighbor effect, and buffering capacity may play a significant role for a particular primer.

Optimization of the annealing temperature should be carried out first with solution-based reactions. It is important to know the optimal temperature before attempting to conduct *in situ* amplification, as *in situ* reactions are simply not as robust a solution-based ones. We hypothesize that this is due to the fact that primers do not have easy access to DNA templates inside cells and tissues, as numerous membranes, folds, and other small structures can keep primers from binding homologous sites as readily as they do in solution-based reactions.

There are two additional ways to determine the real annealing temperatures:

1. To utilize a recently developed thermocycler designed for determination of actual annealing temperature called Robocycler from Stratagene (La Jolla, CA).
2. Another method is to utilize so called Touchdown PCR *(22)*.

The logic of determining the correct annealing temperature for IS-PCR is presented below. During amplification, spurious products often appear in addition to those desired. Therefore, even if the cells do not contain DNA homologous to the primer sequences, many artifactual bands may appear. Many protocols have appeared in the literature to overcome false priming, including HOT-Start *(63,67,75)*, use of DMSO, formamide, and anti-*Taq* antibodies (*75;* i.e., Clonetech's *Taq*Start). An important thing to remember is that false priming will occur if melting temperature *(T_m)* between primer and template is not accurate. Therefore, temperature above the T_m will yield no products, and temperature too far below the T_m often will give unwanted products due to false priming. Therefore determination of optimal annealing temperature is extremely important.

Recently, M. J. Research (Watertown, MA) has devised a thermocycler with the capacity to perform both *in situ* gene amplification in slides and in solution (tubes) simultaneously, in the same block. That kind of thermocyclers can be very useful in determining the optimal amplification of a gene of interest.

3.5. Amplification Protocol, All Types

Prepare an amplification cocktail containing the following: 1.25 μ*M* of each primer, 200 μ*M* (each) dNTP, 10 m*M* Tris-HCl (pH 8.3), 50 m*M* KCl, 2.5 m*M* MgCl$_2$, 0.001% gelatin, and 0.1 U/μL *Taq* polymerase. The following is a convenient recipe that we use in our laboratory to amplify HIV-1 gag gene.

1. 25 μ*M* Forward primer (i.e., SK 38 for HIV-1); 5.0 μL.
2. 25 μ*M* Reverse primer (i.e., SK 39 for HIV-1); 5.0 μL.
3. 10 m*M* each dNTP; 2.5 μL.
4. 1.0*M* Tris-HCl pH 8.3; 1.0 μL.
5. 1.0*M* KCl; 5.0 μL.
6. 100 m*M* MgCl$_2$; 2.5 μL.
7. 0.01% Gelatin; 10.0 μL.
8. *Taq* pol (Ampli-*Taq* 5 U/μL)*; 2.0 μL.
9. H$_2$O; 67.0 μL.

*Note: Other thermostable polymerase enzymes have also been used quite successfully.

1. Layer 8 μL of amplification solution onto each well with a P$_{20}$ micropipet so that the whole surface of the well is covered with the solution. If using a single-well slide for a tissue section, add 12–20 μL of the solution to the well. Be careful—do not touch the surface of the slide with the tip of the pipet.
2. Add a glass coverslip (20 × 60 mm) and carefully seal the edge of the coverslip to the slide with clear nail polish or varnish. If using tissue sections, use a second slide instead of a cover slip (*see* Section 3.6.).
3. Place slides on a heat block at 92°C for 90 s.
4. Place slides in a thermocycling instrument.

5. Run 30 cycles of the following amplification protocol: 94°C, 30 s; 45°C, 1 min; and 72°C, 1 min. These times/temperatures will likely require optimization for the specific thermocycler being used. Furthermore, the annealing temperature should be optimized, as described earlier. These particular incubation parameters work well with SK38 and SK39 primers for the HIV-1 *gag* sequence, when amplified in an MJ Research PTC-100-60 or PTC-100-16MS thermal cycler.

6. After the thermal cycling is complete, dip slides in 100% EtOH for at least 5 min, in order to dissolve the nail polish. Pry off the coverslip using a razor or other fine blade—the coverslip generally pops off quite easily. Scratch off any remaining nail polish on the outer edges of the slide so that fresh coverslips will lay evenly in the subsequent hybridization/detection steps.

7. Place slides on a heat block at 92°C for 1 min—this treatment helps immobilize the intracellular signals.

8. Wash slides with 2X SSC (*see* Section 2., step 9) at room temperature for 5 min.

The amplification protocol is now complete and one can proceed to the labeling/hybridization procedures.

3.6. Comments on Attaching Coverslip/Top Slide

Be certain to carefully paint the polish around the entire periphery of the cover slip or the edges of the dual slide, as the polish must completely seal the coverslip-slide assembly in order to form a small reaction chamber that can contain the water vapor during thermal cycling. For effective sealing, do not use colored polish or any other nail polish which is especially runny—our laboratory prefers to use "Wet & Wild Clear" nail polish. Proper sealing is very important, for this keeps reaction concentrations consistent through the thermal cycling procedure, and concentrations are critical to proper amplification. However, be certain to apply the nail polish very carefully so that none of the polish gets into the actual chamber where the cells or tissues reside. If any nail polish does enter the chamber, discard that slide since the results will be questionable. Please bear in mind that the painting of nail polish is truly a learned skill; therefore, it is strongly recommended that researchers practice this procedure several times with mock slides before attempting an experiment.

In the case of tissue sections, it is best to use another identical blank slide for the cover instead of a coverslip. Apply the amplification cocktail to the appropriate well of the blank slide, place an inverted tissue-containing slide atop the blank slide, and seal the edges as described. Invert the slide once again so that the tissue-containing slide is on the bottom. This technique can be modified to accommodate a hot start (*see* Section 3.6.1.).

3.6.1. Comments on Hot Start Technique

There is much debate as to whether a hot start helps to improve the specificity and sensitivity of amplification reactions. In our laboratory, we find the hot start adds no advantage in this regard; rather, it adds only technical difficulty to the practice of the *in situ* technique. However, recently, a variation of the "hot start" has been reported *(42,75)*. In this procedure, one simply uses anti-*Taq* antibody in the PCR cocktail (containing *Taq*), which keeps the *Taq* enzyme in the cocktail "blocked" until the first cycle of 92°C in which anti-*Taq* antibody gets denatured and restores the full *Taq* activity. This modification essentially serves the same function as the hot start procedure, but without its difficulties *(75)*.

3.6.2. Comments on Thermal Cyclers

Various technologies of thermocycler will work in this application; however, some instruments work much better than others. In our laboratory, we use two types: a standard, block-type thermocycler that normally holds 60 0.5-mL tubes but which can be adapted with aluminum foil, paper towels, and a weight to hold 4–6 slides. We also use dedicated thermocyclers that are specifically designed to hold 12 or 16 slides. We understand that other labs have used stirred-air, oven-type thermocyclers quite successfully; however, we have also heard that there are sometimes problems with the cracking of glass slides during cycling. Thermocyclers dedicated to glass slides are now available from several vendors, including Barnstead Thermolyne of Iowa, Coy Corporation of Minnesota, Hybaid of England, and MJ Research of Massachusetts. Our laboratory has used an MJ Research PTC-100-12MS and a PTC-100-16MS quite successfully. Recently, this company has combined the slide and tubes into a single block, allowing the simultaneous confirmation of *in situ* amplification in a tube. Furthermore, there are newer designs of thermalcyclers which incorporate humidification chambers. However, we do not yet have sufficient experience with this technology to verify whether they can eliminate the need for sealing the slides with nail polish during thermal cycling. Nonetheless, the humidified instruments are especially useful in the reverse transcription and hybridization steps, in which otherwise a humidified incubator is needed.

We suggest that you follow the manufacturer's instructions on the use of your own thermocycler, bearing in mind the following points:

1. Glass does not easily make good thermal contact with the surface on which it rests. Therefore, a weight to press down the slides and/or a thin layer of mineral oil to fill in the interstices will help thermal conduction. If using mineral oil, make certain that the oil is well smeared over the glass surface so that the slide is not merely floating on air bubbles beneath it.
2. The top surfaces of slides lose heat quite rapidly through radiation and convection; therefore, use a thermocycler to envelope the slide in an enclosed chamber (as in some dedicated instruments), or insulate the tops of the slides in some manner. Insulation is particularly critical when using a weight on top of the slides,

for the weight can serve as an unwanted heat sink if it is in direct contact with the slides.

3. Good thermal uniformity is imperative for good results—poor uniformity or irregular thermal change can result in cracked slides, uneven amplification, or completely failed reactions. If adapting a thermocycler that normally holds plastic tubes, use a layer of aluminum foil to spread out the heat.

3.7. One-Step RT-Amplification

If one uses RT enzymes to manufacture cDNA that is subsequently amplified by *Taq* or Vent polymerase, one must use two different buffer systems—one solution for the reverse transcription and another for the DNA amplification. However, one can use a single, recombinant enzyme—rTth—which can do both jobs at once. The typical cocktail for this single-step reaction is the following:

1. 100 μM Forward primer; 0.5 μL.
2. 100 μM Reverse primer; 0.5 μL.
3. 3 mM Nucleotide mix (dNTP); 6.0 μL.
4. 10 mM MnCl$_2$; 2.0 μL.
5. 25 mM MgCl$_2$; 10.0 μL.
6. 10X Transcription buffer (below); 2.0 μL.
7. 10X Chelating buffer (below); 8.0 μL.
8. 1.7 mg/mL BSA; 10.0 μL.
9. 2.5 U/mL rTth enzyme; 2.0 μL.
10. DEPC-treated water; 59.0 μL.

10X transcription buffer: 100 mM Tris-HCl pH 8.3, 900 mM KCl. 10X chelating buffer: 100 mM Tris-HCl pH 8.3, 1M KCl, 7.5 mM EGTA, 0.5% Tween™ 20, 50% (v/v) glycerol.

This reaction requires a slightly variant thermal cycling profile. Our laboratory uses the following amplification protocol: 70°C, 15 min; 92°C, 3 min; 70°C, 15 min; 92°C, 3 min; and 70°C, 15 min.

Then, 29 cycles of the following profile: 93°C, 1 min; 53°C, 1 min; and 72°C, 1 min.

3.8. Direct Incorporation of Nonradioactive Labeled Nucleotides

Several nonradioactive labeled nucleotides are available from various sources (i.e., dCTP-Biotin, digoxin II-dUTP, etc.). These nucleotides can be used to directly label amplification products, and then, the proper secondary agents and chromogens can be used to detect the directly-labeled *in situ* amplification products (*see* Section 2., steps 1 and 12). However, in our opinion—as well as in the opinions of several other laboratory groups—the greatest specificity is *only* achieved by conducting amplification followed by subsequent *in situ* hybridization *(1–16,20,42,44,45,48,57,97,103)*. In the direct labeling protocols, nonspecific incorporation can be significant, and even if this incorporation is minor, it still leads to false-positive signals similar to nonspecific bands in gel electrophoresis following solution-based DNA- or RT-amplification. Therefore, we strongly discourage the direct incorporation of labeled nucleotides as part of an *in situ* amplification protocol.

The only exception to this recommendation is when one is screening a large number of primer pairs for optimization of a specific assay. Then, direct incorporation may be useful. To perform such screenings, add to the amplification cocktail detailed earlier the following: 4.3 μ*M* labeled nucleotide—either 14-biotin dCTP or 14-biotin dATP or 11-digoxigenin dUTP—along with cold nucleotide to achieve a 0.14 m*M* final concentration. Also, if one has worked out the perfect annealing system, either using Robocycler or equivalent system, then one can use direct incorporation without fear of nonspecific labeling, which we have discussed elsewhere in detail *(8)*.

3.9. Special Application of In Situ *Amplification*

3.9.1. In Situ *Amplification and Immunohistochemistry*

Immunohistochemistry and *in situ* amplification can be performed simultaneously in a single cell. For this purpose, we first fix cells or frozen sections of tissue, which are already placed on slides, with 100% methanol for 10 min. Then, slides are washed in PBS. After that, labeling of surface antigen(s) can be carried out by standard immunohistochemistry method (i.e., FITC-conjugated antibody is incubated for 1 h at 37°C, washed, and then cells or tissue section are fixed in 4% paraformaldehyde for 2 h). In various pathology laboratories, many specific surface antigens have been tabulated which can withstand 10% formalin and other routine histopathology procedures and will still bind specific MAbs. If one is using any of these immunohistochemistry panels, then one can routinely use prepared paraffin sections for the detection of cellular antigens. Then, the tissue is prepared for *in situ* amplification, as described earlier.

Following the development of the color of the amplified product in the posthybridization step, one can view the cells or tissue under UV/visible light in an alternate fashion, to detect two signals in a single cell.

3.9.2. Multiple Signals, Multiple Labels in Individual Cells

DNA, mRNA, and protein can all be detected simultaneously in individual cells. As described in the above section, one can label proteins by FITC-labeled antibodies. Then, one can perform both RNA and DNA *in situ* amplification in the cells. If one is using primers for spliced mRNA and if these primers are not going to bind any sequences in DNA, then both DNA and RT amplification can be carried out simultaneously, of course one still needs to perform RT-step but this time without pre-DNAse treatment. In Fig. 1 , we have illustrated how we design such primers. Subsequently, products can be labeled with different kinds of probes, resulting in different colors of signal. For example, proteins can have a FITC-labeled probe, mRNA can show a Rhodamine signal (Rhodamine-conjugated probe, 20 colors are available), and DNA can be labeled with a biotin-peroxidase probe. Each will show a different signal within an individual cell.

3.9.3. In Situ *Amplification on Chromosomal Bands*

Our laboratory has successfully amplified an HIV-1 sequence on chromosomal bands prepared from SUP-T1 infected cell lines. We have modified the chromosomal banding procedure so that it can be use for *tu in situ* gene amplification. The basic principles are the following: carry out chromosomal banding on the specially-designed slides, cover the naked chromosomes with cell ghost membranes so that the signal will remain localized, use fewer cycles of amplification *(10–12)* so the signal will not leak out, and eliminate the heat fixation step . Currently, we are using this technique for peripheral blood mononuclear cells from HIV-1 infected individuals. The following is a detailed procedure.

3.10. Precautions

Sterile technique must be practiced at all times in this procedure. This is necessary in handling cell cultures, both to protect the investigator and to avoid introducing microbial contamination of the cell culture system. Such contamination is often the cause of test failure.

The human peripheral blood used in this procedure may be infectious or hazardous to the investigator. Proper handling, decontamination, and disposal of waste material must be emphasized.

3.10.1. Initial Set-Up

1. Using two culture tubes filled with 5 mL of RPMI media (*see* Section 2.10.), add 0.5 mL of well-mixed whole blood. Rinse pipet three or four times to expel all of the whole blood into the culture tube. (Use three culture tubes for blood from newborns.)
2. With a 1-mL syringe, slowly add 0.1 mL phytohemagglutinin (PHA-C, *see* Section 2., step 10). Gently vortex or invert to ensure complete mixing.
3. Place cultures in a tray with a slight slope upwards towards cap (about an 18° angle). Loosen cap to allow CO_2 penetration.
4. Incubate 66 h at 37°C with 4.5% CO_2 and 90% humidity (as in a water-jacketed incubator).

3.10.2. Arresting Cells at Mitotic Metaphase

1. After 66 h of incubation, remove from incubator and gently resuspend to insure a homogenous mixture.
2. Add 0.1 mL of working velban (*see* Section 2., step 10) to each tube.
3. Mix each culture by gently swirling. Return to CO_2 incubator for 45 min. At this same time, place hypotonic solution sufficient for all cultures in the 37°C incubator to prewarm.

3.10.3. Harvesting

1. After 45 min in velban, centrifuge the culture tubes for 10 min at 800*g*.
2. Aspirate and discard the supernatant leaving 0.25–0.50 mL of liquid on top of the packed cells.
3. Resuspend cells by mixing on a vortex at the lowest setting.
4. *Slowly* add 5–10 mL of prewarmed hypotonic to each tube while vortexing.
5. Gently invert tubes and place in a 37°C water bath for 45–50 min.
6. Centrifuge for 15 min at 800*g*.
7. Aspirate and *save* supernatant, leaving 0.25–0.50 mL of liquid on packed cells.
8. Resuspend cells in remaining supernatant by gently mixing on vortex mixer at the lowest setting.
9. Using a Pasteur pipet with a rubber bulb while continuing to vortex, add the fixative solution (*see* Section 2., step 3) *very slowly* to bring the total volume to 10 mL in each tube.
10. Allow to stand at room temperature for a minimum of 25 min.
11. Repeat steps 6–9, for a total of three changes of fixative.
12. After the third fixation, spin at 800*g*, aspirate supernatant leaving 0.10–0.20 mL on layer of fixed white blood cells.
13. Add 0.10–1.0 mL of fresh fixative to suspension; the amount depends on the density of cell button.

3.10.4. Slide Preparation

1. Resuspend button by bubbling with a fresh Pasteur pipet. Be careful not to draw any liquid into wide portion of pipet—this would make recovery of the mitoses very difficult.
2. Prepare single-well silanized slides, as described earlier. Dip slides in DEPC distilled water. Freeze for 1/2 h before use.
3. Place wet, prechilled slide onto bench while dropping 3–5 drops of specimen 4–5 ft (this may take practice) directly onto the slide. Allow slides to air dry.
4. Add 2–3 drops of supernatant saved in step 7 which contains the red-blood-cell ghost membranes. Allow to air dry once again.
6. Cure slides on a hot plate at 59–60°C for 30–48 h.
7. Heat fix at 105°C for 5–10 s, then fix in 2% PFA for 1–2 h. Next, treat with proteinase K (6 μg/mL) for 3–5 min. Inactivate the proteinase K by placing slides on 95°C heat block for 2 min. Wash in DEPC H_2O2. Air dry.

8. Perform *in situ* amplification as described above with one modification—use only 15–18 cycles instead of 30 cycles.

3.11. In Situ *Hybridization*

Prepare a solution containing: 20 pg/mL of the appropriate probe, 50% deionized formamide, 2X SSC buffer, 10X Denhardts solution, 0.1% sonicated salmon sperm DNA, and 0.1% SDS. The following is a convenient recipe:

1. Probe (^{33}P-labeled, biotinylated, or digoxigenin), 2 μL;
2. Deionized formamide, 50 μL;
3. 20X SSC*, 10 μL;
4. 50X Denhardts solution, 20 μL;
5. 10 mg/mL ssDNA*, 10 μL;
6. 10% SDS, 1 μL; and
7. H$_2$O, 7 μL.

*Note: *See* Section 2., step 8 for preparation of 20X SSC buffer; the salmon sperm should be denatured at 94°C for 10 min before it is added to the hybridization buffer.

Note: 2% BSA can be added if one is observing nonspecific binding. For this purpose, one can add 10 μL of 20% BSA solution and reduce the amount of water.

1. Add 10 μL of hybridization mixture to each well and add coverslips.
2. Heat slides on a block at 95°C for 5 min.
3. Incubate slides at 48°C for 2–4 h in a humidified atmosphere.

Note: The optimal hybridization temperature is a function of the Tm (melting temperature) of the probe. This must be calculated for each probe, as described previously. However, the hybridization temperatures used should not be too high. If that circumstance occurs, then the formula for the hybridization solution should be modified and instead of 50% formamide, 40% formamide should be substituted.

3.11.1. Posthybridization Procedure for 33P Probe

1. Wash slides in 2X SSC for 5 min.
2. Dip slides in 3X nuclear tract emulsion (Kodak NBT-2, diluted 1:1 with water). Note: Steps 3–5 should be carried out in the dark:
3. Slides are air dried, then incubated for 3–10 d in light proof box with a drying agent.
4. Slides are developed for 3 min in Kodak D-19 developer, then rinsed in H$_2$O.
5. Slides fixed for 3 min in Kodak Unifax.
6. Slides are counterstained with May Grunewald Giemsa.

3.11.2. Posthybridization for Peroxidase-Based Color Development

1. Wash slides in 1X PBS twice for 5 min each time.
2. Add 10 µL of streptavidin-peroxidase complex (100 µg/mL in PBS pH 7.2). Gently apply the coverslips.
3. Incubate slides at 37°C for 1 h.
4. Remove coverslip, wash slides with 1X PBS twice for 5 min each time.
5. Add to each well 100 µL of 3'-amino-9-ethylene carbazole in the presence of 0.03% hydrogen peroxide in 50 mM acetate buffer (pH 5.0).
6. Incubate slides at 37°C for 10 min to develop the color—this step should be carried out in the dark. After this period, observe slides under a microscope. If color is not strong, develop for another 10 min.
7. Rinse slides with tap water and allow to dry.
8. Add 1 drop of 50% glycerol in PBS and apply the coverslips.
9. Analyze with optical microscope—positive cells will be stained a brownish red.

3.11.3. Posthybridization for Alkaline-Phosphatase-Based Color Development

1. After hybridization, remove coverslip, wash the slides with two soakings in 2X SSC at room temperature for 15 min.
2. Cover each well with 100 µL of blocking solution *(see below)*, place the slides flat in a humid chamber at room temperature for 15 min.
3. Prepare a working conjugate solution by mixing 10 µL of streptavidin-alkaline phosphatase conjugate (40 µg/mL stock) with 90 µL of conjugate dilution buffer, for each well.
4. Remove the blocking solution from each slide by touching absorbent paper to the edge of the slide.
5. Cover each well with 100 µL of working conjugate solution and incubate in the humid chamber at room temperature for fifteen min. Do not allow the tissue to dry after adding the conjugate.
6. Wash slides by soaking in buffer A for 15 min at room temperature two times.
7. Wash slides once in alkaline substrate buffer at room temperature for 5 min.
8. Prewarm 50 mL of alkaline-substrate buffer to 37°C in a Coplin jar. Just before adding the slides, add 200 µL NBT and 166 µL of BCIP. Mix well.
9. Incubate slides in the NBT/BCIP solution at 37°C until the desired level of signal is achieved (usually from 10 min to 2 h). Check the color development periodically by removing a slide from the NBT/BCIP solution. Be careful not to allow the tissue to dry.
10. Stop the color development by rinsing the slides in several changes of deionized water. The tissue may now be counterstained.

Blocking solution: 50 mg/mL BSA (protein) in 100 mM Tris-HCl (pH 7.8), 150 mM NaCl, and 0.2 mg/mL sodium azide.

Conjugate dilution buffer: 100 mM Tris-HCl, 150 mM MgCl$_2$, 10 mg/mL bovine serum albumin, and 0.2 mg/mL sodium azide.

Buffer A: 100 mM Tris-HCl (pH 7.5), 150 mM NaCl.
Alkaline substrate buffer: 100 mM Tris-HCl (pH 9.5), 150 mM NaCl and 50 mM MgCl$_2$.
NBT (Nitro-blue-tetrazolium): 75 mg/mL NBT in 70% (v/v) dimethly-formamide.
BCIP (4-bromo-5-chloro-3-indolylphosphate): 50 mg/mL in 100% dimethyl-formamide.

3.11.4. Counterstaining and Mounting

1. If you are using red indicator color (like AEC) then use Gill's hematoxalin (Sigma) and if using alkaline phasphotase as indicator color then use nuclear fast red stain as counter stain (stain for 5 min at room temperature).
2. Rinse in several exchanges of tap water.
3. Dehydrate the sections through graded ethanol series (50, 70, 90, 100% v/v for 1 min each).
4. Air dry at room temperature.
5. For permanent mounting, a water-based medium such as CrystalMount™ or GelMount™ or an organic solvent-based medium such as Permount (Fisher Scientific) can be used.
6. Apply one drop of mounting medium per 22-mm coverslip.
7. The slides may be viewed immediately, if you are careful not to disrupt the coverslip. The mounting medium will dry after sitting overnight at room temperature.

4. Validation and Controls

The validity of *in situ* amplification-hybridization should be examined in every run. Attention here is especially necessary in laboratories first using the technique, because occasional technical pitfalls lie on the path to mastery. In an experienced laboratory, it is still necessary to continuously validate the procedure and to confirm the efficiency of amplification. To do this, we routinely run two or three sets of experiments in multi-welled slides simultaneously, for we must not only validate amplification, but we must also confirm the subsequent hybridization/detection steps as well.

In our lab, we frequently work with HIV. A common validation procedure we will conduct is to mix HIV-1 infected cells plus HIV-1 uninfected cells in a known proportion (i.e., 1:10, 1:100, etc.), then we confirm that the results are appropriately proportionate. To examine the efficiency of amplification, we use a cell line which carries a single copy or two copies of cloned HIV-1 virus, then look to see that proper amplification and hybridization has occurred.

In all amplification procedures, we use one slide as a control for nonspecific binding of the probe. Here we hybridize the amplified cells with an unrelated probe. We also use HLA-DQa probes and primers with human peripheral blood mononuclear cells (PBMC) as positive controls, to check various parameters of our system.

In case one is using tissue sections, a cell suspension lacking the gene of interest can be used as a control. These cells can be added on top of the tissue section and then retrieved after the amplification procedure. The cell suspension can then be analyzed with the specific probe to see if the signal from the tissue leaked out and entered the cells floating above.

We suggest that researchers carefully design and employ appropriate positive and negative controls for their specific experiments. In the case of RT-*in situ* amplification, one can use beta-actin, HLA-DQa, and other endogenous-abundant RNAs as the positive markers. Of course, one should always have a RT-negative control for RT *in situ* amplification, as well as DNAse and non-DNAse controls. Controls without *Taq* polymerase plus primers and without primers should always be included.

References

1. Bagasra, O. (1990) Polymerase chain reaction in situ. *Amplifications* 20–21.
2. Bagasra, O., Hauptman, S. P., Lischner, H. W., Sachs, M., and Pomerantz, R. J. (199) Detection of HIV-1 provirus in mononuclear cells by in situ PCR. *N. Engl. J. Med.* **326,** 1385–1391.
3. Bagasra, O., Seshamma, T., and Pomerantz, R. J. (1993) Polymerase chain reaction in situ: intracellular amplification and detection of HIV-1 proviral DNA and other specific genes. *J. Immunol. Methods* **158,** 131–145.
4. Bagasra, O. and Pomerantz, R. J. (1993) HIV-1 provirus is demonstrated in peripheral blood monocytes in vivo: a study utilizing an in situ PCR. *AIDS Res. Hum. Retroviruses* **9,** 69–76.
5. Bagasra, O., Seshamma, T., Oakes, J., and Pomerantz, R. J. (1993) Frequency of cells positive for HIV-1 sequences assessed by in situ polymerase chain reaction. *AIDS* **7,** 82–86.
6. Bagasra, O., Seshamma, T., Oakes, J., and Pomerantz, R. J. (1993) High percentages of CD4-positive lymphocytes harbor the HIV-1 provirus in the blood of certain infected individuals. *AIDS* **7,** 1419–1425.
7. Bagasra O., Qureshi, M. N., Joshi, B., Hewlett, I., Barr, C. E., Henrad, D. (1994) High prevalence of HIV DNA and RNA and localization of HIV-proviral DNA in oral mucosal epithelial cells in saliva from HIV (+) subjects in 1994. Annual Meeting of the Candian Academy of Pathology, 742.
8. Bagasra, O.,T. Seshamma, R. J. Pomerantz. (1993) In situ PCR: a powerful new methodology, in *In Situ Hybridization and Neurology*, Oxford University Press, New York, pp. 143–156.
9. Bagasra, O. and Pomerantz, R. J. (1994) In situ PCR: applications in the pathogenesis of diseases. *Cell Vision* **1,** 13–16.

10. Bagasra, O., Farzadegan, H., Seshamma, T., Oakes, J., Saah, A., and Pomerantz, R. J. (1994) Human immunodeficiency virus type 1 infection of sperm in vivo. *AIDS* **8,** 1669–1674.

11. Bagasra, O. and Pomerantz, R. J. (1994) In situ polymerase chain reaction and HIV-1, in Clinics of North America (Pomerantz, R. J., ed.), W. B. Saunders, Philadelphia, pp. 351–366.

12. Bagasra, O. and Pomerantz, R. J. (1995) Detection of HIV-1 in the brain tissue of individuals who died from AIDS, in *PCR in Neuroscience* (Sarkar, G., ed.), Academic, San Diego, pp. 339–357.

13. Bagasra, O., Seshamma, T., Pastanar, J. P., and Pomerantz, R. (1994) Detection of HIV-1 gene sequences in the brain tissues by in situ polymerase chain reaction, in *Technical Advances in AIDS Research in the Nervous System* (Majors, E., and Levy, J. A., eds.), Plenum, New York, pp. 339–357.

14. Bagasra, O. and Seshamma, T. (1994) *In Situ PCR: Applications in the Pathogenesis of Diseases—A Practical Manual.* (Gu, J., ed.), Eaton, MA, pp. 35–69.

15. Bagasra, O., Seshamma, T., Hansen, J., and Pomerantz, R. (1994) In situ PCR: A manual, in *Current Protocols in Molecular Biology* (Ausubel, F., Brent, R., Kingston, R. E., Moore, D. D., Seidman, J. G., Smitto, J. A., and Steuhl, K., eds.), Wiley, New York, 14.8–14.8.24.

16. Bagasra, O., Seshamma, T., Hansen, J. Bobroski, L., Pomerantz, R. J. (1994) Applications of in situ PCR methods in molecular biology: I. ISPCR protocol for general use. *Cell Vision* **1,** 324–336.

17. Cartun, R. W., Siles, J. F., Li, L. M., Berman, M. M., and Nuovo, G. J. (1994) Detection of hepatitis C virus infection in hepatectomy specimens using immunohistochemistry with reverse transcriptase (RT) in situ polymerase chain reaction (PCR) confirmation. *Cell Vision* **1,** 84.

18. Cary, S. C., Waren, W., Anderson, E., and Giovannoni, S. J. (1993) Identification and localization of bacterial endosymbionts in hydrothermal vent taxa with symbiont-specific polymerase chain reaction amplification and in situ hybridization techniques. *Mol. Marine Biol. Biotechnol.* **2,** 51–62.

19. Chen, R. H. and Fuggle, S. V. (1993) In situ cDNA polymerase chain reaction: a novel technique for detecting mRNA expression. *Am. J. Path.* **143,** 1527–1533.

20. Chieu, K.-P., Cohen, S. H., Morris, D. W. and Jordan, G. W. (1992) Intracellular amplification of proviral DNA in tissue sections using the polymerase chain reaction. *J. Histochem. Cytochem.* **40,** 333–341.

21. Cirocco, R., Careno, M., Gomez, C., Zucker, K., Esquenazi, V., and Miller, J. (1994) Chimerism demonstrated on a cellular level by in situ PCR. *Cell Vision* **1,** 84.

22. Don R. H., Cox, P. T., Wainwright, B. J., Baker, K., and Matiick, J. S. (1994) Touchdown PCR to circumvent spurious priming during gene amplification. *Nucleic Acids Res.* **19,** 4008–4010.

23. Embleton, M. J., Gorochov, G., Jones, P. T., and Winter, G. (1992) In-cell PCR from mRNA amplifying and linking the rearranged immunoglobulin heavy and light chain V-genes within single cells. *Nucleic Acids Res.* **20,** 3831–3837.

24. Embretson, J., Zupanic, M., Beneke, T., Till, M., Wolinsky, S., Ribas, J. L., Burke, A., and Haase, A. T. (1993) Analysis of human immunodeficiency virus-infected tissues by amplification and in situ hybridization reveals latent and permissive infections at single-cell resolution. *Proc. Natl. Acad. Sci. USA* **90,** 357–361.

25. Embretson, J., Zupancic, M., Ribas, J. L., Burke, A., Racz, P., Tennr-Racz, K., and Haase, A. T. (1993) Massive covert infection of helper T lymphocytes and macrophages by HIV during the incubation period of AIDS. *Nature* **62,** 359–362.

26. Gingeras, T. R., Prodanovich, P., Latimer, T., Guatelli, J. C., Richman, D. D., and Baninger, K. J. (1991) Use of self-sustained sequence replication amplification reaction to analyze and detect mutations in zidovudine-resistant human immunodeficiency virus. *J. Infect. Dis.* **164,** 1066–1074.

27. Gingeras, T. R., Richman, D. D., Kwoh, D. Y., and Guatelli, J. C. (1990) Methodologies for in vitro nucleic acid amplification and their applications. *Vet. Microbiol.* **24,** 235–251.

28. Gingeras, T. R., Whitfield, K. M., and Kwoh, D. Y. (1990) Unique features of the self-sustained sequence replication (3SR) reaction in the in vitro amplification of nucleic acids. *Ann. Biol. Clin. Paris* **4X,** 498–501.

29. Gosden, J. and Hanratty, D. (1993) PCR in situ: a rapid alternative to in situ hybridization for mapping short, low copy number sequences without isotopes. *BioTechniques* **5,** 78–80.

30. Greer, C. E., Lund, J. K., and Manos, M. M. (1991) PCR amplification from paraffin-embedded tissues: recommendations on fixatives for long-term storage and prospective studies. *PCR Method Appl.* **95,** 117–124.

31. Greer, C. E., Peterson, S. L., Kiviat, N. B., and Manos, M. M. (1991) PCR amplification from paraffin-embedded tissues: effects of fixative and fixative times. *Am. J. Clin. Pathol.* 95, 117–124.

32. Gressens P., Langston, C., and Martin, J. R. (1994) In situ PCR localization of herpes simplex virus DNA sequences in disseminated neonatal herpes encephalitis. *J. Neuropathol. Exp. Neurol.* **53,** 469–482.

33. Gressens P. and Martin, J. R. (1994) HSV-2 DNA persistence in astrocytes of the trigeminal root entry zone: double labelling by in situ PCR and immunohistochemistry. *J. Neuropathol. Exp. Neurol.* **53,** 127–135.

34. Guatelli, J. C., K. M. Whitfield, D. Y. Kwoh, K. J. Barringer, D. D. Richman and T. R. Gingeras. (1994) Isothermal, in vitro amplification of nucleic acids by a multienzyme reaction modeled after retroviral replication. *Proc. Natl. Acad. Sci. USA* **7,** 1874–1878.

35. Haase, A. T., Retzel, E. F., and Staskus, K. A. (1990) Amplification and detection of lentiviral DNA inside cells. *Proc. Natl. Acad. Sci. USA* **37,** 4971–4975.

36. Heniford, B. W., Shum-Siu, A., Leonberger, M., and Hendler, F. J. (1993) Variation in cellular EGF receptor mRNA expression demonstrated by in situ reverse transcription polymerase chain reaction. *Nucleic Acids Res.* **21,** 3159–3166.

37. Hiort, O., Klauber, G., Cendron, M., Sinnecker, G. H., Keim, L., Schwinger, E., Wolfe, H. J., and Yandell, D. W. (1994) Molecular characterization of the androgen receptor gene in boys with hypospadias. *Eur. J. Ped.* **153,** 317–321.

38. Hacker, G. W., Zehbe, I., Hauser-Kronberger, C., Gu, J., Graf, A.-H., and Dietze, O. (1994) In situ detection of DNA and mRNA sequences by immunogold-silver staining (IGSS). *Cell Vision* **1**, 30–37.
39. Heniford, B. W., Shum-Siu, A., Leonberger, M., and Hendler, F. J. (1993) Variation in cellular EGF receptor mRNA expression demonstrated by in situ reverse transcriptase polymerase chain reaction. *Nucleic Acids Res.* **21**, 3159–3166.
40. Hsu, T. C., Bagasra, O., Seshamma, T., and Walsh, P. N. (1994) Platelet factor XI mRNA amplified from human platelets by reverse transcriptase polymerase chain reaction and detected by in situ amplification and hybridization. *FASEB J.* **8**, 1375.
41. Li, H. H., Gyllensten, U. B., Cui, X. F., Saiki, R. K., Erlich, H. A., and Arnheim, N. (1988) Amplification and analysis of DNA sequences in single human sperm and diploid cells. *Nature* **335**, 414–417.
42. Isaacson, S. H., Asher, D. M., Gibbs, C. J., and Gajdusek, D. C. (1994) In situ RT-PCR amplification in archival brain tissue. *Cell Vision* **1**, 85.
43. Kawasaki, E., Saiki, R., and Erlich, H. (1989) PCR technology, in *Principles and Applications for DNA Amplification* (Ehrlich, H. A., ed.), Stockton, New York.
44. Komminoth, P., Long, A. A., Ray, R., and Wolfe, H. J. (1992) In situ polymerase chain reaction detection of viral DNA. Single copy genes and gene rearrangements in cell suspensions and cytospins. *Diagn. Mol. Pathol.* **1**, 85–97.
45. Komminoth, P., Merk, F. B., Leav, I., Wolfe, H. J., and Roth, J. (1992) Comparison of 32S and digoxigenin-labeled RNA and oligonucleotide probes for in situ hybridization expression of mRNA of the seminal vesicle secretion protein 11 and androgen receptor genes in the rat prostate. *Histochemistry* **93**, 217–228.
46. Kuwata S. (1992) Application of PCR and RT-PCR method to molecular biology study in nephrology. *Nippon Rinsho* **50**, 2868–2873.
47. Larzul, D., Guigue, F., Sninsky, J. J., Mach, D. H., Brechot, C., and Guesdaon, J. L. (1988) Detection of hepatitis B virus sequences in serum by using in vitro enzymatic amplification. *J. Virol. Methods* **20**, 227–237.
48. Long, A. A., Komminoth, P., Lee, E., and Wolfe, H. J. (1993) Comparison of indirect and direct in situ polymerase chain reaction in cell preparations and tissue sections Detection of vira DNA gene reatrangements and chromosomal translocations. *Histochemistry* **99**, 151–162.
49. Lundberg, K. S., Shoemaker, D. D., Adams, M. W., Short, J. M., Sorge, J. A., and Mathur, E. J. (1991) High fidelity amplification using a thermostable DNA polymerase isolated from *Pyrococcus furiosus. Gene* **1**, 81–86.
50. Mahalingham R, Kido, S., Wellish, M., Cohrs, R., and Gilden, D. H. (1994) In situ polymerase chain reaction detection of varicella zoster virus in infected cells in culture. *J. Virol.* **68**, 7900–7908.
51. Mankowski J. L., Spelman, J. P., Reesetar, H. G., Strandberg, J. D., Laterra, J., Carter, D. L., Clements, J. E., Zink, M. C. (1994) Neurovirulent simian immunodeficiency virus replicates productively in endothelial cells of the central nervous system in vivo and in vitro. *J. Virology* **68**, 8202–8208.

52. Mehta, A., Maggioncalda, J., Bagasra, O., Thikkavarapu, S., Saikumari, P., and Block, T. (1994) Detection of herpes simplex sequences in the trigeminal ganglia of latently infected mice by in situ PCR method. *Cell Vision* **2**, 110–115.

53. Mehta A, Maggioncalda, J., Bagasra, O., Thikkavarapu, S., Saikumari, P., Nigel, F. W., and Block, T. (1994) In situ PCR and RNA hybridization detection of herpes simplex virus sequences in trigeminal ganlia of latently infected mice. *Virology* **206**, 633–640.

54. Mitchell, W. J., Gressens, P., Martin, J. R., and DeSanto, R. (1994) Herpes simplex virus type 1 DNA persistence, progressive disease and transgenic immediate early gene promoter activity in chronic corneal infections in mice. *J. Gen. Virol.* **75**, 1201–1210.

55. Murray, G. I. (1993) In situ PCR. *J. Pathol.* **169**, 187,188.

56. Nuovo, G. J. (1990) Human papillomavirus DNA in genital tract lesions histologically negative for condylomata. Analysis by in situ, Southern blot hybridization and the polymerase chain reaction. *Am. J. Surg. Path.* **14**, 643–651.

57. Nuovo, G. J. (1994) Questioning in situ PCR. In situ cDNA polymerase chain reaction: a novel technique for detecting mRNA expression. *Am. J. Path.* **145**, 741.

58. Nuovo, G. J., Becker, J., Margiotta, M., MacConnell, P., Comite, S., and Hochman, H. (1992) Histological distribution of polymerase chain reaction-amplified human papillomavirus 6 and 11 DNA inpenile lesions. *Am. J. Surg. Path.* **16**, 269–275.

59. Nuovo, G. J., Becker, J., Simsir, A., Margiotta, M., Khalife, G., and Shevchuk, M. (1994) HIV-1 nucleic acids localize to the spermatogonia and their progeny. A study by polymerase chain reaction in situ hybridization. *Am. J. Path.* **144**, 1142–1148.

60. Nuovo, G. J., Darfler, M. M., Impraim, C. C., and Bromley, S. E. (1991) Occurrence of multiple types of human papillomavirus in genital tract lesions. Analysis by in situ hybridization and the polymerase chain reaction. *Am. J. Path.* **138**, 53–58.

61. Nuovo, G. J., Delvenne, P., MacConnell, P., Chalas, E., Neto, C., and Mann, M. J. (1991) Correlation of histology and detection of human papillomavirus DNA in vulvar cancers. *Gynecol. Oncol.* **43**, 275–280.

62. Nuovo, G. J., Forde, A., MacConnell, P., and Fahrenwald, R. (1993) In situ detection of PCR-amplified HIV-1 nucleic acids and tumor necrosis factor cDNA in cervical tissues. *Am. J. Path.* **143**, 40–48.

63. Nuovo, G. J., Gallery, F., Hom, R., MacConnell, P., and Bloch, W. (1993) Importance of different variables for enhancing in situ detection of PCR-amplified DNA. PCR *Methods Applications* **2**, 305–312.

64. Nuovo, G. J., Gallery, F., and MacConnell, F. (1992) Detection of amplified HPV 6 and 11 DNA in vulvar lesions by hot start PCR in situ hybridization. *Mod. Path.* 5, 444–448.

65. Nuovo, G. J., Gallery, F., MacConnell, F., and Braun, A. (1994) In situ detection of polymerase chain reaction-amplified HIV-1 nucleic acids and tumor necrosis factor-alpha RNA in the central nervous system. *Am. J. Path.* **144**, 659–666.

66. Nuovo, G. J., Gallery, F., MacConnell, P., Becker, J., and Bloch, W. (1991) An improved technique for the in situ detection of DNA after polymerase chain reaction amplification. *Am. J. Path.* **139,** 1239–1244.

67. Nuovo, G. J., Gorgone, G. A., MacConnell, P., Margiotta, M., and Gorevic, P. D. (1992) In situ localization of PCR-amplified human and viral cDNA. *PCR Methods Applications* **2,** 117–123.

68. Nuovo, G. J., Hocman, H. A., Eliezri, Y. D., Lastarria, D., Comite, S. L., and Silvers, D. N. (1990) Detection of human papillomavirus DNA in penile lesions histologically negative for condylomata. Analysis by in situ hybridization and the polymerase chain reaction. *Am. J. Surg. Path.* **14,** 829–836.

69. Nuovo, G. J., Lidonnici, K., MacConnell, P., and Lane, B. (1993) Intracellular localization of polymerase chain reaction (PCR)-amplified hepatitis C cDNA. *Am. J. Surg. Path.* **17,** 683–690.

70. Nuovo, G. J., MacConnell, P, Forde, A., and Delvenne, P. (1991) Detection of human papillomavirus DNA in formalin-fixed tissues by in situ hybridization after amplification by polymerase chain reaction. *J. Path.* **139,** 847–854.

71. Nuovo, G. J., Margiotta, M., MacConnell, P., and Becker, J. (1992) Rapid in situ detection RT-PCR-amplified HIV-1 DNA. *Diag. Mol. Path.* **1,** 98–102.

72. Nuovo, G. J., Moritz, J., Walsh, L. L., MacConnell, P., and Koulos, J. (1992) Predictive value of human papillomavirus DNA detection by filter hybridization and polymerase chain reaction in women with negative results of colposcopic examination. *Am. J. Clin. Path.* **98,** 489–492.

73. Nuovo, M., Nuovo, G. J., Becker, J., Gallery, F., Delvenne, P., and Kane, P. B. (1993) Correlation of viral infection, histology, and mortality in immuno-compromised patients with pneumonia. Analysis by in situ hybridization and the polymerase chain reaction. *Diag. Mol. Path.* **2,** 0–9.

74. Nuovo, M., Nuovo, G. J., MacConnell, P., Forde, A., and Steiner, G. C. (1992) In situ analysis of Paget's disease of bone for measles-specific PCR-amplified cDNA. *Diag. Mol. Path.* **1,** 256–265.

75. Nuovo, G. J. (1994) *PCR in Situ Hybridization Protocols and Applications,* 2nd ed., Raven, NewYork.

76. O'Leary, J. J., Browne, G., Landers, R. J., Crowley, M., Healy, I. B., Street, J. T., Pollock, A. M., Murphy, J., Johnson, M. I., and Lewis, F. A. (1994) The importance of fixation procedures on DNA template and its suitability for solution-phase polymerase chain reaction and PCR and in situ hybridization. *Histochem. J.* **26,** 337–346.

77. Patterson, B. K., Till, M., Otto, P., Goolsby, C., Furtado, M. R., McBride, L. J., and Woolinsky, S. M. (1993) Detection of HIV-I DNA and messenger RNA in individual cells by PCR-driven in situ hybridization and flow cytometry *Science* **260,** 976–979.

78. Pereira, R. F., Halford, K. W., O'Hara, M. D., Leeper, D. P., Sokolov, B. P., Pollard, M. D., Bagasra, O., and Prockop, D. P. (1995) Cultured stromal cells from marrow serve as stem cells for bone, lung and cartilage in irradiated mice. *Proc. Natl. Acad. Sci. USA* **92,** 4857–4861.

79. Pestaner, J. P., Bibbo, M., Bobroski, L., Seshamma, T., and Bagasra, O. (1994) Potential of in situ polymerase chain reaction in diagnostic cytology. *Acta Cytologia* **38**, 676–680.

80. Pestaner, J. P., Bibbo, M., Bobroski, L., Seshamma, T., and Bagasra, O. (1994) Case report: surfactant protein A mRNA expression utilizing the reverse transcription in situ polymerase chain reaction for metastatic adenocarcinoma. *Cell Vision* **1**, 324–336.

81. Qureshi, M. N., Barr, C. E., Seshamma, T., Pomerantz, R. J., and Bagasra, O. (1994) Localization of HIV-1 proviral DNA in oral mucosal epithelial cells. *J. Inf. Dis.* **171**, 190–193.

82. Qureshi, M. N., Bagasra, O., Joshi, B., Howlett, I., Barr, C. E., and Henrad, D. (1994) High prevalence of HIV DNA and RNA and localization of HIV-provirus DNA in oral mucosal epithelial cells in saliva from HIV+ subjects. *Lab. Invest.* **70**, 127A (#742).

83. Ray, R., Komminoth, P., Machado, M., and Wolfe, H. J. (1991) Combined polymerase chain reaction and in inl hybridization for the detection of single copy genes and viral gnomic sequences in intact cells. *Mod. Pathol.* **4**, 124A.

84. Saito, H., Nishikawa, A., Gu, J., Ihara, Y., Soejima, H., Wada, Y., Sekiya, C., Niikawa, N., and Taniguchi, N. (1994) cDNA cloning and chromosomal mapping of human N-acetylglucosaminyl transferase V+. *Biochem. Biophys. Res. Comm.* **198**, 318–327.

85. Sallstrom, J. F., Zehbe, I., Alemi, M., and Wilander, E. (1993) Pitfalls of in situ polymerase chain reaction (PCR) using direct incorporation of labelled nucleotides. *Anticancer Res.* **13**, 1153.

86. Schwartz, D., Sharma, U., Busch, M., Weinhold, K., Lieberman, J., Birx, D., Farzedagen, H., Margolick, J., Quinn, T., Davis, B., Leitman, S., Bagasra, O., Pomerantz, R. J., and Viscidi, R. (1994) Absence of recoverable infectious virus and unique immune responses in an asymptomatic HIV + long term survivor. *AIDS Res. Hum. Retroviruses* **10**, 1703–1711.

87. Spann, W., Pachmann, K., Zabnienska, H., Pielmeier, A., and Emmerich, B. (1991) In situ amplification of single copy gene segments in individual cells by the polymerase chain reaction. *Infection* **19**, 242–244.

88. Staskus, K. A., Couch, I., Bitterman, P., Retzel, E. F., Zupancic, M., List, J., and Haase, A. T. (1991) Amplification of visna virus DNA in tissue sections reveals a reservoir of latently infected cells. *Microb. Pathog.* **11**, 67–76.

89. Stork, P., Loda, M., Bosari, S., Wiley, B., Poppenhusen, K., and Wolfe, H. J. (1992) Detection of K-ras mutations in pancreatic and hepatic neoplasms by non-isotopic mismatched polymerase chain reaction. *Oncogene* **6**, 857–862.

90. Staecker, H., Cammer, M., Rubenstein, R., and Van De Water, T. (1994) A procedure for RT-PCR amplification of mRNAs on histological specimens. *BioTechniques* **16**, 76–80.

91. Staskus, K. A., Couch, L., Bitterman P., Retzel, E. F., Zupancic, M., List, J., and Haase, A. T. (1991) In situ amplification of visna virus DNA in tissue sections: a reservoir of latently infected cells. *Microbial Pathogenesis* **11**, 67–76.

92. Sukpanichnant, S., Vnencak-Jones, C. L., and McCurley, T. L. (1993) Detection of clonal immunoglobulin in heavy chain gene rearrangements by polymerase chain reaction in scrapings from archival hematoxylin and eosin-stained histologic sections: implications for molecular genetic studies of focal pathologic lesions. *Diagn. Mol. Pathol.* **2,** 168–176.

93. Tsongalis, G. J., McPhail, A. H., Lodge-Rigal, R. D., Chapman, J. F., and Silverman, L. M. (1994) Localized in situ amplification (LISA): a novel approach to in situ PCR. *Clin. Chem.* **40,** 381–384.

94. Walter, M. J., Lehky, T. J., Fox, C. H., and Jacobson, S. (1994) In situ PCR for the detection of HTLV-1 in HAM/TSP patients. *Ann. NY Acad. Sci.* **724,** 404–413.

95. Walboomers, J. M. M., Melchers, W. J. G., Mullink, H., Meijer, C. L. M., Struyk, A., Quint, W. G. J., van der Noorda, J., and ter Schegget, J. (1988) Sensitivity of in situ detection with biotinylated probes of human papillomavirus type 16 DNA in frozen tissue sections of squamous cell carcinoma of the cervix. *Am. J. Pathol.* **139,** 587–594.

96. Winslow, B. J., Pomerantz, R. J., Bagasra, O., and Trono, D. (1993) HIV-1 Latency due to the site of proviral integration. *Virology* **196,** 849–854.

97. Wolfe, H. J., Ross, D., and Wolfe, B. (1990) Detection of infectious agents by molecular methods at the cellular level. *Verhandlungen der Deutschen Gesellschaft fur Pathologie* **74,** 295–300,.

98. Yap, E. P. H. and McGee, J. O'D. (1991) Slide PCR: DNA amplification from cell samples on microscopic glass slides. *Nucleic Acids Res.* **19,** 15.

99. Yin, J., Kaplitt, M. G., and Pfaff, D. W. (1994) In situ PCR and in vivo detection of foreign gene expression in rat brain. *Cell Vision* **1,** 58–59.

100. Zehbe, I., Hacker, G. W., Rylander, E., Sallstrom, J., and Wilander, E. (1992) Detection of single HPV copies in SiHa cells by in situ polymerase chain reaction (in situ PCR) combined with immunoperoxidase and immunogold-silver staining (IGSS) techniques. *Anticancer Res.* **12,** 2165–2168.

101. Zehbe, I., Hacker, G. W., Sallstrom, J. F., Muss, W. H., HauserKronberger, C., Rylander, E., and Wilander, E. (1994) Polymerase chain reaction in situ hybridization (PISH) and in situ self-sustained sequence replication-based amplification (in situ 3SR). *Cell Vision* **1,** 46–47.

102. Zehbe, I., Hacker, G. W., Sallstrom, J. F., Rylander, E., and Wilander, E. (1994) Self sustained sequence replication-based amplification (3SR) for the in situ detection of mRNA in cultured cells *Cell Vision* **1,** 20–24.

103. Zebhe, I., Sallstrom, J. F., Hacker, G. W., Hauser-Kronberger, C., Rylander, E., and Wilander, E. (1994) Indirect and direct in situ polymerase chain reaction for the detection of human papillomavirus An evaluation of two methods and a double staining technique *Cell Vision* **2,** 163–168.

104. Zevallos, E., Bard, E., Anderson, V., and Gu, J. (1994) An in situ (ISPCR) study of HIV-I infection of lymphoid tissues and peripheral lymphocytes. Proceedings 2nd International Workshop. Modern Methods in Analytic Morphology. *Cell Vision* **1,** 87.

105. Zevallos, E., Bard, E., Anderson, V., and Gu, J. (1994) Detection of HIV-I gag sequences in placentas of HIV positive mothers by in situ polymerase chain reaction. *Cell Vision* **2,** 116–121.

24

Expression and Detection
of Green Fluorescent Protein (GFP)

Steven R. Kain and Paul Kitts

1. Introduction

Light is produced from several bioluminescent species through the action of green fluorescent proteins (GFPs) (reviewed in ref. *1*). One such GFP from the jellyfish *Aequorea victoria* fluoresces following the transfer of energy from the Ca^{2+}-activated photoprotein aequorin *(2,3)*. This energy transfer mechanism is radiationless, and proceeds via the interaction of these two proteins in photogenic cells located at the base of the jellyfish. Expression of GFP in vivo does not alter the spectral properties of the protein *(4)*. Full-length GFP appears to be required for fluorescence, however the chromophore responsible for light absorption is located within a hexapeptide at positions 64–69 *(5)*. This region of GFP contains a Ser^{65}-dehydro Tyr^{66}-Gly^{67} cyclic tripeptide which functions as the minimal chromophore.

The cloning and sequencing of the GFP gene from *A. victoria (2,6)* has permitted expression of this protein in heterologous systems. The GFP expressed in both prokaryotic and eukaryotic cells yields green fluorescence when excited by blue or UV light. The GFP fluorescence does not require additional gene products or other factors, and occurs in an apparent species-independent fashion. Studies over the past 1–2 yr have demonstrated the utility of GFP as an in vivo reporter of gene expression in a wide variety of experimental settings (for previous reviews, *see* refs. *7–9*). As examples, this reporter system has been used to address experimental questions such as neuronal development in nematodes *(4)*, protein trafficking *(10,11)*, nuclear transport *(12)*, bacterial and viral infectivity *(13,14)*, and selection of transgenic organisms *(15–17)*. In many cases, GFP is not expressed alone, but rather used to tag other proteins by expression of N- and/or C-terminal GFP chimeras. Such fusion proteins have

From: *Methods in Molecular Biology, vol. 63: Recombinant Protein Protocols:
Detection and Isolation* Edited by: R. Tuan Humana Press Inc., Totowa, NJ

Table 1
Proteins Expressed as Fusions to GFP

Protein	GFP terminus	Host cell or organism	Localization	Refs.
YopE cytotoxin	N	Yersinia		*39*
Histone H2B	N	Yeast	Nucleus	*12,41*
Tubulin	N/C	Yeast	Microtubules	*9*
Nuf2	N	Yeast	Spindle-pole body	*73*
Mitochondrial matrix targeting signal	?	Yeast	Mitochondria	*74*
Nucleoplasmin	C	Yeast	Nucleus	*33*
p93dis1	N	Fission yeast	Spindle-pole body and microtubules	*42*
Exu	N/C	*Drosophila*	Oocytes	*25*
ObTMV movement protein	?	Tobacco		*60*
MAP4	N	Mammalian cells	Microtubules	*48*
Mitochondrial targeting signal	N	Mammalian cells	Mitochondria	*75*
Cyclins	N/C	Mammalian cells	Nucleus, micro-tubules or vesicles	*11*
Chromogranin B	N	HeLa cells	Secreted	*10*
HIV p17 protein	N	HeLa cells	Nucleus	*76*
NMDAR1	N	HEK293 cells	Membrane	*47*
CENP-B	?	Human cells	Nucleus	*77*

been shown to maintain the fluorescence properties of native GFP, as well as the biological activity and subcellular localization patterns of the fusion partner (*see* Table 1).

1.1. Properties of GFP and GFP Variants

1.1.1. The GFP Chromophore

The GFP chromophore consists of a cyclic tripeptide derived from Ser-Tyr-Gly in the primary protein sequence *(5)* and is only fluorescent when embedded within the complete GFP protein. There have been no reports of a truncated GFP (other than a few amino acids from the C-terminus) that is still fluorescent. It may be possible that the whole protein is required to catalyze the formation of the chromophore and to provide the proper environment for the chromophore to fluoresce. Nascent GFP is not fluorescent, since chromophore formation occurs posttranslationally. The chromophore is formed by a cyclization reaction and an oxidation step that requires molecular oxygen

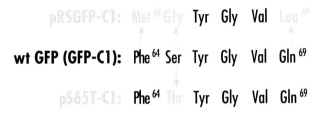

pRSGFP-C1: Met⁵⁴ Gly Tyr Gly Val Leu⁶⁸

wt GFP (GFP-C1): Phe⁶⁴ Ser Tyr Gly Val Gln⁶⁹

pS65T-C1: Phe⁶⁴ Thr Tyr Gly Val Gln⁶⁹

Fig. 1. The mutations that create the GFP-S65T and RSGFP mutants are shown next to wt GFP sequence. RSGFP was originally described as RSGFP4 *(21)*.

(18,19). These steps are either autocatalytic or use factors that are ubiquitous, since fluorescent GFP can be formed in a broad range of organisms. Chromophore formation may be the rate-limiting step in generating the fluorescent protein, especially if oxygen is limiting *(18,19)*. The wt GFP absorbs UV and blue light with a maximum peak of absorbance at 395 nm and a minor peak at 470 nm and emits green light maximally at 509 nm, with a shoulder at 540 nm *(20)*.

1.1.2. GFP Variants

Several variants of *Aequorea* GFP have been described that have altered fluorescence excitation and/or emission spectra. In many cases, the altered spectral properties provide advantages over the wild-type GFP for certain applications. Using a combinatorial mutagenic strategy, a variant of GFP referred to as RSGFP4 was isolated that has a single fluorescence excitation maximum at 490 nm, yet retains an emission maximum similar to wild-type GFP at 505 nm *(21,22)*. A mutant with similar spectral properties was constructed by mutagenesis of Ser[65] in the GFP chromophore to Thr[65] *(23)*. The fluorescence intensity from each of these GFP mutants is 4–6 times greater than wild-type (wt) GFP when excited with light of 450–500 nm owing to the greater amplitude of the excitation peaks. In addition, GFP-S65T acquires fluorescence about four times faster than wt GFP *(23)*. Therefore, this variant should be useful for the analysis of rapid gene induction. (Data on the rate of formation of an active fluorophore are not yet available for RSGFP.) The specific mutations are shown in Fig. 1. The red-shifted excitation spectra of these mutants allow them to be used in combination with wt GFP in double-labeling experiments *(21,22)*. Although the peak positions in the emission spectra of all three proteins are virtually identical, double-labeling can be performed by selective excitation of wt and red-shifted GFP *(14,22)*. Additional variants of GFP include a Tyr[66] to His[66] mutant that produces a blue fluorescent signal *(19)*, and a mutant with a single peak of excitation at 395 nm *(19,24)*.

1.1.3. Acquisition and Stability of GFP Fluorescence

1.1.3.1. PHOTOBLEACHING

The GFP fluorescence is very stable in a fluorometer (Ward, W. W., personal communication). Even under the higher intensity illumination of a fluorescence microscope, GFP is more resistant to photobleaching than is fluorescein *(25)*. The fluorescence of wt GFP, GFP-S65T, and RSGFP is quite stable when illuminated with 450–490 nm light (the major excitation peak for GFP-S65T and RSGFP, but the minor peak for wt GFP). Some photobleaching occurs when wt GFP is illuminated near its major excitation peak with 340–390 nm or 395–440 nm light *(4)*. The rate of photobleaching is less with lower energy lamps such as QTH or mercury lamps. High-energy xenon lamps should be avoided, as these may cause rapid photodestruction of the GFP chromophore. The rate of photobleaching varies with the organism being studied; GFP fluorescence is quite stable in *Drosophila (25)* and zebrafish. In *C. elegans*, 10 mM NaN$_3$ accelerates photobleaching *(4)*.

1.1.3.2. STABILITY TO OXIDATION/REDUCTION

The GFP needs to be in an oxidized state to fluoresce; strong reducing agents, such as 5 mM Na$_2$S$_2$O$_4$ or 2 mM FeSO$_4$, convert GFP into a non-fluorescent form, but fluorescence is fully recovered after exposure to atmospheric oxygen *(26)*. Weaker reducing agents, such as 2% β-mercaptoethanol, 10 mM dithiothreitol (DTT), 10 mM reduced glutathione, or 10 mM L-cysteine, do not affect the fluorescence of GFP *(26)*. GFP fluorescence is not affected by moderate oxidizing agents (Ward, personal communication).

1.1.3.3. STABILITY TO CHEMICAL REAGENTS

The GFP fluorescence is retained in mild denaturants, such as 1% SDS or 8M urea, after fixation with glutaraldehyde or formaldehyde, but fully denatured GFP is not fluorescent. The GFP is very sensitive to some nail polishes used to seal coverslips *(4,25)*; therefore, use molten agarose or rubber cement to seal coverslips on microscope slides. Fluorescence is also quenched by the nematode anesthetic phenoxypropanol *(4)*. The GFP fluorescence is irreversibly destroyed by 1% H$_2$O$_2$ and sulfhydryl reagents such as 1 mM DTNB (5,5'-dithio-bis-[2-nitrobenzoic acid]) *(26)*. Fluorescence survives pH 7–12, but intensity decreases at pH 5.5–7.0 *(27)*. Many organic solvents can be used at moderate concentrations without abolishing fluorescence; however, the absorption maximum may shift *(28)*.

GFP dimerizes via hydrophobic interactions at protein concentrations above 5–10 mg/mL and high salt concentrations with a fourfold reduction in the absorption at 470 nm (Ward, personal communication). Dimer formation is not required for fluorescence, and monomeric GFP is the form of the reporter expressed in most model systems.

1.1.3.4. PROTEIN STABILITY: IN VITRO

GFP is exceptionally resistant to heat (T_m = 70°C), alkaline pH, detergents, chaotropic salts, organic solvents, and most common proteases, except pronase *(27–30)*. Some GFP fluorescence can be observed when nanogram amounts of protein are resolved on native or 1% SDS polyacrylamide gels *(2)*.

Fluorescence is lost if GFP is denatured by high temperature, extremes of pH, or guanidinium chloride but can be partially recovered if the protein is allowed to renature *(27,31)*. A thiol compound may be necessary to renature the protein into the fluorescent form *(32)*.

1.1.3.5. PROTEIN STABILITY: IN VIVO

The GFP appears to be stable when expressed in various organisms. However, no measurement of the half-life of GFP has been reported.

1.1.3.6. TEMPERATURE SENSITIVITY OF GFP CHROMOPHORE FORMATION

Although fluorescent GFP is highly thermostable *(30)*, it appears that the formation of the GFP chromophore is temperature sensitive. In yeast, GFP fluorescence was strongest when the cells were grown at 15°C, decreasing to about 25% of this value as the incubation temperature was raised to 37°C (33). However, GFP and GFP-fusions synthesized in *S. cerevisiae*, at 23°C retain fluorescence despite a later shift to 35°C *(33)*. It has also been noted that *E. coli* expressing GFP show stronger fluorescence when grown at 24°C or 30°C as compared to 37°C (*19*; Ward, personal communication). Mammalian cells expressing GFP have also been seen to exhibit stronger fluorescence when grown at 33°C as compared to 37°C *(11)*.

1.1.4. Sensitivity

1.1.4.1. FLUORESCENCE MICROSCOPY

The GFP, like fluorescein, has a quantum yield of about 80% *(30)*, although the extinction coefficient for GFP is much lower. Wild-type *Aequorea* GFP excited with fluorescein filter sets is approx 10-fold less intense than the same number of molecules of free fluorescein. Nevertheless, in fluorescence microscopy, GFP fusion proteins have been found to give greater sensitivity and resolution than staining with fluorescently labeled antibody *(25)*. The GFP fusions have the advantages of being more resistant to photobleaching and of avoiding background caused by nonspecific binding of the primary and secondary antibodies to targets other than the antigen *(25)*. Although binding of multiple antibody molecules to a single target offers a potential amplification not available for GFP, this is offset because neither labeling of the antibody nor binding of the antibody to the target is 100% efficient. For some applications, the sensitivity of GFP may be limited by autofluorescence or limited penetration of light.

1.1.4.2. As a Quantitative Reporter

The signal from GFP does not have any enzymatic amplification; hence, the sensitivity of GFP will probably be lower than that for reporter proteins that are enzymes, such as β-galactosidase, SEAP, and firefly luciferase. Experiments are currently in progress to determine the minimal amount of GFP that is detectable by fluorescence in transfected mammalian cells and *E. coli*. Purified GFP can be quantified in a fluorometer-based assay in the low nanogram range.

2. Materials
2.1. Reagents for Detection
2.1.1. Fluorescence-Activated Cell Sorting (FACS) Analysis

1. 1X Trypsin/EDTA.
2. PBS/0.5 μg/mL propidium iodide (PI).

2.1.2. Fluorescence Microscopy

1. 95% Ethanol.
2. PBS (pH 7.4).
3. PBS/4% paraformaldehyde (pH 7.4–7.6). Add 4 g of paraformaldehyde to 100 mL of PBS. Heat to dissolve. Store at 4°C.
4. Rubber cement or molten agarose.

2.1.3. SDS PAGE

1. Sample buffer (pH 6.8): 62.5 mM Tris (0.76 g), 2% SDS (w/v) (2.0 g), 10% glycerol (v/v) (10 mL), and 0.1% bromophenol blue (w/v) (0.1 g). Add Tris, SDS, and glycerol to 75 mL ddH$_2$O. Adjust pH to 6.8 with dilute HCl. Add the bromophenol blue, and adjust to a final volume of 100 mL with ddH$_2$O. Store buffer at room temperature.
2. 1.0M Dithiothreitol: Add 1.54 g of dithiothreitol to 8 mL of ddH$_2$O and mix gently until dissolved. Adjust final volume to 10 mL with ddH$_2$O. Aliquot into microcentrifuge tubes and freeze at –20°C.

2.2. Equipment Required for Detection

1. Fluorescence-Activated Cell Sorting (FACS) Machine.
2. Microscope.
3. Filters: Chroma Technology Corporation has developed several filter sets designed for use with GFP; they claim the High Q FITC filter set (#41001) produces the best signal-to-noise ratio for visual work, and the High Q GFP set

(#41014) produces the strongest absolute signal, but with some background. We have also used a Zeiss filter set with a 450–490 nm band-pass excitation filter, 510 nm dichromic reflector and 520–750 nm long-pass emission filter, and the Chroma filter set 31001. Other filter sets may give better performance, and it is necessary to match the filter set to the application (*see* Notes 1 and 2).

2.3. Recombinant GFP (rGFP)

Recombinant green fluorescent protein, a 27-kDa monomer with 238 amino acids purified from transformed *E. coli* is available commercially (Clontech, Palo Alto, CA). The excitation and fluorescence emission spectra for the recombinant protein is identical to GFP purified from *Aequorea victoria (4)*. The purified protein retains its fluorescence capability under many harsh conditions and is suitable as a control reagent for GFP expression studies, or for microinjection as a marker or tracer.

2.4. GFP-Specific Antibodies

A GFP-specific monoclonal and polyclonal antibody is also commercially available (Clontech). Each antibody recognizes rGFP, GFP purified from *Aequorea victoria*, and recombinant wt GFP, GFP-S65T, and RSGFP. The polyclonal antibody is from antiserum raised in rabbits against rGFP isolated from transformed *E. coli*, and then purified by a Protein A column. The MAb is produced from a mouse hybridoma cell line via conventional methods. These antibodies can by used for Western blots and immunoprecipitations, or for *in situ* detection of GFP and GFP chimeras.

3. Methods
3.1. Expression of GFP
3.1.1. Suitable Host Organisms and Cells

Fluorescence has been observed from GFP expressed in the following hosts:

3.1.1.1. Microbes

1. Anabaena *(34)*.
2. Dictyostelium *(35)*.
3. *Escherichia coli (2,4,36–38)*.
4. *Yersinia (39)*.
5. *Saccharomyces cerevisiae (9,12,33,40,41)*.
6. *Schizosaccharomyces pombe (42)*.

3.1.1.2. Invertebrates

1. *Caenorhabditis elegans (4,43,44)*.
2. *Drosophila melanogaster (17,25,45,46)*.
3. Leech (personal communication).

3.1.1.3. VERTEBRATES

1. 293 cells (transformed primary embryonal kidney, human) *(47)*.
2. BHK cells (hamster) *(48)*.
3. CHO-K1 cells (ovary, Chinese hamster) *(14,22,37,48)*.
4. COS-7 cells (kidney, SV40 transformed, African green monkey) *(11,48)*.
5. Ferret neurons *(49)*.
6. GH3 cells (pituitary tumor, rat) *(50)*.
7. HeLa cells (epitheloid carcinoma, cervix, human) *(10,11)*.
8. JEG cells (placenta, human) *(51)*.
9. NIH/3T3 cells (embryo, contact-inhibited, NIH Swiss mouse) *(11,48)*.
10. Myeloma cells (personal communication)
11. PC-12 cells (adrenal pheochromocytoma, rat) (personal communication).
12. Pt K1 cells (kidney, kangaroo rat) *(48)*.
13. Mice *(16)*.
14. Xenopus *(52,53)*.
15. Zebrafish *(15,54)*.

3.1.1.4. PLANTS

1. *Arabidopsis thaliana* (thale cress) (stable) *(55)*.
2. *Citrus sinensis* (L.) Osbeck cv. Hamlin (an embryogenic sweet-orange cell line, H89) *(56)*.
3. *Zea mays* (maize) *(57,58)*.
4. *Nicotiana benthamiana* (tobacco) *(13)*.
5. *Nicotiana clevelandii* (tobacco) *(13)*.
6. *Hordeum vulgar* (barley) *(59)*.
7. *Glycine max* (soybean) *(50)*.
8. Tobacco *(57,59,60)*.

GFP has been expressed in a diverse range of organisms, but GFP may not be an appropriate reporter for some applications (*see* Notes 3–7).

3.1.2. Expression of GFP Fusion Proteins

GFP has been expressed as a fusion to many different proteins. In many cases, chimeric genes encoding either N- or C-terminal fusions to GFP retain the normal biological activity of the heterologous partner, as well as maintaining fluorescent properties similar to native GFP *(9,12,25,47)*. Table 1 indicates the variety of proteins expressed as fusions to GFP. The use of GFP in this capacity provides a fluorescent tag on the protein, which allows for in vivo localization of the fusion protein. The use of GFP fusions can provide enhanced sensitivity and resolution in comparison to standard antibody staining techniques *(25)*, and the GFP tag eliminates the need for fixation, cell permeabilization, and antibody incubation steps normally required when using antibodies tagged with chemical fluorophores. Lastly, use of the GFP tag permits kinetic studies of protein localization and trafficking due to the resistance of the GFP chromophore to photobleaching *(12,25)*.

3.1.3. GFP Expression in Mammalian Cells

Appropriate vectors may be transfected into mammalian cells by a variety of techniques. These methods include calcium phosphate *(61)*, DEAE-dextran *(62)*, various liposome-based transfection procedures *(62)*, and electroporation *(62)*. The efficiency of the transfection procedure is primarily dependent on the host cell line being transfected. Different cell lines may vary by several orders of magnitude in their ability to take up and express exogenous DNA. Moreover, a method that works well for one host cell line may be inferior for another. Therefore, when working with a cell line for the first time, it is advisable to compare the efficiencies of several transfection protocols. This can best be accomplished using vectors, such as those shown in Fig. 2, which have the CMV immediate early promoter for high-level expression in most cell lines. A promoterless vector (such as pGFP-1 available from Clontech) may be used to determine the background autofluorescence in the host cell line.

After a method of transfection is chosen, it must be optimized for parameters such as cell density, the amount and purity of DNA, media conditions, transfection time, and the posttransfection interval required for GFP detection. Once optimized, these parameters should be kept constant in order to obtain reproducible results.

The GFP expression may be detected by fluorescence microscopy, fluorescence-activated cell sorting (FACS) analysis, or fluorometer assays 24–72 h posttransfection, depending on the host cell line used. There is one published report of a stable mammalian cell line expressing GFP *(48)*. For visualizing GFP-expressing cells by fluorescence microscopy, grow the cells on a sterile glass coverslip placed in a 60-mm culture plate. Alternatively, an inverted fluorescence microscope may be used for direct observation of fluorescing cells in the culture plate.

3.1.4. GFP Expression in Plants

Using a variety of methods, GFP has been expressed in several plant species (*see* Table 2). The GFP allows plant scientists to study transformed cells and tissues in vivo, without sacrificing vital cultures. The GFP coding sequences contain a region recognized in *Arabidopsis* as a cryptic plant intron (between bases 400 and 483). The intron is efficiently spliced out in *Arabidopsis* resulting in a nonfunctional protein *(55)*. Functional GFP has been transiently expressed in several other plant species, indicating that the cryptic intron may not be recognized or is recognized less efficiently in other plant species. A

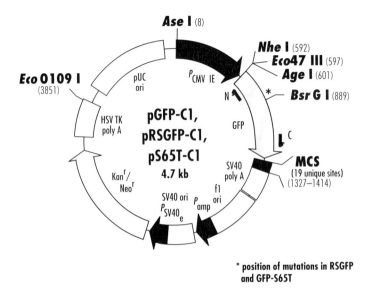

Fig. 2. pGFP-C1, pRSGFP-C1, and pS65T-C1 plasmid map: Each of these vectors can be used to express GFP, or the corresponding GFP variant in mammalian cells. The vectors also permit genes cloned into an MCS at the end of the coding sequences to be expressed as fusions to the C-terminus of GFP or GFP variants. GFP, GFP variants, or fusion proteins are expressed under control of the immediate early promoter of cytomegalovirus ($P_{CMV\ IE}$). An upstream Kozak consensus sequence *(72)* and a downstream SV40 polyadenylation signal direct proper processing of mRNA in eukaryotic cells. The vector backbone also contains an SV40 origin for replication in mammalian cells expressing the SV40 large T antigen. A neomycin-resistance cassette allows for G418 selection of stably transfected eukaryotic cells. A bacterial promoter (P_{amp}) expresses kanamycin resistance in *E. coli*. The backbone also provides a pUC19 origin of replication for propagation in *E. coli*, and an f1 origin for single-stranded DNA production. All sites shown are unique.

modified version of GFP, in which the cryptic sequences have been altered, has been used for GFP expression in *Arabidopsis (55)*.

3.2. GFP Detection

The wt GFP absorbs UV and blue light with a maximum absorbance at 395 nm and a minor peak of absorbance at 470 nm. The GFP-S65T and RSGFP variants have a single excitation peak at approx 490 nm (Fig. 3). The shifted spectra of the variant proteins has advantages for both fluorescence micro–scopy and FACS analysis. The filter sets commonly used with fluorescein isothiocyanate (FITC) and GFP illuminate at 450–500 nm—well above the major excitation peak of wt GFP. These filter sets therefore produce a fluo-

Table 2
Use of GFP in Plants

Species (culture)	Method	Refs.
Arabidopsis thaliana (thale cress roots)	*Agrobacterium* (stable)	*55*
Citrus sinensis (L.) Osbeck cv. Hamlin (embryogenic suspension culture sweet-orange cells H89)	Electroporation	*56*
Zea mays (maize protoplast suspension culture)		*57, 58*
Nicotiana benthamiana (tobacco plant leaves)	PVX infection	*13*
Nicotiana clevelandii (tobacco plant leaves)	PVX infection	*13*
Hordeum vulgar (barley)	Particle bombardment	*59*
Glycine max (soybean suspension culture)	Particle bombardment	*50*
Tobacco	Various	*57,59,60*

rescent signal from wt GFP that is several-fold less than would be obtained with filter sets that excite at 395 nm. However, use of the major peak at 395 nm for excitation of the wt GFP is not advisable owing to rapid photobleaching of the fluorescent signal. For excitation of the wt GFP, we recommend the 470 nm peak to minimize photobleaching. In contrast, the major excitation peak of both GFP-S65T and RSGFP encompasses the excitation wavelength of these commonly used filter sets, so the resulting signal is much brighter. Similarly, the argon ion laser used in most FACS machines emits at 488 nm, so excitation of both GFP protein variants is much more efficient than excitation of wt GFP. In practical terms, this means the detection limits in both microscopy and FACS should be considerably lower with either GFP-S65T or RSGFP. Also, living cells and animals tolerate longer wavelength light better due to the lower energies; therefore, the fluorescent signal obtained with illumination of GFP-S65T and RSGFP at ~490 nm is more stable and less toxic than the fluorescence obtained with wt GFP excited at 395 nm. (*See also* Notes 1, 2, 8, 9, and 10–12.)

3.2.1. Fluorescence-Activated Cell Sorting (FACS)

Several investigators have sorted GFP expressing cells by flow cytometry *(38,63,64)*. In particular, good results have been obtained in FACS analysis with plant protoplasts, yeast, *E. coli*, and mammalian cells infected with bacteria expressing GFP.

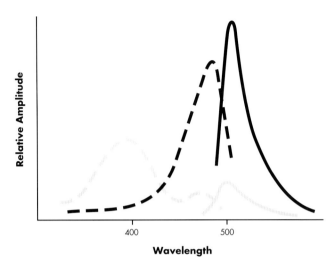

Fig. 3. Excitation (dashed lines) and emission (solid lines) spectra of wt GFP (gray lines) and GFP-S65T (black lines): Although precise data are not available, the excitation and emission spectra of RSGFP appear to be very similar to those of GFP-S65T (*21,* Clontech, unpublished observations). The emission data for wt GFP were obtained with excitation at 475 nm. (The emission peak for wt GFP is several-fold higher when illuminated at 395 nm [the major excitation peak of wt GFP]. However, the filter sets commonly used for GFP fluorescence microscopy and the argon ion lasers used in most FACS machines excite around 450–500 nm). The emission data for GFP-S65T were obtained with excitation at 489 nm, which corresponds closely to the excitation peak of GFP-S65T. (The data in this figure are derived from ref. *23.*)

1. Harvest cells with 1X trypsin/EDTA, 24–72 h posttransfection.
2. Resuspend cells in a small volume of PBS/PI solution.
3. Follow recommended FACS protocol.

3.2.2. Fluorescence Microscopy

In a tissue culture hood:

1. Sterilize glass coverslip.
   a. Place coverslip into 95% ethanol. Shake excess ethanol from coverslip until it is nearly dry.
   b. Flame-sterilize coverslip and allow to completely air-dry in sterile tissue-culture dish.
2. Plate and transfect cells in tissue-culture dish containing coverslip(s).
3. At the end of culture period, remove the tissue culture media and wash cells three times with PBS.

For unfixed cells:

1. Carefully remove the coverslip from the plate with forceps. (Pay close attention to which side of the coverslip contains the cells.)
2. Mount the coverslip onto a glass microscope slide.
   a. Place a tiny drop of PBS on the slide, and allow the coverslip to slowly contact the solution and lie down on the slide.
   b. Carefully aspirate the excess solution around the edge of the coverslip using a Pasteur pipet connected to a vacuum pump.
   c. Attach coverslip to the microscope slides with either molten agarose or rubber cement. Note: Do not use nail polish, as this reagent has been shown to eliminate GFP fluorescence.
   d. Allow to dry for 30 min.
3. Immediately examine slides by fluorescence microscopy, or store up to 2–3 wk at –20°C.

For fixed cells:

1. After cells have been washed with PBS, add 2.0 mL of freshly made PBS/4% paraformaldehyde directly on coverslip.
2. Incubate cells in solution at room temperature for 30 min.
3. Wash cells twice with PBS.
4. Follow steps 1–3 for unfixed cells.

3.3. GFP Controls

3.3.1. Recombinant GFP (rGFP)

Purified GFP can be used as a standard for SDS polyacrylamide *(62,65–67)* or two-dimensional polyacrylamide *(68)* gel electrophoresis, isoelectric focusing *(69)*, Western analysis *(62,65,66,70,71)*, fluorescence microscopy, and microinjection into cells and tissues. When used as a standard for Western blotting applications in conjunction with a GFP antibody, the GFP protein can be used to differentiate problems with detection of GFP fluorescence from expression of GFP protein. Note: Generally, GFP will not fluoresce on a Western blot. Some GFP fluorescence may be observed when the protein is resolved on an SDS-polyacrylamide gel if nanogram quantities of GFP are present.

3.3.1.1. SAMPLE PREPARATION FOR ONE DIMENSIONAL SDS POLYACRYLAMIDE GEL ELECTROPHORESIS

The following procedure is based on the discontinuous polyacrylamide gel system described by Laemmli *(67)*. If an alternate electrophoresis system is employed, the sample preparation should be modified accordingly.

If GFP is added to a total cell/tissue lysate or other crude sample, the amount of total protein loaded per lane must be optimized for the particular application.

1. On a minigel apparatus, 25–75 µg/lane of lysate protein is typically needed for satisfactory separation (i.e., discrete banding throughout the molecular weight range) of a protein mixture derived from a whole cell or tissue homogenate.
2. For use of GFP as an internal standard in electrophoresis experiments, load 500 ng of GFP/lane for gels stained with Coomassie blue staining.
3. For Western blotting applications, load 40–400 pg of GFP/lane for a strong positive signal using either the GFP polyclonal or monoclonal antibody in conjunction with a chemiluminescent detection system *(66)*.
   a. Allow GFP protein to thaw at room temperature. Mix gently until solution is clear, and then place tube on ice.
   b. Dilute the protein solution to the desired final concentration using sample buffer or other suitable buffer for your application. Dilutions are best performed in 1.5-mL microcentrifuge tubes.
   c. For applications requiring the reduction of disulfide bonds (e.g., SDS polyacrylamide reducing gels), add $1.0M$ dithiothreitol to achieve a final concentration of 50 mM (this is a 20-fold dilution of the $1.0M$ stock solution).
   d. Boil the diluted sample for 5 min by placing the tube in a boiling H_2O bath or 100°C heat block.
   e. Load samples on the gel. The final sample volume should be approx 50% of the available space in the well (10–20 µL for a minigel).

3.3.1.2. RECOMBINANT GFP AS A CONTROL FOR FLUORESCENCE MICROSCOPY

rGFP can also be used as a control on microscope slides used in fluorescence microscopy. The purified protein may be used to optimize lamp and filter set conditions for detection of GFP fluorescence, or as a qualitative means to correlate GFP fluorescence with the amount of protein in transfected cells.

1. Unfixed samples: Use this method for live cell fluorescence or other cases where a fixation step is not desired.
   a. Perform 1:10 serial dilutions of the 1.0 mg/mL rGFP stock solution with 10 mM Tris-HCl (pH 8.0) to yield concentrations of 0.1 mg/mL and 0.01 mg/mL. Note: These dilutions should suffice as a positive control. The 1.0 mg/mL solution will give a very bright fluorescent signal by microscopy.
   b. Using a micropipet, spot 1–2 µL of diluted protein onto the microscope slide.
   c. Allow the protein to air-dry for a few seconds, and mark the position of the spot on the other side of the slide to aid in focusing.
   d. Add a coverslip over the spot using a 90% glycerol solution in 100 mM Tris-HCl (pH 7.5).
   e. Fluorescence from the spot is best viewed at low magnifications, using either a 10X or 20X objective.
2. Fixed samples: In some cases it may be necessary to fix the rGFP protein to the microscope slide prior to microscopy. This can be done by dipping the section of the microscope slide, from above, containing the air-dried protein spot into 100% methanol for 1 min. Allow the slide to dry completely and place a coverslip over the sample, as above.

3.3.2. GFP Antibody

The GFP polyclonal and monoclonal antibodies can be used to confirm GFP protein expression by Western blots and to correlate levels of GFP protein expression with fluorescence intensity (detected with microscopy or flow cytometry). In addition to detection of GFP and its variants on Western blots, the GFP antibodies can be used in immunoprecipitation procedures to detect GFP-fusion proteins and *in situ* detection of GFP.

4. Notes

1. Some samples may have a significant background autofluorescence (e.g., worm guts *[4]*). A bandpass emission filter may limit the autofluorescence so that it appears the same color as GFP; using a long-pass emission filter may allow the color of the GFP and autofluorescence to be distinguished. Use of DAPI filters may also allow autofluorescence to be distinguished *(11,46)*.
2. Most autofluorescence in mammalian cells is due to flavin coenzymes (FAD and FMN) that have absorption/emission = 450/515 nm. These values are very similar to those for GFP, so autofluorescence may obscure the GFP signal. The use of DAPI filters may make this autofluorescence appear blue, whereas the GFP signal remains green. In addition, some growth media can cause autofluorescence. When possible, perform microscopy in a clear buffer such as PBS, or medium lacking phenol red. For fixed cells, autofluorescence can also be reduced by washing with 0.1% sodium borohydride in PBS for 30 min after fixation.
3. Variability in the intensity of GFP fluorescence has been noted. This may be due to variability in expression level or in part to the relatively slow formation of the wt GFP chromophore and the requirement for molecular oxygen *(19)*.
4. The slow rate of chromophore formation and the apparent stability of wt GFP may preclude the use of GFP as a reporter to monitor fast changes in promoter activity *(18,19)*. This limitation is reduced by use of GFP-S65T, which acquires fluorescence approximately four times faster than wt GFP *(23)*.
5. Some people have put GFP expression constructs into their system and failed to detect fluorescence. There can be numerous reasons for failure, including use of an inappropriate filter set, expression of GFP below the limit of detection, and failure of GFP to form the chromophore. The rGFP and GFP antibodies may be used as diagnostic tools in these cases.
6. *Escherichia coli* expressing GFP shows stronger fluorescence when grown at 24°C or 30°C compared to 37°C (19; Ward, W. W., personal communication). A similar enhancement of fluorescent signal may be possible by growing yeast expressing GFP at 23°C *(33)*. GFP and GFP-fusions synthesized in S. cerevisiae at 23°C retain brighter fluorescence despite a later shift to higher temperatures (35°C). Hence, incubation at a lower temperature may increase the fluorescence signal.
7. GFP chromophore formation requires molecular oxygen *(18,19)*; therefore, cells must be grown under aerobic conditions.

8. Excite at 470 nm for wt GFP and 490 nm for the GFP-S65T and RSGFP variants. Excitation at the 395 nm peak for wt GFP may result in rapid loss of signal.
9. A tungsten-QTH, mercury, or argon light source is preferable. Xenon lamps are too high in energy and will rapidly destroy the chromophore.
10. In microscopy, the primary issue is the choice of filter sets. In general, conditions used for fluorescein should give some signal, although autofluorescence may be a problem in many cell types/organisms. A simple control for the microscope setup is to spot a small volume of purified rGFP on a microscope slide.
11. Exciting GFP intensely for extended periods may generate free radicals that are toxic to the cell. This problem can be minimized by excitation at 450–490 nm.
12. Fusion of a protein to wt GFP may cause a red-shift in the absorption spectrum. Excitation at longer wavelengths (450–490 nm) may give improved fluorescence.

References

1. Prasher, D. C. (1995) Using GFP to see the light. *Trends Genet.* **11**, 320–323.
2. Inouye, S. and Tsuji, F. I. (1994) Aequorea green fluorescent protein: expression of the gene and fluorescent characteristics of the recombinant protein. *FEBS Lett.* **341**, 277–280.
3. Ward, W. W. (1979) Energy transfer processes in bioluminescence, in *Photochemical and Photobiological Reviews* (Smith, K. C., ed.), Plenum, New York, pp. 1–57.
4. Chalfie, M., Tu, Y., Euskirchen, G., Ward, W. W., and Prasher, D. C. (1994) Green fluorescent protein as a marker for gene expression. *Science* **263**, 802–805.
5. Cody, C. W., Prasher, D. C., Westler, W. M., Prendergast, F. G., and Ward, W. W. (1993) Chemical structure of the hexapeptide chromophore of *Aequorea* green-fluorescent protein. *Biochemistry* **32**, 1212–1218.
6. Prasher, D. C., Eckenrode, V. K., Ward, W. W., Prendergast, F. G., and Cormier, M. J. (1992) Primary structure of the *Aequorea victoria* green fluorescent protein. *Gene* **111**, 229–233.
7. Coxon, A. and Bestor, T. H. (1995) Proteins that glow in green and blue. *Chem. Biol.* **2**, 119–121.
8. Cubitt, A. B., Heim, R., Adams, S. R., Boyd, A. E., Gross, L. A., and Tsien, R. Y. (1995) Understanding, improving and using green fluorescent proteins. *Trends Biochem.* **20**, 448–455.
9. Stearns, T. (1995) The green revolution. *Current Biol.* **5**, 262–264.
10. Kaether, C. and Gerdes, H.-H. (1995) Visualization of protein transport along the secretory pathway using green fluorescent protein. *FEBS Lett.* **369**, 267–271.
11. Pines, J. (1995) GFP in mammalian cells. *Trends Genet.* **11**, 326,327.
12. Flach, J., Bossie, M., Vogel, J., Corbett, A., Jinks, T., Willins, D. A., and Silver, P. A. (1994) A yeast RNA-binding protein shuttles between the nucleus and the cytoplasm. *Mol. Cell. Biol.* **14**, 8399–8407.
13. Baulcombe, D. C., Chapman, S., and Santa Cruz, S. (1995) Jellyfish green fluorescent protein as a reporter for virus infections. *Plant J.* **7**, 1045–1053.

14. Kain, S. R., Adams, M., Kondepudi, A., Yang, T. T., Ward, W. W., and Kitts, P. (1995) The green fluorescent protein as a reporter of gene expression and protein localization. *BioTechniques* **19**, 650–655.

15. Amsterdam, A. and Hopkins, N. (1995) Transient and transgenic expression of green fluorescent protein (GFP) in living zebrafish embryos. CLONTECHniques **X(3)**, 30.

16. Ikawa, M., Kominami, K., Yoshimura, Y., Tanaka, K., Nishimune, Y., and Okabe, M. (1995) Green fluorescent protein as a marker in transgenic mice. *Devel. Growth Differ.* **37**, 455–459.

17. Yeh, E., Gustafson, K., and Boulianne, G. L. (1995) Green fluorescent protein as a vital marker and reporter of gene expression in *Drosophila. Proc. Natl. Acad. Sci. USA* **92**, 7036–7040.

18. Davis, D. F., Ward, W. W., and Cutler, M. W. (1995) Posttranslational chromophore formation in recombinant GFP from *E. coli* requires oxygen. Proceedings of the 8th International Symposium on Bioluminescence and Chemiluminescence (in press).

19. Heim, R., Prasher, D. C., and Tsien, R. Y. (1994) Wavelength mutations and posttranslational autoxidation of green fluorescent protein. *Proc. Natl. Acad. Sci. USA* **91**, 12,501–12,504.

20. Ward, W. W., Cody, C. W., Hart, R. C., and Cormier, M. J. (1980) Spectrophotometric identity of the energy transfer chromophores in *Renilla* and *Aequorea* green-fluorescent proteins. *Photochem. Photobiol.* **31**, 611–615.

21. Delagrave, S., Hawtin, R. E., Silva, C. M., Yang, M. M., and Youvan, D. C. (1995) Red-shifted excitation mutants of the green fluorescent protein. *Bio/Technology* **13**, 151–154.

22. Yang, T. T., Kain, S. R., Kitts, P., Kondepudi, A., Yang, M. M., and Youvan, D. C. (1996) Dual color microscopic imagery of cells expressing the green fluorescent protein and a red-shifted variant. *Gene* (in press).

23. Heim, R., Cubitt, A. B., and Tsien, R. Y. (1995) Improved green fluorescence. *Nature* **373**, 663,664.

24. Ehrig, T., O'Kane, D. J., and Prendergast, F. G. (1995) Green fluorescent protein mutants with altered excitation spectra. *FEBS Lett.* **367**, 163–166.

25. Wang, S. and Hazelrigg, T. (1994) Implications for bcd mRNA localization from spatial distribution of exu protein in Drosophila oogenesis. *Nature* **369**, 400–403.

26. Inouye, S. and Tsuji, F. I. (1994) Evidence for redox forms of the Aequorea green fluorescent protein. *FEBS Lett.* **351**, 211–214.

27. Bokman, S. H., and Ward, W. W. (1981) Renaturation of *Aequorea* green-fluorescent protein. *Biochem. Biophys. Res. Comm.* **101**, 1372–1380.

28. Robart, F. D., and Ward, W. W. (1990) Solvent perturbations of *Aequorea* green fluorescent protein. *Photochem. Photobiol.* **51**, 92s.

29. Roth, A. (1985) Purification and protease susceptibility of the green-fluorescent protein of Aequorea victoria with a note on Halistra ura. Ph.D. thesis, Rutgers University, New Brunswick, NJ.

30. Ward, W. W. (1981) Properties of the Coelenterate green-fluorescent proteins, in *Bioluminescence and Chemiluminescence, Basic Chemistry and Analytical Applications* (DeLuca, M., and McElroy, W. D., ed.), Academic, New York, pp. 235–242.

31. Ward, W. W. and Bokman, S. H. (1982) Reversible denaturation of *Aequorea* green-fluorescent protein, physical separation and characterization of the renatured protein. *Biochemistry* **21,** 4535–4540.

32. Surpin, M. A. and Ward, W. W. (1989) Reversible denaturation of Aequorea green fluorescent protein—thiol requirement. *Photochem. Photobiol.* **49,** WPM-B2.

33. Lim, C. R., Kimata, Y., Oka, M., Nomaguchi, K., and Kohno, K. (1995) Thermosensitivity of green fluorescent protein fluorescence utilized to reveal novel nuclear-like compartments in a mutant nucleoporin Nsp1. *J. Biochem.* **118,** 13–17.

34. Buikema, W. (1995) Oral presentation at Fluorescent Proteins and Applications Meeting, Palo Alto, CA, March 6–7.

35. Hodgkinson, S. (1995) GFP in Dictyostelium. *Trends Genet.* **11,** 327,328.

36. Burlage, R., Yang, Z., and Mehlhorn, T. (1995) Fluorescent microorganisms are detected in bacterial transport experiments. Poster presentation at Fluorescent Proteins and Applications Meeting, Palo Alto, CA, March 6–7.

37. Kitts, P., Adams, M., Kondepudi, A., Gallagher, D., and Kain, S. (1995) Green fluorescent protein (GFP): a novel reporter for monitoring gene expression in living organisms. *CLONTECHniques* **X(1),** 1–3.

38. Yu, J. and van den Engh, G. (1995) Flow-sort and growth of single bacterial cells transformed with cosmid and plasmid vectors that include the gene for green-fluorescent protein as a visible marker. Abstracts of papers presented at the 1995 meeting on "Genome Mapping and Sequencing," Cold Spring Harbor, NY, p. 293.

39. Jacobi, C. A., Roggenkamp, A., Rakin, A., and Heesemann, J. (1995) Yersinia enterocolitica Yop E-GFP hybrid protein expression in vitro and in vivo. Poster presentation at Fluorescent Proteins and Applications Meeting, Palo Alto, CA, March 6–7.

40. Riles, L., Waddle, J., and Johnston, M. (1995) GFP as a reporter of gene expression in yeast. Poster presentation at Fluorescent Proteins and Applications Meeting, Palo Alto, CA, March 6–7.

41. Schlenstedt, G., Saavedra, C., Loeb, J. D. J., Cole, C. N., and Silver, P. A. (1995) The GTP-bound form of the yeast Ran/TC4 homologue blocks nuclear protein import and appearance of poly(a)$^+$ RNA in the cytoplasm. *Proc. Natl. Acad. Sci. USA* **92,** 225–229.

42. Nabeshima, K., Kurooka, H., Takeuchi, M., Kinoshita, K., Nakaseko, Y., and Yanagida, M. (1995) p93dis, which is required for sister chromatid separation, is a novel microtubule and spindle pole body-associating protein phosphorylated at the Cdc2 target sites. *Genes Devel.* **9,** 1572–1585.

43. Sengupta, P., Colbert, H. A., and Bargmann, C. I. (1994) The C. elegans gene odr-7 encodes an olfactory-specific member of the nuclear receptor superfamily. *Cell* **79,** 971–980.

44. Treinin, M. and Chalfie, M. (1995) A mutated acetylcholine receptor subunit causes neuronal degeneration in *C. elegans. Neuron* **14,** 871–877.
45. Barthmaier, P. and Fyrberg, E. (1995) Monitoring development and pathology of Drosophila indirect flight muscles using green fluorescent protein. *Devel. Biol.* **169,** 770–774.
46. Brand, A. (1995) GFP in *Drosophila. Trends Genet.* **11,** 324,325.
47. Marshall, J., Molloy, R., Moss, G. W. J., Howe, J. R., and Hughes, T. E. (1995) The jellyfish green fluorescent protein, a new tool for studying ion channel expression and function. *Neuron* **14,** 211–215.
48. Olson, K. R., McIntosh, J. R., and Olmsted, J. B. (1995) Analysis of MAP4 function in living cells using green fluorescent protein (GFP) chimeras. *J. Cell. Biochem.* **130,** 639–650.
49. Lo, D. C., McAllister, A. K., and Katz, L. C. (1994) Neuronal transfection in brain slices using particle-mediated gene transfer. *Neuron* **13,** 1263–1268.
50. Plautz, J. D., Day, R. N., Dailey, G. M., Welsh, S. B., Hall, J. C., Halpain, S., and Kay, S. A. (1996) Green fluorescent protein and its derivatives as a versatile marker for gene expression in living *Drosophila melanogaster,* plant and mammalian cells. *Gene* **173,** 83–87.
51. Yu, K. L. and Dong, K. W. (1995) Application of luciferase and green fluorescent protein as reporter for analysis of human gonadotropin-releasing hormone gene promoters. Poster presentation at Fluorescent Proteins and Applications Meeting, Palo Alto, CA, March 6–7.
52. Tannahill, D., Bray, S., and Harris, W. A. (1995) A *Drosophila* E(spl) gene is "neurogenic" in *Xenopus.* A green fluorescent protein study. *Devel. Biol.* **168,** 694–697.
53. Wu, G.-I., Zou, D.-J., Koothan, T., and Cline, H. T. (1995) Infection of frog neurons with vaccinia virus permits in vivo expression of foreign proteins. *Neuron* **14,** 681–684.
54. Amsterdam, A., Lin, S., and Hopkins, N. (1995) The Aequorea victoria green fluorescent protein can be used as a reporter in live zebrafish embryos. *Devel. Biol.* **171,** 123–129.
55. Haseloff, J. and Amos, B. (1995) GFP in plants. *Trends Genet.* **11,** 328,329.
56. Niedz, R. P., Sussman, M. R., and Satterlee, J. S. (1995) Green fluorescent protein: an in vivo reporter of plant gene expression. *Plant Cell Reports* **14,** 403–406.
57. Galbraith, D. W., Sheen, J., Lambert, G. M., and Grebenok, R. J. (1995) Flow cytometric analysis of transgene expression in higher plants: green fluorescent protein. *Methods Cell Biol.* **50,** 1–12.
58. Hu, W. and Chein, C.-L. (1995) Expression of Aequorea green fluorescent protein in plant cells. *FEBS Lett.* **369,** 331.
59. Törmäkangas, K., Ritala, A., and Teeri, T. H. (1995) Expression of GFP in plant cells. Poster presentation, at Fluorescent Proteins and Applications Meeting, Palo Alto, CA, March 6–7.
60. Beachy, R. (1995) Oral presentation at Fluorescent Proteins and Applications Meeting, Palo Alto, CA, March 6–7.

61. Chen, C. and Okayama, H. (1988) Calcium phosphate-mediated gene transfer, a highly efficient transfection system for stably transforming cells with plasmid DNA. *BioTechniques* **6**, 632–638.
62. Sambrook, J., Fritsch, E. F., and Maniatis, T. (1989) Molecular Cloning: A Laboratory Manual, Cold Spring Harbor Laboratory, Cold Spring Harbor, NY.
63. Cheng, L. and Kain, S. (1995) Analysis of GFP and RSGFP expression in mammalian cells by flow cytometry. *CLONTECHniques* **X(4)**, 20.
64. Ropp, J. D., Donahue, C. J., Wolfgang-Kinball, D., Hooley, J. J., Chin, J. Y. W., Hoffman, R. A., Cuthbertson, R. A., and Bauer, K. D. (1995) *Aequorea* green fluorescent protein (GFP) analysis by flow cytometry. *Cytometry* **21**, 309–317.
65. Harlow, E. and Lane, E. (1988) Antibodies, A Laboratory Manual, Cold Spring Harbor Laboratory, Cold Spring Harbor, NY.
66. Kain, S. R., Mai, K., and Sinai, P. (1994) Human multiple tissue western blots, a new immunological tool for the analysis of tissue-specific protein expression. *BioTechniques* **17**, 982–987.
67. Laemmli, U. K. (1970) Cleavage of structural proteins during the assembly of the head of bacteriophage T4. *Nature* **227**, 680–685.
68. O'Farrell, P. H. (1970) High resolution two-dimensional electrophoresis of proteins. *J. Biol. Chem.* **250**, 4007–4111.
69. Scopes, R. K. (1987) Protein Purification, Principles and Practice, 2nd ed., Springer-Verlag, New York.
70. Bjerrum, O. J., and Heegaard, N. H. H., eds. (1988) *CRC Handbook of Immunoblotting of Proteins.* CRC, Boca Raton, FL.
71. Towbin, H., Staehelin, T., and Gordon, J. (1979) Electrophoretic transfer of proteins from polyacrylamide gels to nitrocellulose sheets: procedure and some applications. *Proc. Natl. Acad. Sci. USA* **76**, 4350–4356.
72. Kozak, M. (1987) An analysis of 5'-noncoding sequences from 699 vertebrate messenger RNAs. *Nucleic Acids* **15**, 8125–8148.
73. Silver, P. (1995) Oral presentation at Fluorescent Proteins and Applications Meeting, Palo Alto, CA, March 6–7.
74. Cox, J. (1995) Oral presentation at Fluorescent Proteins and Applications Meeting, Palo Alto, CA, March 6–7.
75. Rizzuto, R., Brini, M., Pizzo, P., Murgia, M., and Pozzan, T. (1995) Chimeric green fluorescent protein as a tool for visualizing subcellular organelles in living cells. *Current Biol.* **5**, 635–642.
76. Bian, J., Lin, X., and Tang, J. (1995) Nuclear localization of HIV-1 matrix protein P17, The use of A. victoria GFP in protein tagging and tracing. *FASEB J.* **9**, A1279 #132.
77. Sullivan, K. F., Hahn, K., and Shelby, R. D. (1995) Bioluminescent labeling of human centromere DNA sequences *in vivo.* Poster presentation at Fluorescent Proteins and Applications Meeting, Palo Alto, CA, March 6–7.

IV

APPLICATIONS OF RECOMBINANT GENE EXPRESSION

25

Transgenic Animals as Bioreactors for Expression of Recombinant Proteins

Jaspal S. Khillan

1. Introduction

Transfer of gene into the early embryo forms the basis of a powerful tool to impart new and useful characteristics into existing breeds of animals *(1)*. Recombinant DNA techniques have made it possible to isolate and clone individual genes for specific functions. Isolated and characterized genes can be introduced into the genome of the animal by microinjection of DNA at early stage embryo. Animals that contain exogenous genes integrated into their genome are called transgenic animals. By micromanipulating the preimplantation embryo, genetic material from any source can be introduced into the genome of any species. The gene is usually transmitted to the progeny in normal Mendelian fashion. Several mammalian species including mouse, rabbit, goat, sheep, pig, and cow have been made transgenic by the microinjection of genes into early embryo *(1–8)*.

The techniques of transgenic technology have contributed significantly in understanding the molecular basis of tissue and stage-specific regulation of gene expression during development and differentiation. Several copies of a well characterized gene are injected into one of the pronuclei of a newly fertilized embryo. The DNA integrates, apparently randomly, into the host chromosome usually at one locus as multicopy head to tail concatemer. The expression of the gene is regulated by promoter/regulatory sequences present at the 5'-end of the gene. However the flanking chromosomal sequences at the site of insertion also influence the expression of the gene. The expression of gene can also be regulated by external stimuli by using promoters of inducible genes such as metallothionein, a gene inducible by heavy metal ions such as cadmium and zinc. Transgenic animals with a gene construct in which the promoter/regula-

From: *Methods in Molecular Biology, vol. 63: Recombinant Protein Protocols: Detection and Isolation* Edited by: R. Tuan Humana Press Inc., Totowa, NJ

tory sequence of metallothionein was fused with the gene for growth hormone, expressed high levels of hormone in their blood when a diet containing cadmium chloride or zinc sulfate was given. The increased level of growth hormone in blood caused increase in their size to twice that of their normal littermates *(9)*.

Transgenic animals prepared with the gene constructs in which the promoter/ regulatory sequences from one gene are fused with the coding sequences of another gene, express protein in the tissues specific for the promoter function, thus suggesting that the product of a gene can be targeted to any organ or tissue by using specific promoter/regulatory sequences. This targeted expression of genes laid down the foundation of a very important application of this technology to use transgenic animals as bioreactors for the proteins of biomedical importance *(10–12)*. Transgenic livestock may serve as biofactories for large scale production of valuable proteins in their body fluids such as blood or milk. Isolation of proteins from fluids has advantages over the tissues because the body fluids are renewable and proteins are easy to recover. The proteins therefore may be directly harvested by standard chromatographic techniques.

Transgenic pigs prepared with a gene construct in which the gene for human β-globin was fused with the promoter of porcine β-globin gene, expressed moderate to high levels of human β-globin in blood *(5)*. Though it may be convenient to draw blood from large species, repeated blood collections may adversely affect the health of the animal. Also, a continuous circulation of biologically active proteins in the body may be harmful.

Mammary gland on the other hand has several advantages over the blood system. For example, milk can be collected in large quantities, milk proteins do not circulate in the body, and proteins such as caseins and β-lactoglobin are expressed abundantly and exclusively in the mammary gland. Proteins such as tissue plasminogen activator (tPA), human chorionic gonadotropin (HCG), blood clotting factor IX, and α1-antitrypsin (ATT) have been expressed in the milk of transgenic goat, sheep, and rabbit prepared with the gene constructs in which the promoter/regulatory sequences of β-casein, whey acidic protein and β-lactoglobin were fused with the coding sequences of the gene *(2,3,6,7,10,11)*. The proteins expressed in the transgenic system are processed properly for the glycosylation and other posttranslational modifications. A high expressing line of transgenic animals may be selected and propagated to harvest sufficient amount of protein on a daily basis. It is estimated that if the expression of factor IX is achieved to the level of endogenous proteins, at an efficiency of 10% for the protein processing, a herd of 10 ewes can meet the demand of factor IX for the whole world.

The following methods describe the preparation of transgenic animals. Since the mammalian embryos from different species are very similar, the method will be explained for the preparation of transgenic mice.

Table 1
Embryo Culture Medium

Compound	mM	g/L
NaCl	94.66	5.533
KCl	4.78	0.356
CaCl$_2$ 2H$_2$O	1.71	0.252
KH$_2$PO$_4$	1.19	0.162
MgSO$_4$ 7H$_2$O	1.19	0.293
NaHCO$_3$	25.0	2.101
Sodium lactate	23.28	2.610
Sodium pyruvate	0.33	0.036
Glucose	5.56	1.000
Bovine serum albumin		4.000
Penicillin G (100 U/mL)		0.060
Streptomycin sulphate (50 µg/mL)		0.050
Phenol red		0.010

2. Materials

2.1. Animals

1. Four- to five-week-old C57BL6 X CBA hybrid F1 female mice or FVB/N female mice as embryo donors and 8–10-week-old stud males of the same strains.
2. Female mice of CD1 and NIH-GP strains as pseudopregnant recipient mothers.
3. 8–10-wk-old FVB/N or CD1 males to obtain pseudopregnant recipients.

2.2. Hormones

1. Pregnant mare serum gonadotropin (PMSG).
2. Human chorionic gonadotropin (HCG).

A solution of 50 IU/mL for each hormone is prepared in 0.9% NaCl. Aliquots of 1 mL are stored frozen protected from light up to 4–6 wk.

2.3. Anesthesia

Avertin is the most commonly used anesthesia in mice. A stock solution of 100% Avertin is prepared by mixing 10 g of tribromoethanol with tertiary amyl alcohol. The stock solution is diluted to 2.5% in water before use and stored under dark at room temperature for prolonged periods. About 0.015–0.017 mL/gm body weight is administered to anesthetize adult mice.

2.4. Embryo Culture Medium

A modified Whitten's medium, M16 medium *(13)*, is used to culture early embryos. The composition of M16 medium is given in Table 1. Dissolve chemi-

Table 2
Size of Microtools

	Holding pipet		
	Outer diameter, μM	Inner diameter, μM	Microinjection needle, μM
Mouse	80–100	30–50	0.5–1.0
Higher Species	100–120	40–50	0.5–1.0

cals in 2X glass distilled water in a final volume of 1 L. Filter through a millipore filter into small containers and gas the air space with 5% CO_2 in air to maintain pH 7.2–7.4. Adjust osmolarity to 288–292 m osmoles. The medium can be stored up to 2 wk at 4°C.

2.5. Microtools

For the micromanipulation of embryos three types of needles and micropipets are required. The size of various needles and micropipets is described in Table 2. Capillary tubing for all types of needles should be washed overnight in 100% ethanol and rinsed with dH_2O followed by heating at 180°C.

1. Isolation and transfer pipet: For the collection of embryos, an isolation pipet with an opening of about 140–180 µm id is prepared. Glass capillaries (World Precision Instruments [Sarasota, FL], 1-mm id) are prepulled on a gas burner. The prepulled capillary is cut with a diamond pencil and the tip is polished on a microforge. Same pipet may be used for the transfer of embryo into pseudopregnant females.
2. Holding pipet: Holding pipet is prepared in the same way as the isolation needle. The opening of the holding pipet is about 40–60 µm.
3. Microinjection needle: Microinjection needle with an opening of 0.5–1.0 mm is pulled on an automatic needle puller, e.g., Flaming/Brown micropipet puller, Sutter Instrument Co. The microinjection needle is generally prepared from a capillary that contains filament throughout its length that helps to fill DNA solution from the base of the needle.

2.6. DNA for Microinjection

DNA for microinjection should be highly pure (*see* Notes 5–7). The gene is first excised from the vector sequences as a linearized fragment and purified by either polyacrylamide gel or by ethanol precipitation. A solution of 1–2 µg DNA/mL in TE buffer is prepared for microinjection.

2.7. Buffers and Solutions

1. A 10 mg/mL proteinase K solution is stored in 200-µL aliquots at –20°C.
2. Tris-EDTA (TE) buffer: 10 mM Tris-HCl and 0.1 mM EDTA pH 7.2. The buffer can be stored at room temperature indefinitely.

3. 20X SSC: 175.3 g sodium chloride and 88.2 g sodium citrate/L pH 7.0. The solution is stored at room temperature.

2.8. Major Equipments

1. High power microscope equipped with Nomarski objectives.
2. Microforge.
3. Vibration free table.
4. Dissection stereo microscope.
5. Surgical instruments.
6. A pair of Leitz micromanipulators.
7. Micrometer syringe.

2.9. Radiolabeled Compounds

α-P^{32} labeled deoxyribonucleotide triphosphate to prepare gene specific probes. Radiolabeled compounds are health hazards therefore must be handled behind a thick protective plastic shield.

3. Methods

There are three different methods to prepare transgenic animals:

1. Microinjection of DNA into pronucleus of zygote.
2. Retroviral vectors and retroviral infection of embryo (*see* Note 1).
3. Embryonic stem cell injection into blastocysts (*see* Note 2).

Transgenic animals are generally prepared by the micromanipulation of preimplantation embryos. Fig. 1 shows various stages of preimplantation embryo. The most commonly used method is direct microinjection of DNA into one of the pronuclei of a newly fertilized embryo (Fig. 1A). The overall strategy is shown in Fig. 2. Although this method is the most efficient to introduce foreign genes into the animal genome it has certain limitations as well (*see* Note 3).

3.1. Microinjection of DNA into Zygote

Direct microinjection of DNA into pronucleus of a zygotic embryo is the most efficient and reliable method to generate transgenic animals. Several hundred copies (600–800) of the DNA are injected into one of the pronuclei of a newly fertilized embryo and the embryos are transferred into a pseudopregnant female. About 4–40% of the newborns contain exogenous DNA stably integrated into their genome. When the transgenic animal is crossed with a normal animal, the exogenous DNA is transmitted to the subsequent generations in normal Mendelian fashion. Following are the steps for the microinjection of DNA into embryos:

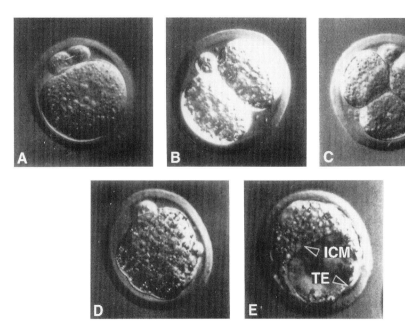

Fig. 1. Different stages of a preimplantation embryo.

3.1.1. Preparation of Pseudopregnant Females

To obtain pseudopregnant recipients, female mice are mated with the vasectomized males. The vasectomized males are prepared as follows:

1. Make a horizontal incision in the abdominal region of 6–8-wk-old stud males at the level of hind legs.
2. Pull out testis on each side along with the vas deference.
3. Cut about half cm of the vas deference and push testis back into the scrotal sac followed by closing the skin with the wound clips.
4. Allow 5–7 d for mice to recover and then mix with the females.
5. Check females daily for the presence of vaginal plug, which is formed from the coagulation of proteins secreted in the semen. The plugs generally fall off after 12–14 h of mating. About 20 vasectomized males are sufficient to provide 3–4 pseudopregnant females every day.

3.1.2. Collection of Embryos

Female mice are superovulated to obtain a large number of synchronized embryos. Usually 4–5 FVB/N or C57BL6 superovulated female mice are sufficient to provide 50–70 embryos.

1. Inject 5 IU PMSG intraperitoneally to females around 2:00 PM PMSG mimics the effect of follicle stimulating hormone (FSH).

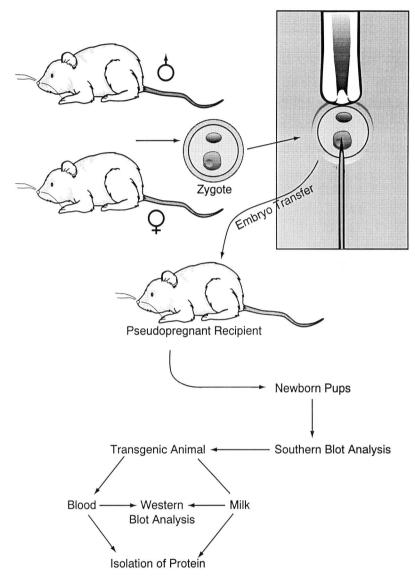

Fig. 2. General strategy to prepare transgenic animals.

2. After 48 h inject 5 IU HCG, which has the same effect as the lutenizing hormone (LH).
3. Mix female mice with the stud males overnight.
4. Check females for vaginal plugs next morning, only the plugged mice are used for embryo isolation. The females that do not mate can be recycled after 3–4 wk. Each superovulated female can produce 15–25 embryos.

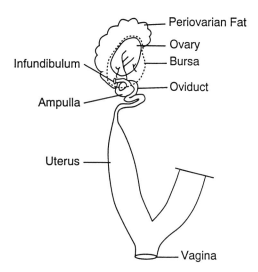

Fig. 3. Reproductive organs of female mouse. The ovary and oviduct is surrounded by a thin membrane called the ovarian bursa. The embryos along with the cumulus mass are present at the swollen portion of the oviduct called ampulla. The embryos are collected by teasing the ampulla. After the microinjection, the embryos are transferred by inserting a needle from the infundibulum by making an incision in the bursa. The embryos are delivered in the ampulla.

5. On the same morning, separate plugged females from the vasectomized males for foster mothers.
6. Prepare following 60-mm Petri dishes:
   a. Approx 5 mL of M16 medium.
   b. Approx 8–10 mL of M16 medium.
   c. Place a drop of M16 medium (~100 µL) in the center of the Petri dish. Cover medium with light mineral oil equilibrated with the same medium.
7. Incubate at 37°C for about 1 h.
8. Sacrifice superovulated and plugged females by cervical dislocation and flush abdominal area with 70% ethanol.
9. Open peritoneal cavity, cut out ovary and oviduct together (Fig. 3) and transfer into first Petri dish.
10. Release embryos by teasing the ampulla, a swollen portion of the oviduct, with tweezers under a dissecting microscope. The embryos at this stage are surrounded by a mass of cumulus cells that provide nutrition to the egg.
11. Add one to two drops of hyaluronidase solution. The cumulus cells will dissociate in about 3–5 min of incubation.
12. Collect embryos in an isolation pipet attached to a rubber tubing as shown in Fig. 4 and transfer into second Petri dish (a few bubbles in the needle are helpful to control the flow of embryos).

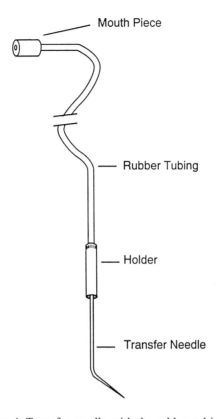

Fig. 4. Transfer needle with the rubber tubing.

13. Wash embryos 3–5 times in picking and releasing manner at different areas of Petri dish, picking only embryos each time, and leaving cumulus cells behind.
14. Transfer clean embryos to the drop of medium in third Petri dish and store at 37°C in CO_2 incubator until microinjection.

3.1.3. Microinjection of DNA

1. Transfer 20–25 embryos into a drop of M16 medium on a glass-bottomed Petri dish (Nomarski lens can be used only with the glass surfaces). Cover the drop with paraffin oil. Transfer Petri dish to microscope stage and focus onto embryos. Alternatively, depression glass slides may be used for this purpose.
2. Set up holding pipet first. Fill holding pipet half-way with the medium and insert into the holder of the left micromanipulator. Bring pipet close to an embryo and hold it at the tip by suction via a 30-mL glass syringe or a micrometer syringe.

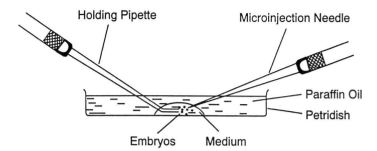

Fig. 5. Microinjection setup: The embryos are placed in a drop of medium covered with paraffin oil. The embryo is held at the tip of holding pipet (left) and the DNA is injected by inserting microinjection needle (right).

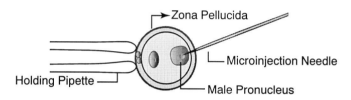

Fig. 6. Microinjection of DNA into male pronucleus at high magnification. Embryo is held at the tip of holding pipet by suction (left), so that the male pronucleus is opposite to the microinjection needle filled with DNA (right). The DNA is injected by inserting the needle and pushing the solution by a syringe. A successful injection is judged by a clear swelling of the pronucleus.

3. Fill microinjection needle with the DNA by dipping its base into the DNA solution and insert into the holder on right manipulator as shown in Fig. 5. At 200X magnification, focus simultaneously on the tip of the needle and embryo (Fig. 6). Position embryo by sucking and releasing so that the male pronucleus is directly opposite to the injection needle. Usually the DNA is injected into male pronucleus, as it is bigger in size and is not obstructed by the polar bodies (*see* Note 8).

4. Insert needle into the pronucleus and inject DNA with a micrometer syringe attached to the other end of the tube until a clear swelling of the pronucleus is observed. Avoid touching needle to the nucleoli. Pull out needle carefully and repeat the process for rest of the embryos.

5. After injection, remove dead embryos and store healthy embryos at 37°C until ready for transfer into foster mothers. About 60–70% of the embryos survive microinjection.

3.1.4. Transfer of Embryos into Pseudopregnant Mothers

1. Anesthetize a pseudopregnant female by injecting 0.4–0.5 mL 2.5% Avertin and wipe the back with 70% ethanol. Shave off the fur in the middle and make an

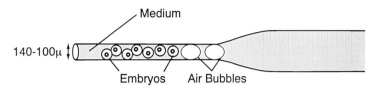

Fig. 7. Transfer needle with embryos: A few air bubbles behind the embryos serve as an indicator of successful transfer when the air bubbles are transferred along with the embryos.

incision of about 1 cm in the mid region. Oviducts on both sides can be reached from the same incision. Place mouse under a dissecting microscope.

2. Cut peritoneal membrane on one side and locate the pink-colored ovary. Pull out reproductive organs by pulling the periovarian fat that surrounds the ovary (Fig. 3). Place a cotton applicator under the oviduct. Locate the opening of the oviduct through ovarian bursa, a membrane that surrounds the ovary. Tear open the ovarian bursa with a pair of tweezers just above the infundibulum.

3. Pick up ~10–15 embryos in a transfer pipet (Fig. 7) and introduce into the ampulla through infundibulum (Fig. 3). A few air bubbles behind the embryos help to control the flow of embryos and also serve as an indicator of a successful transfer when the air bubbles are transferred along with the embryos.

4. Push organs back into the peritoneal cavity with the help of a cotton applicator. Close the opening with a suture and repeat the procedure for the other side. Clamp skin with wound clips and keep females under a heat lamp till recovery. The pups are born after 19–21 d.

3.1.5. Analysis of Transgenic Animals

Microinjected DNA usually integrates into the host chromosome at a single locus as a multicopy head to tail unit (*see* Note 3). The number of copies varies between one to several hundred per haploid genome. Since the integration of DNA occurs at one-cell stage of the embryo, every cell including germline cells of the newborn carry the transgene and therefore is transmitted to the subsequent generations like other hereditary traits.

To analyze animals for the presence of exogenous DNA, chromosomal DNA is isolated from a piece of tail of 2–3-wk-old animals. Alternatively the placental or ear tissue may also be used. The tissue is minced and digested in a lysis buffer (50 mM Tris-HCl, pH 8.0, 100 mM EDTA, 100 mM NaCl, 1.0% sodium dodesyl sulfate (SDS), and 35 µg/mL proteinase K) for 8–10 h and the DNA is extracted by phenol:chloroform extraction procedure followed by precipitation with ethanol. DNA is dissolved in TE buffer and analyzed either by the dot blot or by the Southern blot analysis *(13)*.

3.1.5.1. Dot Blot Analysis

Initial screening of the newborn pups is carried out by the dot blot procedure. About 5–10 μg DNA in 200 μL 10X SSC is denatured by heating in boiling water bath for 5–10 min followed by transfer onto a nitrocellulose membrane. The membrane is hybridized to a radiolabeled DNA probe. Transgenic animals are identified by exposing the nitrocellulose membrane to X-ray films.

3.1.5.2. Southern Blot Analysis

After the initial screening, transgenic animals are confirmed by Southern blot analysis. About 10 μg of DNA is digested with specific restriction enzyme(s) and the digested DNA is separated by electrophoresis on agarose gel. The separated DNA is then transferred onto a nitrocellulose membrane and probed with radiolabeled probes. The presence of unique DNA fragments on the X-ray film represent integration of gene into chromosomal DNA. The number of copies of the transgene is determined by including known amount of microinjected DNA on the same gels in parallel lanes.

3.1.5.3. Analysis of RNA

To analyze animals for the expression of gene, total RNA is extracted from the target tissue such as blood or mammary gland. For transgenic animals with mammary gland specific promoter/regulatory sequences, the tissue is taken from lactating animals as the gene will be active only during lactation. Total RNA is then subjected to Northern blot analysis. A known amount of RNA, 25–40 μg, is electrophoresed on denaturing agarose gels containing formaldehyde. The separated RNAs are transferred onto a nylon membrane and the membrane is hybridized with radiolabeled probes specific for the gene.

3.1.5.4. Analysis of Protein

The presence of protein is generally checked by the Western blot analysis. Proteins from blood or milk samples are separated on 6–10% polyacrylamide gels and electroblotted onto a nylon membrane. The membrane is then treated with antibodies specific for the protein and the amount of protein is estimated by comparing with the proteins of known concentration run on the same gel. Once the presence of protein is confirmed, it can be isolated by standard chromatographic procedures.

3.1.6. Propagation of Transgenic Lines

Transgenic animals, that are positive for the gene, are mated with normal animals to propagate the transgenic lines (*see* Note 4). Most of the transgenic animals prepared by this method transmit gene to their next generation in a typical Mendelian fashion, i.e., approx 50% of the progeny is positive for the gene. However, 5–10% percent of the founder transgenic animals are mosaic for the gene integration as the integration in some embryos occurs at late cleavage stages. Consequently, a small percent of the progeny of these animals will be positive for the gene. The progeny that are positive, however, will transmit the gene to the subsequent generations.

4. Notes

1. Four cell stage to early gestation embryo can be infected with retroviruses or recombinant retroviruses *(1)* that integrate into the genome after the viral RNA is reverse transcribed into DNA. Since the infection of embryo is carried out at the late cleavage stages, the newborns from these embryos are mostly mosaic for the gene integration. Transmission of DNA to subsequent generations is obtained if the viral DNA integrates into cells that contribute to the germline such as cells of the ovary and testis. The limitations of the procedure are that the construction of retroviral vectors is difficult. Genes of larger size can not be integrated because only 8–10 kb DNA can be packaged into the viral coat.

2. Embryonic stem (ES) cell lines are established by culture of inner cell mass (ICM) of a blastocyst (Fig. 1E). These cells are pleuripotent and can be maintained in undifferentiated state indefinitely. When introduced into a host blastocyst, ES cells have potential to contribute to any organ of the developing embryo including testis and ovary. The gene can be introduced into ES cells either by electroporation or by microinjection and the cells are selected for the integration. After selection, the cells are directly microinjected into the cavity of a blastocyst. The cells integrate into ICM of the host blastocyst and contribute to the formation of different tissues. The animal born from such embryo is chimeric, i.e., contains microinjected cells as well as cells from the host blastocyst. Transmission of gene to the next generation occurs only if the microinjected cells contribute to the germline. The limitations of this procedure are that many of the chimeric animals fail to transmit the gene to their progeny and also the culture and maintenance of ES cells is quite difficult.

3. Though most of the gene constructs used in transgenic animals express their product, a proper regulation of expression has not been achieved. The level of expression of the gene depends upon the site of integration within the chromosome. It is not very unusual to have transgenic animals that do not express the gene. The design and construction of gene construct is very critical. Some of the limitations associated with the technology are:
   a. Site of insertion and the number of copies of the gene cannot be controlled.
   b. Multiple copies of the gene in tandem represent chromosomal instability.
   c. The level of expression of the gene is not related to the number of copies integrated into the genome. A large number of copies does not represent higher expression.
   d. Production of transgenic animals of higher species is very expensive and their maintenance is also difficult.

e. The zygotes of pig, sheep, and cattle are not transparent, therefore microinjection of DNA in these species is relatively difficult. However this can be overcome by slow centrifugation of embryos before microinjection.

f. The response of animals to gonadotropins for superovulation is highly variable.

g. The efficiency of gene integration in embryos of higher species is far lower than in mouse, e.g., in pig about 10%, sheep approx 1.0%, and cattle less than 1%.

h. Gene integration may also interrupt the function of some vital gene or cause chromosomal rearrangements that may be detrimental to the survival of embryo.

4. In spite of several limitations, the transgenic technology has a strong potential for its use in the field of agriculture and biological sciences. In transgenic animals, mostly the gene integration occurs at a single locus in the genome that is unique for each transgenic line. Occasionally, the gene integrates at more than one locus. Since the integration of foreign DNA occurs at one chromosome, the founder animals are heterozygous for the transgene.

To create homozygous lines, heterozygous progeny are intercrossed. As expected from the Mendelian transmission of the gene, about a quarter of the progeny from this cross will be homozygous for the gene. These animals contain double the number of copies and may have twice the level of expression. Homozygous animals can be identified by dot blot or Southern blot analysis in which the homozygous animal will display a stronger signal on X-ray film. Homozygosity is further confirmed by mating with the normal animals. All progeny from this cross should be positive for the transgene. On the average it takes about 6–7 mo to create a homozygous line of mice. This period is much longer in species that have long gestation periods. Once the homozygosity is established these animals can be crossed with each other to maintain homozygous lines.

5. The DNA for microinjection should be absolutely free from all contaminants such as traces of agarose, phenol, or ethanol that are toxic to the embryo. The DNA may be further purified by filtration with microfilters. The vector sequences in the gene construct should be removed as much as possible. Tissue specificity and the level of expression of transgene is greatly influenced by the presence of vector sequences. A 100 to 1000-fold increase in the level of expression was observed when the vector sequences were deleted from the human β-globin gene construct *(14)*. Similar observations were made with the other constructs that contained metallothionein and elastase promoter upstream of growth hormone gene.

6. Both linear and circular DNA have been shown to integrate into the host chromosome, however linear molecules integrate with at least five times higher frequency than circular molecules. There seems to be a little effect of structure at the ends of DNA. In general, DNA with single-stranded ends has higher efficiency of integration. Concentration of DNA plays a significant role in the viability of embryos. Usually 1–2 μg/mL DNA is optimum concentration that is equivalent to about 600–800 copies/pl of 5 kb DNA. At lower concentrations the frequency of integration is also low, whereas at a concentration of 10 μg/mL or more the survival rate of embryos is drastically reduced.

7. In general genomic clones express better than the cDNA clones. A size of 20 kb or less is optimum for the microinjection, however DNA of up to 50 kb and dissected pieces of chromosome *(15)* have been used successfully to create transgenic animals. A DNA of larger size may be nicked during micromanipulations.

8. Though the DNA can be microinjected into either of the pronuclei, the microinjection into male pronucleus is more convenient because of its larger size and it is not obstructed by the polar bodies. On the other hand DNA microinjected into the cytoplasm usually does not integrate into the genome of the embryo *(16)*.

References

1. Jaenisch R. (1988) Transgenic animals. *Science* **240,** 1468–1474.
2. Bühler, Th. A., Bruyere, Th., Went, D. F., Stranzinger, G., and Bürki, K. (1990). Rabbit β-casein promoter directs secretion of human interleukin-2 into the milk of transgenic rabbits. *Biotechnology* **8,** 140–143.
3. Ebert, K. M., Selgrath, J. P., DiTullio, P., Denman, J., Smith, T. E., Memon, M. A., Schindler, J. E., Monastersky, G. M., Vitale, J. A., and Gordon, K. (1991). Transgenic production of a variant of human tissue-type plasminogen activator in goat milk, generation of transgenic goats and analysis of expression. *Biotechnology* **9,** 835–838.
4. Carver, A. S., Dalrymple, M. A., Wright, G., Cottom, D. S., Reeves, D. B., Gibson, Y. H., Keenan, J. L., Barrass, J. D., Scott, A. R., Colman, A., and Garner, I. (1993). Transgenic livestock as bioreactors: stable expression of human alpha-1-antitrypsin by a flock of sheep. *Biotechnology* **11,** 1263–1270.
5. Swanson, M. E., Martin, M. J., O'Donnell, J. K., Hoover, K., Lago, W., Huntress, V., Parsons, C. T., Pinkert, C. A., Pilder, S., and Logan, J. S. (1992). Production of functional human hemoglobin in transgenic swine. *Biotechnology* **10,** 557–560.
6. Sharma, A., Martin, M. J., Okabe, J. F., Truglio, R. A., Dhanjal, N. K., Logan, J. S., and Kumar, R. (1994). An isologous porcine promoter permits high level expression of human hemoglobin in transgenic swine. *Biotechnology* **12,** 55–59.
7. Clark, A. J., Bessos, H., Bishop, J. O., Borwn, P., Harris, B. S., Lathe, R., McClenaghan, M., Prowse, C., Simons, J. P., Whitelaw, C. B. A., and Wilmut, I. (1989). Expression of human anti-hemophilic factor IX in the milk of transgenic sheep. *Biotechnology* **7,** 487–492.
8. Krimpenfort, P., Rademakers, A., Eyestone, W., van der Schans, A., van den Broek, S., Kooiman, P., Kootwijk, E., Platenburg, G., Pieper, F., Strijker, R., and de Boer, H. (1991). Generation of transgenic diary cattle using "in vitro" embryo production. *Biotechnology* **9,** 844–847.
9. Palmiter, R. D., Brinster, R. L., Hammer, R. E., Trunbauer, M. E., Rosenfeld, M. G., Birnberg, N. C., and Evans, R. M. (1982). Dramatic growth of mice that develop from eggs microinjected with metallothionein-growth hormone fusion genes. *Nature* **300,** 611–615.
10 Van Brunt, J. (1988). Molecular farming, transgenic animals as bioreactors. *Biotechnology* **6,** 1149–1154.

11. Moffat, A. S. (1991). Transgenic animals may be down on the pharm. *Science* **254,** 35,36.
12. Spalding, B. J. (1992). Transgenic pharming advances. *Biotechnology* **10,** 498,499.
13. Hogan, B.,Constantini, F., and Lacy, E. (1986) *Manipulating the Mouse Embryo: A Laboratory Manual.* Cold Spring Harbor Laboratory, Cold Spring Harbor, NY.
14. Townes, T. M., Chen, H. Y., Lingrel, J. B., Palmiter, R. D., and Brinster, R. L. (1985). Expression of human β-globin genes in transgenic mice: effects of a flanking metallothionein-human growth hormone fusion gene. *Mol. Cell. Biol.* **5,** 1977–1983.
15. Richa J. and Lo C. W. (1989) Introduction of human DNA into mouse eggs by injection of dissected chromosome fragments. *Science* **245,** 175–177.
16. Brinster, R. L., Chen, H. Y., Trumbauer, M. E., Yagle, M. K., and Palmiter, R. D. (1995). Factors affecting the efficiency of introducing foreign DNA into mice by microinjecting eggs. *Proc. Natl. Acad. Sci. USA* **82,** 4438–4442.

26

Artificial Cells for Bioencapsulation of Cells and Genetically Engineered *E. coli*

For Cell Therapy, Gene Therapy, and Removal of Urea and Ammonia

Thomas M. S. Chang and Satya Prakash

1. Introduction

Enzymes, proteins, cells, microorganisms, adsorbents, magnetic materials and other biologically active materials can be encapsulated within artificial cells with artificial polymer membranes *(1–9)*. This type of artificial cells are not liposomes. The artificial cells protect the encapsulated biological materials from immunological reactions. At the same time, the enclosed materials continue to act in the immunologically isolated environment. The use of artificial cells containing adsorbent for hemoperfusion is a routine procedure used in patients for the removal of toxic substances or unwanted metabolites. Clinical studies are ongoing in patients using artificial cells as red blood cell substitutes, for enzyme therapy, and for delivery of biotechnological agents. Chang first developed a drop method for the bioencapsulation of cells and proposed its use in cell therapy *(4,5)*. "Microencapsulation of intact cells … the enclosed material might be protected from destruction and from participation in immunological processes, whereas the enclosing membrane would be permeable to small molecules of specific cellular product which could then enter the general extracellular compartment of the recipient. … The situation is comparable to that of a graft placed in an immunologically favorable site" *(4)*. This was not explored by others for sometime. However, with increasing interests in biotechnology, many groups are now actively studying this for bioencapsulation of cells or genetically engineered microorganisms in cell and gene therapy. This chapter describes only two examples: bioencapsulation of hepatocytes or

From: *Methods in Molecular Biology, vol. 63: Recombinant Protein Protocols: Detection and Isolation* Edited by: R. Tuan Humana Press Inc., Totowa, NJ

genetically engineered microorganisms. These examples are used to demonstrate the methods of preparations and their potential applications in cell and gene therapy.

2. Materials

2.1. Hepatocytes

Hepatocytes were isolated from Wistar rat liver using the method of Seglen *(10)*. The hepatocytes obtained are contaminated with nonparenchymal cells, damaged cells, and connective tissues. In order to recover only the hepatocytes, the following procedure is carried out at −4°C. The cell suspension is filtered stepwise through nylon meshes of 250, 100, and 60 μm. The suspension is centrifuged at 50g for 5 min. The pellet is then dispersed in fresh William's E medium. The number of cells and viability were determined. Viability is measured by the Trypan blue exclusion test. The cell suspension is held in ice-cold William's E medium until the bioencapsulation procedure. For the bioencapsulation procedure, isolated hepatocytes were suspended with William's E medium, 100 μg/mL of streptomycin and penicillin in ice-cold autoclaved 4% sodium alginate in 0.09% sodium chloride solution.

2.2. Genetically Engineered *E. coli DH5 Cells*

Genetically engineered *E. coli* DH5 cells which contain urease gene from *K. aerogens*, was the generous gift form Prof. R. P. Haussinger *(11)*. The bacteria were grown in Luria Bertani (LB) medium. The composition of the LB medium is consisted of 10 g/L bacto tryptone, 5 g/L bacto yeast extract, and 10 g/L sodium chloride. The pH of the LB media is adjusted to 7.50 by adding about 1.00 mL/L of 1N NaCl. Log phase bacterial cells were harvested by centrifuging at 10,000g for 20 min at 4°C. The supernatant was discarded. The cell mass is then washed and centrifuged at 10,000g for 10 min at 4°C for five times with sterile cold water to remove media components. Bacterial cells were suspended in an autoclaved sodium alginate in ice cold 0.90% sodium chloride solution.

2.3. Sodium Alginate

Alginates are heteropolymer carboxylic acids, coupled by 1 → 4 glycosidic bonds of (β-ᴅ-mannuronic (M) and (α-ʟ-gluronic acid unit (G). Alkali and magnesium alginate are soluble in water, whereas alginic acids and the salts of polyvalent metal cations are insoluble. Thus, when a drop of sodium alginate solution enters a calcium chloride solution, rigid spherical gels are formed by ionoirotpic gelation. Sodium alginate is usually prepared as a 2% solution in 0.90% sodium chloride solution. Sodium alginate is from Kelco Gel® low viscosity alginate. Before use, it is either filter sterilized or heat sterilized for 5 min.

2.4. Solutions

All the solutions are kept in an ice bath before use and during the process of bioencapsulation. The pH of the solutions is kept at 7.40 by buffering with N-2-Hydroxylethyl piperazine-N-2 enthanesulfonic acid (HEPES). Except for sodium alginate, the solutions are sterilized by filtering through a sterile 0.2-μm Millipore filter.

3. Methods

3.1. General Procedure

The original drop method for cell encapsulation involves chemically crosslinking the surface of aqueous droplets that contains cells *(4,5)*. This is further modified into the following drop technique using a milder physical crosslinking agent *(12,13)*. Alginate-polylysine-alginate (APA) microcapsules containing hepatocytes or genetically engineered *E. coli* is prepared using the apparatus setup shown in Fig. 1A. The steps are as follows (Fig. 2).

1. The viscous alginate-bacterial suspension or alginate hepatocyte solution is passed through a 23-gage stainless steel needle using a syringe pump (compact infusion pump model-975 Harvard Instruments, Cambridge, MA). Sterile compressed air, through a 16-gage coaxial stainless steel needle, are used to shear the droplets coming out of the tip of the 23-gage needle. The two needles in combination make up the droplet needle assembly that is shown in more detail in the Fig. 1B. Each droplet falls into the sterile, ice-cold solution of calcium chloride (1.40%, pH 7.20, heat sterilized). Upon contact with the calcium chloride buffer, alginate gelation is immediate. The droplets are then allowed to gel for 15 min in the gently stirred mixture of ice-cold, sterile solution of calcium chloride (1.40%).
2. After gelation in the calcium chloride solution, alginate gel beads are suspended for 10 min in a 0.05% polyl-L-ysine (mol wt 22,500 for hepatocytes and 16,100 for *E. coli*) in HEPES buffer saline, pH 7.20. The positively charged poly-L-lysine forms a complex with surface alginate to form a semipermeable membrane.
3. The beads are then washed with HEPES buffer saline (pH 7.20), and placed in an alginate solution (0.10%) for 4 min. The alginate neutralizes any excess poly-L-lysine on the surface. The alginate-poly-L-lysine-alginate (APA) capsules are then washed in a 3% citrate bath (3% in 1:1 HEPES-buffer saline, pH 7.20) to liquefy the gel in the microcapsules. The APA microcapsules form, which either contain a suspension of hepatocytes or genetically engineered bacteria *E. coli*, and are stored at 4°C for use in experiments. All the conditions are kept sterile during the microencapsulation process.

3.2. Procedure Specific for Encapsulating High Concentrations of Smaller Cells like Hepatocytes or E. coli

The general procedure described above was originally designed for the bioencapsulation of one or two islets *(12–14)*. When this general procedure is

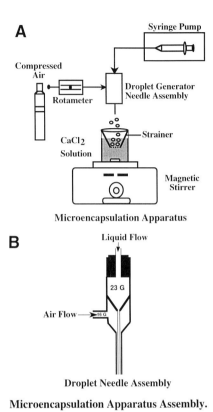

Fig. 1. Microencapsulation apparatus assembly. Reprinted with permission from ref. *27.*

used to bioencapsulate a large number of dispersed cells like hepatocytes or *E. coli,* the following problems can occur *(15)* (Fig. 2). In step 1, when alginate gel is formed, some cells are observed to protrude out of the surface of the gel spheres. In step 2 of membrane formation with polylysine, the protruding cells are entrapped in the membrane matrix, with some protruding outside of the membrane. Cellular entrapment into the capsular membrane will increase with increasing concentration of cells encapsulated. Therefore, with increasing concentration of encapsulated cells a greater number of microcapsules with membrane imperfection will be expected. Severe problems as described above are observed with hepatocyte concentrations of $10–20 \times 10^6$. When these microcapsules are implanted, the hosts immediately recognize the protruding cells on the surface resulting in acute cell-mediated host immune response and rejection. In other cases, where the cells have not protruded but only integrated into the membrane matrix, resulting in a weak and poorly formed area in the

STANDARD METHOD FOR
BIOENCAPSULATION OF LIVING CELLS

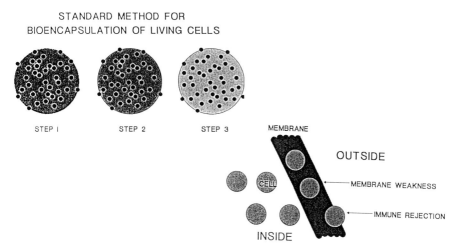

Fig. 2. Standard method using polylysine-alginate for encapsulation of cells. Some entrapped cells protrude on the surface of the alginate gel spheres. When the membrane is formed, a few cells may be incorporated into the membrane matrix. At these sites, the membranes are weakened, some sites may be perforated after implantation. Immune rejection occurs at these sites because there is no immunoisolation at these sites. One exposed cell on the surface can result in rejection by the body of the whole microcapsule. Reprinted with permission from ref. *16*.

membrane. When these cells are implanted into mice, macrophage and lymphocytes appear to be able to perforate the capsular membrane at these sites and infiltrate the microcapsule at these sites.

To prevent the above problem, the following novel approach is devised *(16,17)*. The steps are summarized in Fig. 3, using a modification of the micrpencapsulation apparatus setup shown in Fig. 1.

1. Small calcium alginate gel microspheres containing entrapped cells are first allowed to form. Like in the general procedure, there are cells protruding out of the surface of the smaller calcium alginate gel microspheres. These gel microcapsules are then resuspended in alginate solution to go through same droplet formation as before.
2. The small microspheres are entrapped within larger calcium alginate gel microspheres. When these smaller microspheres are entrapped within the large microspheres, the larger ones do not have cell extruding on the surface. Up to five small microspheres can be entrapped within each larger microsphere.
3. The next two steps are the same as described before. In the last step, the entire content of the microcapsule is liquefied by citrate. This also liquefies the small calcium alginate gel microspheres inside the microcapsule. This way, the hepatocytes in the smaller gel microsphere are released to float freely in the final microcapsule.

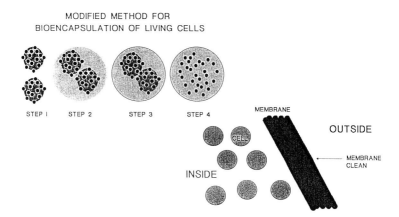

Fig. 3. Novel two-step method of cell encapsulation. Cells are entrapped in the smaller alginate gel microspheres. The smaller microspheres are then entrapped in larger microspheres. This way, no cells are near the surface of the larger alginate gel microsphere. When membrane is formed on the surface of the larger microsphere, no cells are incorporated into the membrane matrix. The gel in the large microcapsules including the smaller microspheres inside is then liquefied. This way, the encapsulated cells are freely dispersed in the microcapsule. Reprinted with permission from ref. *16*.

Microscopic studies show that the encapsulated cells are not embedded within the walls of the microcapsular membrane during the process of encapsulation *(16,17)*. Implantation results in better immunoisolation.

3.3. Applications

3.3.1. Bioencapsulated Cells in Cell Therapy

The present example is to use bioencapsulated hepatocytes in basic investigations to study the feasibility of encapsulated cells in cell and gene therapy. Hepatocytes or hepatocytes with genetic induction can be used. One can immunoisolate the cells by using the bioencapsulation concept discovered by Chang *(2)*. Hepatocytes and other types of cells can be obtained from humans or animals for implantation.

3.3.1.1. Bioencapsulation of Hepatocytes

Typically, hepatocytes were enclosed within alginate-polylysine-alginate (APA) microcapsules of 300 µm mean diameters *(18)*. Different concentrations of hepatocytes can be microencapsulated into the artificial cells. In a typical case, each 300-µm diameter artificial cell contains 120 ± 20 SD hepatocytes. Thus, the 4 mL of 62,000 artificial cells injected contained a total of about 7.40

$\times 10^6$ hepatocytes. Higher concentrations of hepatocytes can also be enclosed. Thus, 1.1 mL of microcapsules can contain 15×10^6 hepatocytes *(19,20)*. The 300-μm diameter microcapsules are flexible. They can be easily injected using syringes with 20-gage needles. Permeability of the membrane can be adjusted. Detailed analysis has been carried out using HPLC of a large spectrum of mol wt dextran *(21)*. The permeability can be adjusted to have different cutoff mol wt depending on the applications. Thus, for hepatocytes, it can be adjusted to allow albumin to pass through but not immunoglobulins.

3.3.1.2. THE VIABILITY OF RAT HEPATOCYTES AFTER ENCAPSULATION

Wong and Chang *(22)* noted that after isolation from the liver, the percentage of viable hepatocytes as determined by trypan blue stain exclusion was about 80%. After bioencapsulation of hepatocyes within alginate-polylysine microcapsules, the percent of viable cells was 63.40%. This is mainly due to the effect of the procedure on the hepatocyte membrane integrity. As will be shown later, the membrane can recover in certain in vivo conditions.

3.3.1.3. EXPERIMENTAL CELL THERAPY IN FULMINANT HEPATIC FAILURE RAT

Wong and Chang showed that galactosamine-induced fulminant hepatic failure rats that received control artificial cells died 66.1 ± 18.6 h after galactosamine induction *(18)*. The survival time of the group that received one peritoneal injection of 4 mL of microcapsules containing 7.40×10^6 hepatocytes was 117.3 ± 52.7 h SD. Paired analysis showed that this is significantly (P < 0.025) higher than that of the control group. The total number of hepatocytes injected in this initial study was very small, later study by another group using higher concentrations of hepatocytes resulted in increase in long-term survival rates *(23)*.

3.3.1.4. EXPERIMENTAL CELL THERAPY IN GUNN RATS—AN ANIMAL MODEL FOR HUMAN NONHEMOLYTIC HYPERBILIRUBINEMIA (CRIGLER-NAJJAR TYPE I)

Bruni and Chang investigated the use of artificial cells containing hepatocytes as cell therapy to lower bilirubin levels in Gunn rats (Fig. 4) *(19,20)*. In the first experiment, 3.5-mo-old Gunn rats weighing 258 ± 12 g were used. During the 16 d control period, the serum bilirubin increased at a rate of 0.32 ± 0.07 mg/100 mL/d. This reached 14.00 ± 1 mg/100 mL at the end of the control period. Each animal then received an ip injection of 1.10 mL of microcapsules containing 15×10^6 viable Wistar rat hepatocytes. Twenty days after implantation of the encapsulated hepatocytes, the serum bilirubin decreased to a level of 6.00 ± 1 mg/100 mL. The level remained low 90 d after the implantation. In the second experiment, control groups of Gunn rats were compared to those receiving cell therapy. The bilirubin levels did not decrease in the control group

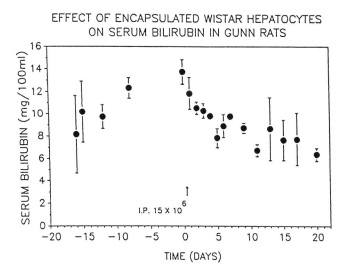

Fig. 4. Implantation of artificial cells containing hepatocytes: Effects on hyper-bilirubinemia in Gunn Rats. Reprinted with permission from ref. *19*.

and the group which received control microcapsules contained no hepatocytes. In the group receiving encapsulated hepatocytes there was significant decreases in the plasma bilirubin level. Significant decreases in the plasma bilirubin level were observed in the rat group receiving encapsulated hepatocytes. Analysis showed that implanted encapsulated hepatocytes lowered bilirubin by carrying out the function of the liver in the conjugation of bilirubin.

3.3.1.5. IMMUNOISOLATION OF BIOENCAPSULATED RAT HEPATOCYTES WHEN IMPLANTED INTO MICE

Wong and Chang reported studies on the ip implantation of free or bioencapsulated rat hepatocytes into 20–22 g male normal CD-1 Swiss mice or CD-1 Swiss mice with galactosamine-induced fulminant hepatic failure (FHF) *(22)*. This is a basic study to see if rat hepatocytes can remain viable and be immunoisolated inside freely floating artificial cells in mice. Therefore, aggregated microcapsules were not analyzed since hepatocytes do not have good viability under this condition.

As expected, free rat hepatocytes implanted into normal CD-1 Swiss mice were rapidly rejected. By day 14, there were no intact hepatocytes detected in the mice (Fig. 5). Rat hepatocytes after implantation into CD-1 Swiss mice with galactosamine-induced FHF (FHF) were rejected completely after 4–5 d. In the case of bioencapsulated hepatocytes, not only did they stay viable, there was also a significant increase ($p < 0.001$) in the percentage of viable hepato-

RAT XENOGRAFT IN NORMAL MICE

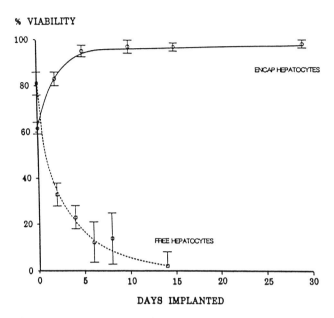

Fig. 5. Rat hepatocytes transplanted into mice. The viability and regeneration of artificial cell microencapsulated rat hepatocyte xenograft transplants in mice. This is compared to rat hepatocytes without bioencapsulation. Reprinted with permission from ref. *22*.

cytes within the microcapsules after 2 d of implantation (Fig. 5). The percentage of viable cells increased with time so that 29 d after implantation, the viability increased from the original 62% to nearly 100%. There was no significant changes in the total number of hepatocytes in the microcapsules. The viability of encapsulated rat hepatocytes implanted into galactosamine induced FHF mice also increased to nearly 100%.

In conclusion, rat hepatocytes in free floating microcapsules can be immunoisolated. As a result, xenograft of rat hepatocytes are not immunologically rejected in mice. Instead, we had have the unexpected findings of improvement in cell viability when followed for up to 29 d.

3.3.1.6. HEPATOCYTES-SECRETED HEPATIC STIMULATORY FACTOR IS RETAINED INSIDE THE MICROCAPSULE ARTIFICIAL CELLS

Studies by Kashani and Chang show that hepatocytes in the microcapsule secrete factor(s) capable of stimulating liver regeneration *(25)*. This factor is retained inside the microcapsules after secretion. Using Sephacryl gel chromatography, they showed that this factor has a mol wt >110,000. The hepatic stimulating factor accumulating in the microencapsulated hepatocyte suspension, helps to increase the viability and recovery of the membrane integrity of hepatocytes inside the artificial cells.

3.3.1.7. SUMMARY

Basic research using bioencapsulated hepatocytes shows the feasibility of this technology for cell therapy. Further improvements in biocompatibility may allow this approach to be used for cell and gene therapy in humans. This is becoming increasingly feasible because of the increasing progress in genetic engineering and molecular biology.

3.3.2. Bioencapsulation of Genetically Engineered E. coli Cells for Urea and Ammonia Removal

Urea and ammonia removal are needed in kidney failure, liver failure, environmental decontamination and regeneration of water supply in space travel. Standard dialysis machines are usually complex and expensive. Several alternatives have not been sufficiently effective. Prakash and Chang therefore studied the use of bioencapsulated genetically engineered *E. coli* DH5 cells containing *K. aerogens* urease gene *(26,27)*. The bioencapsulation of the genetically engineered bacteria was prepared by the general procedure using the apparatus assembly shown in Fig. 1. The details of the bioencapsulation process parameters are described in Section 5.

3.3.2.1. EFFICIENCY OF REMOVING UREA AND AMMONIA FROM PLASMA IN VITRO

Results show that log phase APA microencapsulated bacteria lowered 87.89 ± 2.25% of the plasma urea within 20 min and 99.99% of urea in 30 min (Fig. 6). Encapsulated bacteria are slightly more effective in removing urea from the plasma than from the modified reaction media. The bacteria did not produce ammonia during urea utilization. Furthermore, the encapsulated bacteria decrease plasma ammonia concentrations from 975 ± 70.15 to $81.15 \pm 7.37 \, \mu M$ in 30 min *(27)*. This ammonia removal efficiency of encapsulated bacteria in plasma is not significantly different than in the aqueous media. This efficiency of ammonia removal is better than currently used methods for urea or ammonia removal *(27)*.

3.3.2.2. Operational Stability of APA Microcapsules Containing Genetically Engineered Bacteria

Maximum operational stability is required to minimize the associated expense. Results show that one can use APA encapsulated bacteria for up to three cycles (Fig. 7). The APA-encapsulated bacteria plasma urea removal rate

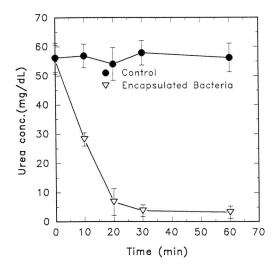

Fig. 6. APA-encapsulated bacteria on plasma urea removal. Reprinted with permission from ref. *27*.

is greater in the second and third cycles than in the first. This is probably owing to an increase in total biomass inside microcapsules with time. There is no leakage of encapsulated bacteria in the first, second, and third cycles. The cumulative urea removal capacity of encapsulated bacteria is determined for the three cycles and the results are shown in Fig. 7 *(27)*. The more recent procedure designed specifically for microencapsulation of high concentrations of smaller cells or microorganisms (*see* Section 3.2.) may allow encapsulated bacteria to be used for even more cycles.

3.3.2.3. Model Analysis of Efficacy for Urea Removal in Kidney Failure Patients

Using a single pool model, urea removal efficiency by the encapsulated bacteria from the total body fluid compartment of 40 L with a urea concentration of 100 mg/dL is analyzed *(27)*. A quantity of 40.00 ± 8.60 g of APA-encapsulated bacteria can remove 87.89 ± 2.25% of the total body urea (40 g) within 20 min and 99.99% in 30 min. This is compared as follows to other urea removal systems. It requires 388.34 g of oxystarch to remove the same amount of urea under the same conditions. It requires 1212.12 g of microcapsule containing urease-zirconium-phosphate to remove 40 g urea from the total body water. Overall, urea removal efficiency of microencapsulated genetically engineered bacteria is found to be 10–30 times higher than the best available urea removal systems available at present (Fig. 8).

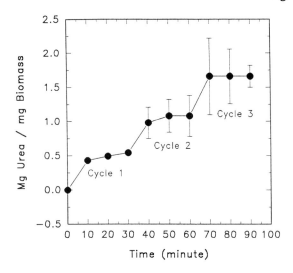

Fig. 7. Cumulative removal of plasma urea by APA encapsulated bacteria. Reprinted with permission from ref. *27*.

3.3.3. Encapsulation of Microorganism for Removing Cholesterol from Plasma

Encapsulation of other microorganisms for other applications is also possible *(28–31)*. For example, Garofalo and Chang *(28,29)* selected *Pseudomonas pictorum* (ATCC #23328) as another model system because of its ability to degrade cholesterol.

3.3.3.1. METHOD OF PREPARATION

Alginate-polylysine-alginate microcapsules do not have sufficient permeability for the lipoprotein-cholesterol macromolecules (50–1000 Å) in plasma. A new method was devised by Garofalo and Chang *(28,29)*. Briefly, A 2% agar (Difco, E. Molesley, Surrey, UK) and 2% sodium alginate (Kelco) solution was autoclaved for 15 min and cooled to 45–50°C. Bacteria Pseudomonas pictorum suspended in 0.40 mL of 0.90% NaCl was added drop by drop to 3.6 mL of agar alginate solution at 45°C, while being vigorously stirred. Three mL of the mixture obtained was extruded through the syringe, keeping the temperature at 45°C. The drops were collected in cold (4°C) 2% calcium chloride and allowed to harden. These agar-alginate beads were about 2 mm in diameter. After 15 min, the supernatant was discarded and the beads were resuspended in 2% sodium citrate for 15 min. This procedure removed alginate in the alginate-agar matrix. This resulted in a very porous agar matrix. They were then washed and stored in 0.90% saline at 4°C. Other compositions and variations in procedures were also studied.

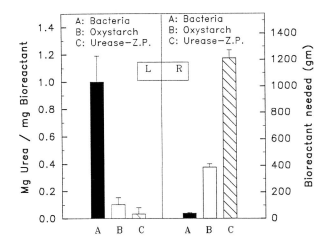

Fig. 8. Comparative study. (L) Urea removal capacity of APA-encapsulated bacteria, oxystarch, and urease-Z.P. (R) Amount of bioreactant needed to lower urea concentration from 100 mg/dL to 6.86 mg/dL from the total 40 L body fluid compartment in a 70 kg adult Men. Reprinted with permission from ref. *27*.

3.3.3.2. In Vitro Properties Including Removal of Cholesterol from Plasma

These high porosity agar beads stored at 4°C did not show any signs of deterioration. The beads retained their activity even after 9 mo of storage. There was no evidence of leakage of the enclosed bacteria. Open-pore agar beads were incubated in serum and their cholesterol depletion activity was compared to controls and nonimmobilized bacteria. The bacterial action was not significantly different between the immobilized and nonimmobilized forms. Bacterial reaction was found to be the limiting step in the overall reaction of immobilized bacteria. Other methods to remove cholesterol (e.g., LDL immunosorbents) are capacity limited. Immobilized microorganism as shown here, have almost unlimited capacity to deplete cholesterol levels. However, for practical applications, a suitable bacteria with a higher rate of cholesterol removal is needed. There is no doubt this will become available in the future with the help of genetic engineering.

4. Notes

1. The effect of alginate concentration: The concentration of alginate is very crucial in the microencapsulation process. Alginate concentration in the tested range of 1.00–2.25% (w/v), does not affect the viability of encapsulated bacteria *(27)*. Bacterial growths inside the microcapsules are independent of alginate concentration in the tested range. However, the quality of microcapsules improves with

increasing alginate concentration from 1–2% (w/v). Overall, using 2% (w/v) alginate resulted in perfect spherical shape and sturdy microcapsules with maximum number of encapsulated bacterial cells *(27)*.

2. The effect of air flow rate: Air flow rate has very important role in determining quality and size of the microcapsules. Microcapsule diameters decrease with increasing air flow rate. At an air flow rate of 2 L/min, the microcapsules had an average diameter of 500 ± 45 µm diameter. At air flow rates above 3 L/min, microcapsules are found to be irregular in shape.

3. Effect of liquid flow rate: The effects of changes in the rates of pumping the liquid (bacterial cells suspended in sodium alginate solution) and on the microencapsulation process has been determined. Flow rates from 0.00264–0.0369 mL/min resulted in an increase in microcapsule diameter. Increasing the liquid flow rate of 0.00724 to 0.278 mL/min resulted in microcapsules with spherical shape. At a low liquid flow rate, there is a tendency for the alginate to gel and clog the needle exit *(27)*.

4. Mechanical stability: We have evaluated the mechanical strength of the alginate microbeads and APA microcapsules as a function of cell leakage. This study is designed to evaluate the effective and safe use of the microencapsulated bacteria. For this, 100 microcapsules are thoroughly washed and placed in 250-mL flasks containing L. B. medium at 30°C on a orbital shaker for 7 h. The bacterial cell released is then evaluated using a coulter counter (Coulter Electronics, Luto Beds, UK). Results show that APA microcapsules are stable up to an agitation of 210 rpm in terms of absence of bacterial leakage *(27)*. For the assessment of bacterial cells released, one can also use the standard plate counting technique.

5. Overall optimal conditions for microencapsulation of genetically engineered bacteria: Based on the above studies, the optimal conditions are: 2% (w/v) alginate concentration, 0.0724 mL/min liquid flow rate, and 2 L/min air flow rate. This results in microcapsules of desired characteristics for the experimental use for the removal of urea and ammonia.

Acknowledgments

The support of the Medical Research Council of Canada and the Quebec MESST Virage "Centre of Excellence in Biotechnology" award to TMSC is gratefully acknowledged.

References

1. Chang, T. M. S. (1991) Artificial cells, in *Encyclopedia of Human Biology* (Dulbecco R., ed.), Academic, San Diego, pp. 377–383.
2. Chang, T. M. S. (1964) Semipermeable microcapsules. *Science* **146,** 524,525.
3. Chang, T. M. S. (1972) *Artificial Cells.* C. C. Thomas Publisher, Springfield, IL.
4. Chang, T. M. S. (1965) Semipermeable aqueous microcapsules. PhD Thesis, McGill University.

5. Chang, T. M. S., MacIntosh, F. C., and Mason, S. G. (1966) Semipermeable aqueous microcapsules: I. preparation and properties. *Cancer J. Physio. Pharmacal.* **44,** 115–128.

6. Chang, T. M. S., MacIntosh, F. C., and Mason, S. G. (1971) Encapsulated hydrophilic compositions and methods of making them. *Canadian Patent* **873,** 815.

7. Chang, T. M. S. (1993) Living cells and microorganisms immobilized by microencapsulation inside artificial cells, in *Fundamentals of Animal Cell Encapsulation and Immobilization* (Goosen, M. F. A., ed.), CRC, pp. 143–182.

8. Chang, T. M. S. (1995) Artificial cells with emphasis on bioencapsulation in biotechnology. *Biotechnol. Annu. Rev.* **1,** 267–293.

9. Chang, T. M. S. (1992) Recent advances in artificial cells with emphasis on biotechnological and medical approaches based on microencapsulation, in *Microcapsules and Nanoparticles in Medicine and Pharmacology* (Donbrow, M., ed.), CRC, Boca Raton, FL, pp. 323–339.

10. Seglen, P. O. (1976) Preparation of isolated rat liver cells. *Methods Cell Biol.* **13,** 29–83.

11. Mulrooney, S. B., Sturart, P., and Haussinger, R. P. (1989) *J. Gen. Microbiol.* **135,** 1769–1776.

12. Lim, F. and Sun, A. M. (1980) Microencapsulated islets as bioartificial endocrine pancreas. *Science* **210,** 908,909.

13. Goosen, M. F. A., O'Shea, G. M., Gharapetian, H. M, Chou S., and Sun, A. M. (1985) Optimization of microencapsulation parameters: semipermeable microcapsules as a bioartificial pancreas. *Biotechnol. Bioeng.* **27,** 146–150.

14. Soon-Shiong, P., Heintz, R. E., Merideth, N., Yao, Q. X., Yao, Z., Zheng, T., Murphy, M., Moloney, M. K., Schmehll, M., Harris, M., Mendez, R., Mendez, R., and Sandford, P. A. (1994) Insulin independence in a type 1 diabetic patient after encapsulated islet transplantation. *Lancet* **343,** 950,951.

15. Wong, H. and Chang, T. M. S. (1991) Microencapsulation of cells within alginate poly-L-lysine microcapsules prepared with standard single step drop technique: histologically identified membrane imperfections and the associated graft rejection. *J. Biomater. Artif. Cells Immobilization Biotechnol.* **182,** 675–686.

16. Wong, H. and Chang, T. M. S. (1991) A novel two step procedure for immobilizing living cells in microcapsules for improving xenograft survival. *J. Biomater. Artif. Cells Immobilization Biotechnol.* **19,** 687–698.

17. Chang, T. M. S. and Wong, H. (1992) A novel method for cell encapsulation in artificial cells. United State Patent No. 5,084,350.

18. Wong, H., and Chang, T. M. S. (1986) Bioartificial liver: implanted artificial cells microencapsulated living hepatocytes increases survival of liver failure rats. *Int. J. Artif. Organs* **9,** 335–346.

19. Bruni, S., and Chang, T. M. S. (1989) Hepatocytes immobilized by microencapsulation in artificial cells: Effects on hyperbilirubinemia in Gunn Rats. *J. Biomat. Artif. Cells Artif. Organs* **17,** 403–412.

20. Bruni, S. and Chang, T. M. S. (1991) Encapsulated hepatocytes for controlling hyperbilirubinemia in Gunn Rats. *Int. J. Artif. Organs* **14,** 239–241.

21. Coromili, V. and Chang, T. M. S. (1993) Polydisperse dextran as a diffusing test solute to study the membrane permeability of alginate polylysine microcapsules. *Biomater. Artif. Cells Immobilization Biotechnol.* **21,** 323–335.
22. Wong, H. and Chang, T. M. S. (1988) The viability and regeneration of artificial cell microencapsulated rat hepatocyte xenograft transplants in mice. *Biomater. Artif. Cells Artif. Organs* **16,** 731–740.
23. Dixit, V., Gordon, V. P., Pappas, S. C., and Fisher, M. M. (1989) Increased survival in galactosamine induced fulminant hepatic failure in rats following intraperitoneal transplantation of isolated encapsulated hepatocytes, in *Hybrid Artificial Organs* (Baquey, C. and Dupuy, B., eds.), Colloque ISERM, Paris, France, **177,** pp. 257–264.
24. Bruni, S. and Chang, T. M. S. (1995) Kinetic analysis of UDP-glucouronosyltransferase in bilirubin conjugation by encapsulated hepatocytes for transplantation into Gunn Rats. *Artif. Organs* **19,** 449–457.
25. Kashani, S. and Chang, T. M. S. (1991) Physical chemical characteristics of hepatic stimulatory factor prepared from cell free supernatant of hepatocyte cultures. *Biomater. Artif. Cells Immobilization Biotechnol.* **19,** 565–578.
26. Prakash, S. and Chang, T. M. S. (1993) Genetically engineered *E. coli* cells containing *K. aerogenes* gene, microencapsulated in artificial cells for urea and ammonia removal. *Biomater. Artif. Cells Immobilization Biotechnol.* **21,** 629–636.
27. Prakash, S. and Chang, T. M. S. (1995) Preparation and in vitro analysis of microencapsulated genetically engineered *E. coli* for urea and ammonia removal. *Biotechnol. Bioengineering* **46,** 621–626.
28. Garofalo, F. and Chang, T. M. S. (1989) Immobilization of *P. pictorum* in open pore agar, alginate polylysine-alginate microcapsules for serum cholesterol depletion. *Biomater. Artif. Cells Artif. Organs* **17,** 271–90.
29. Garofalo, F. and Chang, T. M. S. (1991) Effects of mass transfer and reaction kinetics on serum cholesterol depletion rates of free and immobilized Pseudomonas pictorum. *Appl. Biochem. Biotechnol.* **27,** 75–91.
30. Lloyd-George, I. and Chang, T. M. S. (1993) Free and microencapsulated *Erwinia herbicola* for the production tyrosine. *J. Biomater. Artif. Cells Immobilization Biotechnol.* **21,** 323–335.
31. Chang, T. M. S. (ed.) *Artificial Cells, Blood Substitutes & Immobilization Biotechnology, An International Journal.* Marcel Decker, NY.

27

Hyperexpression
of a Synthetic Protein-Based Polymer Gene

**Henry Daniell, Chittibabu Guda, David T. McPherson,
Xiaorong Zhang, Jie Xu, and Dan W. Urry**

1. Introduction

Environmental problems require the development of biodegradable plastics of benign production that can be synthesized from renewable resources without the use of toxic and hazardous chemicals and will help in solving the increasing global disposal burden. Protein-based polymers offer a wide range of materials similar to that of petroleum-based polymers such as elastomers and plastics. Protein-based polymers can be prepared of varied design and composition through genetic engineering without the use of hazardous and noxious solvents and can be made biodegradable with chemical clocks to set their half-lives such that they can be environmentally friendly over their complete life cycles of production and disposal. Compositions tested to date have been shown to be extraordinarily biocompatible, allowing for medical applications ranging from the prevention of postsurgical adhesions and tissue reconstruction to programmed drug delivery *(1)*. Among nonmedical applications, there are transducers, molecular machines, super absorbents, biodegradable plastics, and controlled release of agricultural crop enhancement agents like herbicides, pesticides, and growth factors.

Bioelastic materials are based on elastomeric and related polypeptides comprised of repeating peptide sequences *(2)*; they may also be called elastic and plastic protein-based polymers. The parent polymer, $(Val^1\text{-}Pro^2\text{-}Gly^3\text{-}Val^4\text{-}Gly^5)_n$ or poly(VPGVG), derives from sequences that occur in all sequenced mammalian elastin proteins *(3)*. In the most striking example, the sequence $(VPGVG)_n$ occurs in bovine elastin with $n = 11$ without a single substitution *(3)*. A particularly interesting analog is poly (AVGVP) as it reversibly forms a plastic on raising the temperature.

From: *Methods in Molecular Biology, vol. 63: Recombinant Protein Protocols:
Detection and Isolation* Edited by: R. Tuan Humana Press Inc., Totowa, NJ

What is required for the commercial viability of protein-based polymers is a cost of production that would begin to rival that of petroleum-based polymers. The potential to do so resides in low cost bioproduction. We have recently demonstrated a dramatic hyperexpression of an elastin protein-based polymer, (Gly-Val-Gly-Val-Pro)$_n$ or poly(GVGVP), which is a parent polymer for a diverse set of polymers that exhibit inverse temperature transitions of hydro-phobic folding, and assembly as the temperature is raised through a transition range and which can exist in hydrogel, elastic, and plastic states. Electron micrographs revealed formation of inclusion bodies in *E. coli* cells occupying up to 80–90% of the cell volume under optimal growth conditions *(3a)*. The beauty of this approach is the lack of any need for extraneous sequences for the purposes of purification *(4)* or adequate expression. The usual strategy for expression of a foreign protein or protein-based polymer in an organism such as *E. coli* antici-pates that the foreign protein will be injurious to the organism. Accordingly, the transformed cells are grown up to an appropriate stage before expression of the foreign protein is begun and expression is generally considered viable for only a few hours. The situation is quite different for the elastic protein-based polymer considered here. This may result in part due to the extraordinary biocompatibility exhibited by (GVGVP)$_n$ and its related polymers. The elastic protein-based polymer, (GVGVP)$_n$ and its γ-irradiation crosslinked matrix as well as related polymers and matrices appear to be ignored by a range of animal cells and by tissues of the whole animal *(5–7)*. This chapter describes in detail meth-odologies to accomplish hyperexpression of a protein based polymer in *E. coli*.

2. Materials

2.1. Partial Purification of Polymer Protein

PBS buffer (pH 7.4): 10 mM NaHPO$_4$, 2 mM KH$_2$PO$_4$, 137 mM NaCl, 3 mM KCl.

2.2. SDS-Gel Electrophoresis and Copper Staining

1. Solution 1: 30% acrylamide (dissolve 58.4 g acrylamide and 1.6 g bis-acrylamide in 200 mL distilled water).
2. Solution 2: 1.5M Tris-HCl, pH 8.8.
3. Solution 3: 10% sodium dodecyl sulfate (SDS).
4. Solution 4: 0.5M Tris-HCl, pH 6.8.
5. Solution 5: 10% ammonium persulfate (freshly prepared).
6. TEMED (tetramethylethylenediamine).
7. Electrophoresis buffer: Dissolve 12 g Tris, 57.6 g glycine, and 4 g SDS in 4 L of water.
8. 2X Gel loading buffer: 100 mM Tris-HCl, pH 6.8, 200 mM dithiothreitol (DTT), 4% SDS, 0.2% bromophenol blue, 20% glycerol.

9. High range protein marker (Bio-Rad, Hercules, CA).
10. Staining buffer: $0.19M$ Tris-HCl, pH 8.8 and 0.1% SDS.
11. $0.3M$ CuCl$_2$.

All the stock solutions should be prepared in distilled water. Store solutions 1, 2, 4, and 6 at 4°C and 9 at –20°C. All others solutions may be stored at room temperature.

2.3. Transmission Electron Microscopy

1. $0.2M$ Cacodylate buffer: Dissolve 42.8 g sodium cacodylate in 1 L distilled water, adjust pH to 7.2 by adding concentrated NaOH or HCl.
2. 3% Glutaraldehyde (final concentration) in $0.05M$ cacodylate buffer.
3. 1% Osmium tetroxide (final concentration) in $0.05M$ cacodylate buffer.
4. 2% Agarose: Add 0.2 g agarose in 10 mL water and dissolve by boiling in a microwave.
5. Graded series of ethanol (30, 50, 70, 80, 90, 100%).
6. Propylene oxide.
7. Spurr's low viscosity embedding resin *(8)*.
   a. Vinyl cyclohexane dioxide (VCD [ERL 4206])—20 g.
   b. Diglycidyl ether of polypropylene glycol (DER 736)—12 g.
   c. Nonenyl succinic anhydride (NSA)—52 g.
   d. Dimethyl amino ethanol (DMAE)—0.8 g.
   Add and mix the first three components thoroughly before adding DMAE, then mix again. This mixture can be stored in a refrigerator inside a desiccator.
8. Super glue (Devcon, Wooddale, IL).

Store all solutions at 4°C except propylene oxide.

2.4. Staining of Grids

1. 1% Uranyl acetate, pH 4.0 in double distilled water.
2. Lead citrate, pH 12.0.
3. $0.1N$ NaOH.
4. NaOH electrolytic pellets.

Store solutions 1 and 2 at 4°C.

3. Methods
3.1. Partial Purification of Polymer Protein

1. Pellet cells by centrifugation ($5000g$, 10 min, 4°C) from 48 h TB grown cultures without IPTG induction. Discard supernatant.
2. Wash the pellet twice with 10 mL PBS buffer (pH 7.4).
3. Resuspend pellet in 4 mL PBS buffer.
4. Lyse cells by sonication twice, 15 min each (50% amplitude) or by French Press (1200 psi).
5. Centrifuge to remove cell debris ($14,000g$, 10 min, 4°C).
6. Collect supernatant and distribute into microfuge tubes, 1.5 mL each.

7. Incubate tubes at 37°C for 20 min to allow phase transition of polymer to insoluble form
8. Spin at 5000g for 3 min at room temperature to allow settling down of polymer.
9. Discard supernatant and resuspend the pellet in 100 μL PBS buffer.
10. Store tubes on ice for 15 min to allow reverse phase transition to soluble form.
11. Centrifuge at 14,000g for 10 min at 4°C.
12. Collect supernatants as partially purified polymer sample.

3.2. SDS-Gel Electrophoresis of the Polymer Protein

Prepare and carry out electrophoresis of SDS-polyacrylamide gels according to Laemmli *(9)*.

1. Lower gel preparation (10%): Add 10 mL solution 1, 7.5 mL solution 2, 0.3 mL, solution 3, 12 mL distilled water, 100 μL solution 5, and 10 μL TEMED into a 100-mL flask and mix well using a transfer pipet. Immediately pour the solution in between gel plates and allow it to polymerize for 30 min.
2. Stacking gel preparation: Add 1.4 mL solution 1, 2.6 mL solution 4, 100 μL solution 3, 5.9 mL distilled water, 50 μL solution 5, and 10 μL TEMED into a 100-mL flask and mix well using a transfer pipet. Immediately pour the solution on the lower gel in between the plates and place the comb in place avoiding air bubbles. Allow it to polymerize for 30 min.
3. Centrifuge 1.5 mL of culture in a microfuge tube at 12,000g for 45 s.
4. Remove supernatant and wash the pellet in 500 μL of Tris-HCl (50 mM, pH 7.6).
5. Centrifuge again to pellet cells, remove supernatant, and resuspend pellet in 100 μL of water.
6. To a fresh microfuge tube, add 20 μL of above sample and an equal volume of 2X SDS gel loading buffer and boil for 5 min. Also boil the high-range protein marker (2 μL) and partially purified polymer (2 μL) after adding 2X loading buffer.
7. Immediately load 40 μL of each sample into individual wells along with high-range protein marker (Bio-Rad) and partially purified polymer protein.
8. Run the gel at 26 mA for 5 h (the current and the run time can be adjusted according to the gel size and convenience).
9. After electrophoresis, take out the gel and soak in Tris-HCl (0.19M, pH 8.8 + 0.1% SDS) for 10 min with gentle shaking.
10. Rinse in distilled water once and soak again in 0.3M CuCl$_2$ solution for 5 min on a shaker *(10)*.
11. Observe polymer polypeptides as negatively stained bands against a dark background.

3.3. Transmission Electron Microscopy

1. Wash *E. coli* cells twice with distilled water by centrifuging and resuspending the pellet.
2. Resuspend the pellet after second wash in 5 mL of 3% glutaraldehyde (final concentration) in 0.05M cacodylate buffer for 3 h at 4°C.
3. Pellet cells, remove supernatant, and resuspend the pellet in 0.05M cacodylate buffer for 12 h to remove excess glutaraldehyde.

4. Pellet cells and resuspend pellet in 5 mL of 1% osmium tetroxide (final concentration) in cacodylate buffered solution ($0.05M$) for 1.5 h.
5. Wash the pellet again with $0.05M$ cacodylate buffer to remove any unbound Os.
6. Pellet cells, remove supernatant and, keep the pellet as dry as possible.
7. Boil 2% agarose solution in a microwave and cool it to 50°C.
8. Add 2 mL of 2% agarose solution to the pellet and mix the pellet and agar with a toothpick, which should form a slurry and solidify in a few minutes.
9. From the solidified pellet, mince about 1 mM^3 size blocks with razor blade and place them in glass vials (~10 pieces per vial).
10. Carry out dehydration steps in glass vials. Add 2 mL of each grade of ethanol (30, 50, 70, 80, 90, and 100%) at an interval of 20 min. Remove the previous grade completely using a transfer pipet before adding the next higher grade of ethanol.
11. Similarly, treat samples with three changes of propylene oxide in 20 min intervals.
12. Add a mixture of propylene oxide:Spurr's resin in the proportion of 3:1. Repeat this step three times by gradually increasing resin content over propylene oxide (1:1, 3:1, and pure resin) in intervals of 2 h each on a gyrator. Finally, add pure resin and allow to infiltrate for 12 h on the gyrator with 2–3 changes of pure resin in between. After 12 h, remove the old resin and add 2 mL of fresh resin.
13. Embed blocks in plastic molds and allow the resin to cure by incubating at 65°C for 8 h.
14. Cut the sample out of the mold using a saw, mount on the resin block using super glue, and incubate at 65°C for 40 min.
15. Trim blocks with glass knives and section using a microtome.
16. Pickup silver sections with a thickness of ~60 nm on copper grids for staining. Steps 1–10 should be done at 4°C.

3.4. Staining of Grids

1. Put a piece of parafilm in a Petri plate. Add drops of uranyl acetate (pH ~4.0) separately on parafilm according to the number of grids to be stained (one drop per grid).
2. Carefully place one grid in each drop and wait for 40 min.
3. Wash each grid in double distilled water by dipping it at least 20X using a pointed forceps and dry on a filter paper.
4. Add a few electrolytic pellets of NaOH in a Petri dish containing parafilm.
5. Add drops of lead citrate (pH ~12.0), place one grid in each drop and wait for 2 min.
6. Wash each grid first in $0.1N$ NaOH, then in double distilled water, and dry the grids on filter paper.

Observe specimens under a transmission electron microscope at 60 kV accelerating voltage.

3.5. Evaluation of Results

Construction of a synthetic protein-based polymer gene: As an illustration of an uninduced hyper-expression of a protein-based polymer in *E. coli*, we have chosen a gene encoding 121 repeats of the elastomeric pentapeptide

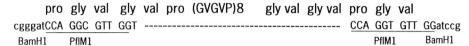

Fig. 1. Amino acid sequence and flanking restriction endonuclease sites of the basic polymer building block coding for (GVGVP)$_{10}$. Using synthetic oligonucleotides and PCR, (GVGVP)$_{10}$ was amplified with flanking BamHI and pflM1 ends and the 121-mer gene was inserted into pUC118 as a BamHI fragment. For expression under control of the T7 polymerase gene promoter, a 121-mer gene was created by concatenation of the PflMI 10-mer fragment with terminal cloning adaptors and subsequently inserted into the expression vector pET-11d.

GVGVP. This gene, (GVGVP)$_{121}$, was constructed by ligase concatenation of a DNA sequence encoding (GVGVP)$_{10}$ and isolation of a concatemer having 12 repeats of this monomer gene plus an additional C-terminal GVGVP sequence encoded by a 3' cloning adaptor *(10a)*. The gene encoding (GVGVP)$_{10}$ was synthesized and cloned into a multipurpose cloning plasmid from which it was then excised by digestion at flanking sites with the restriction endonuclease PflM1 (Fig. 1). A substantial amount of the PflM1 gene fragment was purified and self-ligated in the presence of limited amounts of synthetic double-stranded oligonucleotide adapters that provided the additional restriction sites needed for cloning the resulting concatemers. PflM1 cleaves at its recognition site in the DNA to leave two single-stranded extensions that are not self-complementary (i.e., nonpalindromic) but are only complementary to each other; therefore, proper translational polarity is maintained by head-to-tail tandem coupling of the monomer gene units by ligase during the concatenation reaction.

3.5.1. Vector Construction and Polymer Gene Expression

Concatemer genes recovered by the above procedure, including the (GVGVP)$_{121}$, were ultimately placed into the expression vector pET-11d (Novagen, Madison, WI) immediately adjacent to the initiator ATG codon. This vector is part of the T7 expression system *(11)* that utilizes the coliphage T7 RNA polymerase to drive expression from the T7 promoter on the plasmid. In this case, the polymerase is provided by the host strain HMS 174 (DE3), a lambda lysogen carrying the T7 RNA polymerase gene on the stably integrated phage genome *(12)*. Expression of the RNA polymerase gene is under control of the *lac*UV5 promoter and is therefore inducible by addition of IPTG to the growth medium. Expression of an inserted foreign gene in pET-11d is regulated by two *lacI* repressor genes, one located in the plasmid pET-11d and the other in the genome of the host strain HMS 174 (DE3).

Gene expression was studied in samples each grown in either Luria broth (LB) or in Terrific broth (TB, ref. *13*) in the presence or absence of ampicillin (100 µg/mL) at 37°C. After 3 h of growth (at an OD of 0.8), cultures were induced with 1 mm isopropylthio-β-D-galactoside (IPTG) and continued to grow for different durations; cells were stored at 4°C at the end of the time-course.

Cell lysates of both TB and LB grown cultures separated on SDS-polyacrylamide gels are shown in Fig. 2. Polymer protein can be seen by negative staining around 60 kDa (Fig 2). The pattern of polymer production is observed to be the same in both gels, although the quantity of polymer is several-fold more in TB grown cultures (uninduced). The amount of polymer in uninduced 6 h sample (lane 3) is approximately comparable to that of the induced 6 h sample (lane 4). However, there is a dramatic increase in the expression of polymer in uninduced cultures grown for 24 h (lane 5) over induced cultures of the same age (lane 6). This increase is more pronounced in TB grown cultures compared to LB grown cultures which is not surprising because it is known that in TB grown cultures, copy number of the plasmid increases by four- to sevenfold and the cell density increases by 10-fold over those of the LB grown cultures *(13)*. In contrast, the amount of polymer produced in induced cultures is negligible (lanes 6, 8, and 10) accompanied by irregular shapes of cells (*see* the electron micrograph in Fig. 3C). Decrease in polymer production in induced cells could be directly correlated with loss of the introduced plasmid and reduced cell growth. No plasmid DNA was found in cells induced with IPTG beyond 6 h of growth (data not shown). Reduced cell growth after IPTG induction has been reported earlier. For example, Brosius *(14)* reported that induction of trp/lac (tac) hybrid promoter with 1–5 mM IPTG in *E. coli* strain RB 791 (lac repressor overproducing strain) caused reduced cell growth, ultimately leading to cell lysis. Masui *(15)* also reported that the growth rate of *E. coli* T19 cells induced with IPTG was reduced after 5–6 h. In our studies, the highest expression of protein is observed in 24 h uninduced cultures (lane 5), followed by a gradual reduction in cultures grown beyond 24 h (lanes 7 and 9). This may be owing to cell lysis as well as decrease in plasmid copy number after 24 h as evident from light microscopic observations and plasmid DNA isolation studies (data not shown). However, it should be pointed in this context that this polymer protein, $(GVGVP)_{121}$, is extraordinarily stable in *E. coli* cells as seen in Fig. 2 (lane 12) that shows polymer present in a 48 h uninduced culture stored for over three months at 4°C.

The polypeptide observed at the same mol wt as β-galactosidase (116.3 kDa), after 24 h of growth (Fig. 2, lanes 6, 8, 10) in induced cells but not in uninduced cells (Fig. 2, lanes 3, 5, 7, 9) may be β-galactosidase (a chromosomal gene) induced by addition of gratuitous inducer (IPTG). This result is in accordance with earlier reports that bacteria produce β-galactosidase only when

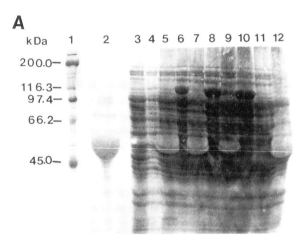

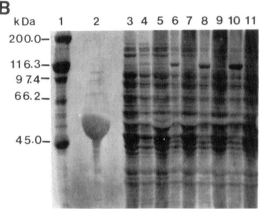

Fig. 2. Crude protein extracts from *E. coli* strain HMS 174 (DE3) transformed with pET 11d-120 mer separated on SDS-PAGE gels. **(A)** Cultures grown in Terrific Broth (TB). **(B)** Cultures grown in Luria Broth (LB). Lane 1: high range protein marker showing (top to bottom) myosin, β-galactosidase, phosphorylase b, bovine serum albumin, and ovalbumin; lane 2: partially purified polymer standard; lane 3: uninduced-6 h; lane 4: induced-6 h; lane 5: uninduced-24 h; lane 6: induced-24 h; lane 7: uninduced-48 h; lane 8: induced-48 h; lane 9: uninduced-72 h; lane 10: induced-72 h; lane 11: host strain without plasmid; and lane 12: uninduced-48 h culture stored for 3 mo at 4°C. For induction, 1 m*M* IPTG was added when the cultures reached 0.8 OD.Reprinted from Guda et al. *(3a)*.

its substrate is added to the medium *(16)*. Excess production of polymer in 24 h uninduced cells may also be attributed to reduction or dilution of repressor protein as evidenced by increase in β-galactosidase production in induced cells

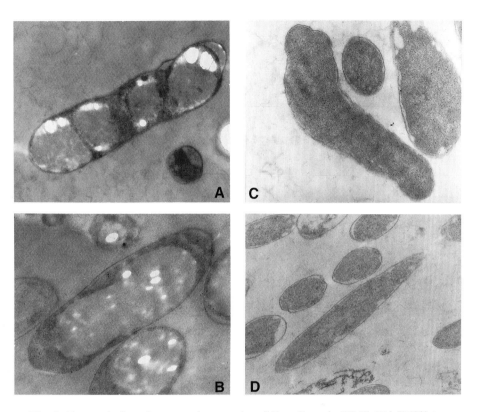

Fig. 3. Transmission electron micrographs of *E. coli* strain HMS 174 (DE3) transformed with pET 11d-121 mer showing polymer production in uninduced and induced cells. **(A,B)** uninduced cells, 24 h; **(C)** an induced cell, 24 h; **(D)** an uninduced cell without plasmid.

of the same age (Fig. 2). Dilution of the repressor protein in rapidly growing cells should have enabled the RNA polymerase to efficiently bind and initiate transcription of the T7 promoter that drives the polymer gene. It is known that the amount of repressor protein produced by DE3 on host chromosome is not sufficient to block the T7 polymerase transcription; BL21 (DE3) cells transformed with the pT7-5 plasmid that carries a *cryIIA* gene driven by the T7 promoter produced large quantities of *cryIIA* crystals without any need for induction with IPTG *(17)*.

Light micoscopic studies using oil immersion lens (Carl Zeiss, plan 100/1, 25 oil ph3) showed distinct intracellular inclusion bodies both in induced and uninduced cells at 6 h growth period (data not shown). The first inclusion body in a cell is found generally but not necessarily at one end of the cell. But, as the cell growth progressed, the number and size of inclusion bodies increased only in uninduced cells up to 24 h. In contrast, induced cells grown beyond 6 h attain irregular shapes without any inclusion body. These results correlate well with SDS-PAGE results., where maximum polymer production is observed in 24 h TB grown culture. Hence, we restricted the TEM studies to 24 h in induced and uninduced cells including the host strain without plasmid as control.

Electron micrographs of induced and uninduced *E. coli* cells are shown in Fig. 3. At 24 h, uninduced cells show a number of inclusion bodies pushing the cytoplasm aside and occupying the cell volume to a maximum extent possible (Fig. 3A). The inclusion bodies are well separated from the cytoplasm although a definite membranous boundary is lacking. Such lack of membrane boundaries around inclusion bodies has been reported in the past *(18)*. Polymer inclusion bodies appear as glittering bodies amidst a dense dark background of the cell cytoplasm; these structures are distinct from the cell cytoplasm by their lighter stain and round to oval shape with poorly infiltrated regions showing dense reflecting matrix. In some cells there are single large inclusion bodies occupying about 70–80% of the cell volume (Fig. 3B) whereas in others several bodies occupy nearly 80–90% of the cell volume (Fig. 3A). In contrast, induced cells have irregular cell wall with dense cytoplasm and no inclusion body (Fig. 3C). Host cells not transformed with plasmid DNA show normal cell growth with no inclusion body (Fig. 3D).

The progression in the number and size of inclusion bodies up to 24 h grown cells (Fig. 3A,B) correlate well with increase in the amount of polymer protein produced (Fig. 2A, lanes 3,5) in SDS-polyacrylamide gels. Similarly, absence of inclusion bodies in induced cells at 24 h (Fig. 3C) and in plasmid-minus host cells (Fig. 3D) is accompanied by lack of polymer production. From this we correlate the appearance of inclusion bodies in transformed *E. coli* cells to production of polymer protein. Similar correlations were made between the amount of inclusion product within the cells to the quantity of chimeric product seen in SDS-polyacrylamide gels in *E. coli* strains overproducing insulin chains A and B *(18)*. They estimated that at peak production level, the inclusion bodies could occupy as much as 20% of the *E. coli* cellular volume. In our studies, by quantitative analyses of electron micrographs using BioScan Optimas version 3.10, the volume occupied by these bodies were estimated to be as much as 80–90% of the cell volume under optimal conditions. The mean area occupied by the polymer in fully grown cells was about 65–75% of the cell volume. Inclusion bodies are formed when foreign proteins are synthesized to levels above their solubility or when precipitation of the native protein takes place *(19)*. The protein based polymer, $(GVGVP)_{121}$, self associates at 37°C, at which point fermentations were carried out *(1)*. In many instances, high level expression of not only heterologous proteins, but also native *E. Coli* proteins results in formation of inclusion bodies as evident in the a subunit of RNA polymerase *(20)*.

4. Notes

1. Culture conditions: Flask volume should be at least four times larger than the culture volume. The shaker speed should not be more than 200 rpm. Too much shaking increases cell density but it has a negative effect on polymer yield.
2. SDS-gel electrophoresis: Load the protein samples immediately after boiling. Do not store samples on ice, since SDS would precipitate.
3. Transmission electron microscopy: If unbound osmium tetroxide is not washed properly in subsequent buffer washes, Os will deposit all over the specimen causing Os pepper on sections. Proper fixation of tissues is essential for good quality sections on microtome. Hence, follow fixation steps carefully, especially the duration of fixation. Super glue instantly bonds skin; therefore, wear protective gloves.
4. Staining for TEM: Uranyl acetate is photosensitive. Try to avoid exposure to light as much as possible and store in the dark at 4°C. Keep the Petri dish covered with aluminum foil during 40 min staining period. Lead citrate solution should be prepared on freshly boiled, double-distilled water since it is extremely sensitive to CO_2. Use a narrow-necked bottle to avoid exposure to CO_2 in air and store at 4°C. Exposure to CO_2 in air results in formation of lead carbonate resulting in deposition of lead carbonate on the sections. While staining, do not let grids float on lead citrate. Immerse grids completely inside the stain drop, to avoid exposure to CO_2 in air. You should also hold your breath while handling the grids to prevent CO_2 exposure to grids. NaOH is highly corrosive. Do not use near instruments that do not have any protective cover on them.

Acknowledgments

Research presented in this chapter was supported in part by a contract from the U. S. Army, Natick RD & E center (#DAAK60-93-C-0094 to DWU), NIH, NIAID grant P30-AI-27767 to the UAB AIDS center, National Institute of Health grant (#GM16551-01), and USDA-NRICGP grant (95-02770) to HD. The authors are thankful to Roland R. Dute and William J. Moar of Auburn University for critical discussions.

References

1. Urry, D. W., Nicol, A., Gowda, D. C., Hoban, L. D., McKee, A., Williams, T., Olsen, D. B., and Cox, B. A. (1993) Medical applications of bioelastic materials, in *Biotechnological Polymers: Medical, Pharmaceutical and Industrial Applications* (Gebelein, C. G., ed.), Technomic, Atlanta, GA, pp. 82–103.
2. Urry, D. W. (1995) Elastic biomolecular machines: synthetic chains of amino acids patterned after those in connective tissue can transform heat and chemical energy into motion. *Scientific American* **272,** 64–69.

3. Yeh, H., Ornstein-Goldstein, N., Indik, Z., Sheppard, P., Anderson, N., Rosenbloom, J., Cicilia, G., Yoon, K., and Rosenbloom, J. (1987) Sequence variation of bovine elastin mRNA due to alternative splicing. *Collagen Related Res.* **7,** 235–247.

3a. Guda, C., Zhang, X., McPherson, D. T., Xu, J., Cherry, J. H., Urry, D. W., and Daniell, H. (1995) Hyperexpression of an environmentally friendly synthetic polymer gene. *Biotechnol. Lett.* **7,** 745–750.

4. McPherson, D. T., Morrow, C., Mineham, D. S., Wu, J., Hunter, E., and Urry, D. W. (1992) Production and purification of a recombinant elastomeric polypeptide, G-(VPGVG)19-VPGV from *Escherichia coli. Biotechnol. Prog.* **8,** 347–352.

5. Urry, D. W., Parker, T. M., Reid, M. C., and Gowda, D. C. (1991) Biocompatibility of the bioelastic materials, poly(GVGVP) and its γ-irradiation cross-linked matrix: surgery of generic biological test results. *J. Bioactive Compatible Polym.* **6,** 263–282.

6. Urry, D. W., Nicol, A., McPherson, D. T., Xu, J., Shewry, P. R., Harris, C. M., Parker, T. M., and Gowda, D. C. (1995) Properties, preparations and applications of bioelastic materials, in *Handbook of Biomaterials and Applications*, vol. 2, Marcel Dekker, New York, pp. 1619–1673.

7. Nicol, A., Gowda, D. C., Parker, T. M., and Urry D. W. (1993) Elastomeric polytetrapeptide matrices: Hydrophobicity dependence of cell attachment from adhesive, (GGIP)n, to non-adhesive, (GGAP)n, even in serum. *J. Biomed. Mater. Res.* **27,** 801–810.

8. Spurr, A. R. (1969) A low viscosity epoxy resin embedding medium for electron microscopy. *J. Ultrastruct. Res.* **26,** 31–36.

9. Laemmli, U. K. (1970) Cleavage of structural proteins during the assembly of the head of bacteriophage T4. *Nature* **227,** 680–685.

10. Lee, C., Levin, A., and Branton, D. (1987) Copper staining: a five-minute protein stain for sodium dodecyl sulfate-polyacrylamide gels. *Anal Biochem.* **166,** 308–312.

10a. McPherson, D. T., Xu, J., and Urry, D. W. (1996) Product purification by reversible phase transition following *Escherichia coli* expression of genes encoding up to 251 repeats of the elastomeric pentapeptide GVGVP. *Protein Expression and Purification* **7,** 51–57.

11. Studier, F. W., Rosenberg, A. H., Dunn, J. J., and Dubendorff, J. W. (1990) Use of T7 RNA polymerase to direct expression of cloned genes. *Methods Enzymol.* **185,** 60–89.

12. Studier, F. W. and Moffat, B. A. (1986) Use of Bacteriophage T7 RNA polymerase to direct selective high level expression of cloned genes. *J. Mol. Biol.* **189,** 113–130.

13. Tartof, K. D. and Hobbs, C. A. (1987) Improved media for growing plasmid and cosmid clones. *Bethesda Res. Lab. Focus* **9,** 12.

14. Brosius, J. (1984) Toxicity of an overproduced foreign gene product in *Escherichia coli* and its use in plasmid vectors for the selection of transcription terminators. *Gene* **27,** 161–172.

15. Masui, Y., Mizuno, T., and Inouye, M. (1984) Novel high-level expression cloning vehicles: 10^4-fold amplification of *Escherichia coli* minor protein. *Bio/Technology* **2**, 81–85.
16. Lewin, B. (1990) A panoply of operons: the lactose paradigm and others, in *Genes IV*, Oxford University Press, New York, pp. 240–263.
17. Daniell, H., PoroboDessai, A., Prakash, C. S., and Moar, W. J. (1994) Engineering plants for stress tolerance via organelle genomes, in *Biochemical and Cellular Mechanisms of Stress Tolerance in Plants* (Cherry, J. H., ed.), NATO ASI Series in Cell Biology, Springer-Verlag, Berlin Heidelberg, pp. 589–604.
18. Williams, D. C., Van Frank, R. M., Muth, W. L., and Burnett, J. P. (1982) Cytoplasmic inclusion bodies in *Escherichia coli* producing biosynthetic human insulin proteins. *Science* **215**, 687,688.
19. Kane, J. F. and Hartley, D. L. (1988) Formation of recombinant protein inclusion bodies in *Escherichia coli*. *Trends Biotechnol.* **6**, 95–101.
20. Gribskov, M. and Burgess, R. R. (1983) Overexpression and purification of the σ subunit of *Escherichia coli* RNA polymerase. *Gene* **26**, 109–118.

28

Transplantation of Cells in an Immunoisolation Device for Gene Therapy

Victoria E. Carr-Brendel, Robin L. Geller,
Tracy J. Thomas, Daniel R. Boggs, Susan K. Young,
Joanne Crudele, Laura A. Martinson, David A. Maryanov,
Robert C. Johnson, and James H. Brauker

1. Introduction

Treatment of genetic deficiency diseases, the so-called inborn errors of metabolism has been fairly limited in scope. Whereas there has been significant success with injections of the missing gene product in cases such as diabetes and hemophilia, management of most genetic diseases has been limited to treatment of various clinical symptoms and not the underlying disease state. For the past decade, many groups have been working toward the goal of gene therapy, i.e., the correction of genetic deficiency diseases by the introduction of a correct copy of the defective gene into affected individuals. Many different diseases have been targeted (*see* ref. *1* for review) utilizing a variety of cells and animal models *(1–14)*. Although many groups have been able to obtain high levels of gene expression in vitro, in vivo expression has been disappointing. In most cases, while the gene product could be detected within hours after introduction into the host, high level expression was not detectable after 30–45 d *(4,5,11,12,14–19)*.

Several factors could be responsible for the loss of expression including death of the engineered cells after introduction into the host, host immune responses to the newly introduced gene products, and/or downregulation of the exogenous promoter. We have focused on approaches for cell transplantation that could address some of these issues. Our work entails the use of an implantable device (immunoisolation device) that encapsulates the cells in such a way that the grafted tissues are protected from contact with host immune cells. In

From: *Methods in Molecular Biology, vol. 63: Recombinant Protein Protocols: Detection and Isolation* Edited by: R. Tuan Humana Press Inc., Totowa, NJ

373

addition, host-derived vascularization occurs at the device/soft tissue interface thereby enhancing cell survival. The vital role of vascularization in implanted cell viability was first addressed several years ago by Thompson et al. who demonstrated that improved bilirubin metabolism was achieved in Gunn rats by implantation of cells expressing the appropriate enzyme, if new blood vessels were induced at the site of implantation *(20)*. They achieved increased vascularization by implanting the cells in a matrix of polytetraflouroethylene (PTFE) fibers coated with type 1 collagen and heparin-binding growth factor. This neovascularization appeared to be necessary to create a supportive environment for the transplanted cells.

To date, most approaches to gene therapy have utilized the patient's own cells which are removed, genetically engineered, and then returned to the patient (autologous somatic cell gene therapy). This requires the development of systems that can be tailored to the individual patient. An alternative to autologous somatic cell gene therapy could be the use of allogeneic cell lines as the source of engineered cells. Use of allogeneic cells eliminates the need for complicated cellular isolation and engineering procedures that are specific to each patient. However, it is limited by host immune rejection of allogeneic tissue. The use of an immunoisolation device, when applied to gene therapy overcomes this problem. Through the use of an immunoisolation device it is possible to maintain allogeneic cells such that the graft tissue remains functional but is protected from the host immune response without the need for immunosuppressive drugs. An additional advantage to the use of allogeneic cells within a device is they would be rejected if they escaped from the device. Moreover, uniformity could be introduced into the process because allogeneic cells, potentially useful for many or all patients, could be prepared at a central industrial-scale location. The quality control issues for the cells would then be met at the same levels as are currently used for recombinant protein production.

We have developed an immunoisolation device that is composed of a bilayer flat sheet membrane. One of the components of the bilayer membrane is a PTFE membrane (5 μm pore size) that is able to alter the host foreign body response such that blood vessels are formed at the membrane interface *(21)*. The other (inner) layer is a 0.45 μm hydrophilized PTFE that eliminates contact between the host tissue and the implanted cells. This membrane configuration can immunoprotect allogeneic graft tissue; Sprague Dawley rat fetal lung tissue has survived for one year when implanted into male Lewis rats *(22)*.

Optimizing the combination of immunoisolation and gene therapy is a multistage process. In this chapter, we will describe the use of a flat-sheet diffusion chamber design (Fig. 1), first proposed by Algire et al. *(23)*, as the implantable chamber. The specific method by which the membranes are sealed is not important. We will describe one approach using a proprietary device and

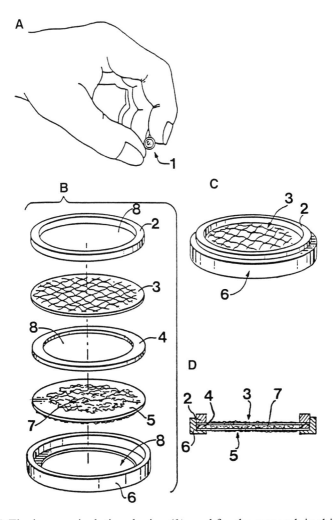

Fig. 1. (A) The immunoisolation device *(1)* used for the research in this chapter was approximately a centimeter in diameter. **(B)** Schematic diagram of the components of a flat-sheet, friction-sealed immunoisolation chamber. Tissue *(7)* is loaded between two cell impermeable membranes *(3,5)* in the presence of a spacer *(4)* and this sandwich is then held together by two, friction-sealed titanium metal rings *(2,6)* whose lumen is indicated *(8)*. **(C)** Diagram of a top view loaded, friction-sealed device. The titanium housing *(6)* and insert *(2)* hold the membrane *(5)* immobile. **(D)** A cross-section through a tissue-loaded, friction sealed device. The titanium insert *(2)*, housing *(6)*, and silicone spacer *(4)* are represented as hatched areas. The flat sheet membranes *(3,5)* are held together by the housing and insert, encapsulating the tissue *(7)*.

accessories, but the same results could be achieved with other sealing designs. More important is the choice of membranes. We will describe the use of Biopore™ membranes (Millipore, Bedford, MA), which can be obtained as a cell culture insert (Millicell-CM). The membranes can be cut from the inserts and used as they are. These membranes will support the viability of the cells contained within. However, the viability of the cells can be significantly enhanced with the use of an outer membrane laminated to the Biopore membrane. This outer membrane should be chosen based on its abilities to enhance vascularization at the membrane tissue interface as has been described elsewhere *(21)*.

As a model system we have chosen to examine the delivery of factor IX (FIX) from engineered cells implanted into host animals within an immunoisolation device. FIX is a blood clotting component that is nonfunctional in patients with Hemophilia B *(24,25)*. Hemophilia B is an X-linked genetic disorder and the incidence of the disease is about one in 30,000 male births per year *(2)*. The most common clinical manifestation is spontaneous bleeding into joints or hemarthrosis, which results in chronic pain and joint deformities. Historically, the treatment of spontaneous bleeding episodes has employed concentrated plasma factors that contain vitamin-K dependent proteins and their activation products; more recently the treatment of choice is highly purified factor IX.

In order to assess the usefulness of any particular gene therapy approach, it is necessary to address the following issues: The gene product must be delivered to the correct body compartment (in this case the blood), and the engineered cells must be capable of sustained high level expression sufficient to correct the disorder in vivo. Whereas we have chosen to examine the expression of FIX, the immunoisolation device could potentially be adapted for use with any secreted gene product of a size appropriate to transit the membrane. As a model system, the device is particularly useful in demonstrating that any correction observed is due to the specific presence of the engineered cells. By explanting the device it is possible to quantitatively remove all of the cells at any point and then follow the loss of the gene product over time.

Canine fibroblasts were transduced with a retroviral vector encoding the gene for human Factor IX (hFIX). The fibroblasts expressed around 200 ng hFIX/10^6 cells/24 h when assayed in vitro. The engineered fibroblasts were implanted within the device into athymic rodents that were monitored for systemic delivery of hFIX as measured by the presence of FIX in their serum. No hFIX was detected in serum collected from these implanted rats when assayed by ELISA. However, when the devices were explanted, frozen fixed, and directly stained for hFIX expression, the transduced cells did stain positive for hFIX (*see* red-brown staining within the device, Fig. 2A). No staining was detected from devices loaded with nontransduced cells (Fig. 2B). Thus, by

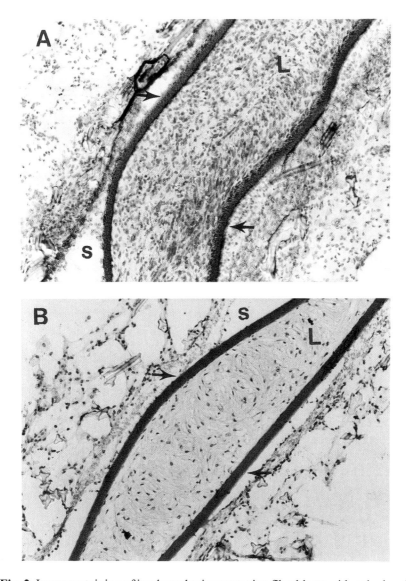

Fig. 2. Immunostaining of implanted primary canine fibroblasts with polyclonal anti-bodies specific to hFIX. Cells shown in **(A)** were transduced with a recombinant retrovirus encoding hFIX. Nontransduced cells are shown in **(B)**. Approximately 1 × 10^5 cells were loaded into a flat-sheet, friction-sealed immunoisolation device, and implanted within the epididymal fat-pad of a male athymic rat. Devices were explanted after 7 d, frozen fixed, and stained for hFIX. Cells which express hFIX stain red-brown within the lumen (L) of the device. Arrows indicate location of the device membranes. (S) Indicates ortifactual spaces created by histological processing.

using an immunoisolation device it may be possible to detect by immuno-staining, in vivo expression too low to be detectable in the serum. This demonstrates that gene products may be expressed at levels too low to be detectable by ELISA, but may be detectable by immunostaining if the cells can be recovered. Analysis of histological sections stained for FIX revealed graft tissue survival and expression of FIX by many of the graft cells (Fig. 2A), whereas nontransduced fibroblasts did not stain positive for FIX (Fig. 2B).

In a separate series of experiments, primary canine fibroblasts were transduced with a retrovirus encoding canine factor IX (cFIX). The transduced cells expressed 1 µg of cFIX/ 10^6 cells/ 24 h. The transduced cells were loaded into devices at 10^5 cells/device, and 10 devices were implanted into an athymic rat. Devices were placed in the epididymal fat pads and subcutaneously. We were able to demonstrate expression of cFIX in plasma from this animal by ELISA at levels of 4–16 ng/mL for 60 d after implantation *(26)*.

2. Materials

1. Membranes for immunoisolation device.
2. 70% EtOH (EtOH).
3. 95% EtOH.
4. 0.9% Saline.
5. Binocular dissecting scope.
6. Sterile jeweler's forceps, gauze, towels.
7. Surgical medium: DMEM supplemented with 20% fetal bovine serum, 1% penicillin G (10,000 U/mL)/streptomycin (10,000 µg/mL) and 1% L-glutamine (200 m*M*).
8. 2% Glutaraldeyde in 0.2*M* Sorenson's phosphate buffer (stable at 4°C for 2–3 wk).
9. Biopore membranes, Millicell-CM.
10. Silicone for spacer (o ring).
11. Laboratory animals for implants.
12. Surgical instruments.
13. ELISA materials.
14. Histology materials: 10% neutral buffered formalin, 80, 95, and 100% EtOH, absolute EtOH/xylene, absolute xylene, paraffin, tissue cassettes, microtome (such as Leica 2035), hematoxylin (Polyscientific, Bay Shore, NY), eosin (formulation obtained from Polyscientific), 0.5% acid alcohol, mounting medium).

3. Methods
3.1. Cell Preparation

The choice of cell should be determined by screening various types of cells within the device for cell survival (by analyzing histological sections of the device and cells, *see* Section 3.5.) before genetically engineering the cell to express the desired gene. Survival of different cell types within the device will vary. Implanting cells within a device does limit access to oxygen and nutrients, and some cells may be more susceptible to those stresses. The following protocol should be followed for preparing cells for implantation (these steps should be performed within a laminar flow hood, using sterile cell growth medium, and so on).

1. Culture cells using normal tissue culture procedures.
2. On day of implant, remove cells from tissue culture flasks.
3. Pellet cells and resuspend in surgical medium to a concentration of between 1×10^3 cells/10 μL and 1×10^7 cells/10 μL. Store cells on ice until loading.
4. Load the cell suspension onto the bottom membrane as detailed below. A 20 μL pipetman is useful for the loading procedure.

3.2. Sterilization and Loading of an Immunoisolation Device

To prepare an immunoisolation device, obtain flat-sheet membranes of the appropriate permeabilities as specified in ref. *21*, or use the Biopore membranes from a cell culture insert (Millicell-CM). When using the Biopore membranes, remove the membrane from the 1 cm plastic ring with a sterile scalpel blade. We target rodents as animal models, so our device is about a centimeter in diameter, which is also the size of the Biopore membranes from cell culture inserts. Biopore membranes arrive sterile. Therefore, if care is taken and proper sterile technique is used, the membrane should remain sterile. However, for those membranes with compromised sterility or that are obtained nonsterile, the following technique should be followed. Immerse the membranes in 95% EtOH for 20–30 s to wet the membrane (for Biopore membranes they will change from opague to translucent when wetted). Incubate the membranes overnight in 70% EtOH to sterilize the membranes. EtOH wetting and sterilization is not appropriate for all membrane compositions, so it is important to investigate which sterilization method is appropriate if non-PTFE compositions are used. The membranes should be handled with sterile instruments, and every effort should be made to touch the membranes only at their extreme edges, so that the membrane integrity is not compromised (*see* Note 1).

We use a proprietary bilayer membrane having a large pore, "vascularizing" membrane as the outer layer to improve biocompatibility *(21)*. The membranes are friction-sealed between a titanium ring and housing to sandwich the grafted tissue between the membranes (Fig. 1). Using titanium as the housing material offers the advantage of implanting a metal that is relatively biologically inert, with no known leachables. The titanium rings and housing are particularly useful because they are easily sterilized by autoclaving. However, it is up to the discretion of the researcher how the device should be sealed. A spacer between the membranes may be helpful in preventing damage to the tissue when the membranes are sealed together. The following assembly steps should be performed within a laminar flow hood, using sterile towels to cover the surface and sterile instruments to handle the housings and membranes (*see* Note 2).

1. Place a titanium ring and housing on the lid of a sterile Petri plate.
2. Observe the membrane and housing through a binocular microscope to assure proper placement of the membrane. Place the membrane into the housing.
3. Place a silicone spacer onto the outer edge of the membrane, being careful not to touch or puncture the membrane.
4. Add a volume of cell suspension (no greater than 10 μL) that contains the appropriate number of cells (*see* Section 3.1. for cell preparation; also *see* Note 3).
5. Gently place the other membrane on top of the cell suspension, being careful not to disturb the cells or wick them to the outside of the device.
6. Place the titanium ring onto the outer membrane.
7. Friction seal the insert to the housing. Do not allow tissue to ooze out the device.
8. Wash device three times in 0.9% saline.
9. Place the device in surgical medium at 37°C, 5% CO_2, humidified incubator until surgery.

3.3. Implantation of the Immunoisolation Device

Identifying a suitable device implant site is an important component of immunoisolation and gene therapy. The choice of implant site is based on a variety of factors, including cell survival at that site, the kinetics of gene product delivery, the site at which the gene product is needed, and so on. We have implanted the FIX-secreting cells into two sites in rodents, the epididymal or ovarian fat-pads, and ventral or dorsal subcutaneous sites (*see* Note 4).

1. Fat-pad surgical implantation.
    a. Anesthetize the rodent with the appropriate anesthetizing agent *(27)*.
    b. Shave and swab the ventral abdominal surface with providone iodine prep solution (Baxter, Deerfield, IL).
    c. Make a 1–2-cm incision just cranial to the anus (females) or penis (males) at the midline with a sterile scalpel.
    d. Cut the abdominal muscle layer to the same size as the outer incision.
    e. Locate the fat-pad, remove it from the peritoneum, and spread it out over saline-wetted sterile gauze.
    f. Center the device on the fat-pad.
    g. Wrap the device in the fat and secure it by gluing the fat over it with small amounts of methylmethacrylate vetting glue (VETBond™, 3M, St. Paul, MN), taking care so that no glue is placed on the membranes. The glue is very difficult to histologically section, so the less used, the better. In addition, the glue is bioresorbable and can increase the level of inflammation around the device.
    h. Return the fat-pad with device to the peritoneum.
    i. Suture the abdominal wall together using a suitable suture material with a running stitch.

j. Seal the surface incision with wound clips.

k. Reswab the entire abdominal area with iodine solution.

2. Surgical implantation into a subcutaneous site.

a. Anesthetize the rodent with the appropriate anesthetizing agent. Shave and swab the dorsal or ventral area of the rodent with above iodine solution.

b. Make a 1–2-cm incision through the skin, but not through muscle facie.

c. Blunt dissect either side of the incision, by separating the skin from the muscle to create pockets slightly larger than the device.

d. Place chamber within pocket, taking care no to touch the device to rodent's fur.

e. Seal and reswab incision as detailed above.

3.4. Removal of the Immunoisolation Device

Removing the immunoisolation device, allows the quantitative retrieval of all of the implanted cells. This permits demonstration that any gene expression observed was due to the implanted cells. The removal of the device also allows the researcher to assess cell survival within the device and to utilize immunostaining to detect gene product that is being expressed at very low levels (Fig. 2). The following details the protocol that we follow to remove devices from rodents:

1. Euthanize rodents according to the appropriate guidelines.
2. Saturate device explant area with 70% ETOH to wet fur.
3. Locate device by palpating implant area. Perform incision at appropriate spot to remove device, being careful to avoid cutting the device itself. Avoid excess bleeding and under no circumstances tug at the device.
4. Cut device from surrounding tissue.
5. Trim device of excess tissue.
6. Place device in 2% glutaraldehyde in Sorenson's buffer.
7. Store fixed devices at 4°C for a minimum of 12 h before histological processing.

3.5. Histological Processing and Staining

By analyzing tissue survival within the device, the researcher is able to assess cell survival and host response to the implant. This is especially important when analyzing gene product expression, because histological analysis may generate clues about the expression (e.g., in cases where the gene product could not be detected systemically and the tissue did not survive and/or the host recognized the gene product as foreign and mounted an immune response). The types of histological processing that are employed will again depend on device composition, but for Biopore membrane-containing devices, the following histological protocols are followed:

1. Remove the membranes from the titanium ring and housing, being careful to preserve the host tissue associated with the membranes. Do not allow the host tissue to slide off when the housing is removed. Place within a tissue cassette.

2. Immerse cassette as follows:
   a. 10% Neutral-buffered formalin: 1 h.
   b. 10% Neutral-buffered formalin: 1 h.
   c. 80% EtOH: 1 h.
   d. 95% EtOH: 1 h.
   e. 95% EtOH: 1 h.
   f. 100% EtOH: 1 h.
   g. 100% EtOH: 1 h.
   h. Absolute EtOH/xylene: 1 h.
   i. Xylene: 1 h.
   j. Xylene: 1 h.
   k. Paraffin: 1 h, 60°C.
   m. Paraffin: 1–2 h, 60°C.
3. Embed the membrane sandwich on edge in a deep paraffin mold and allow it to solidify.
4. Prepare a water bath for standard sectioning and cut sections through the device using a microtome. Frequently change the disposable microtome blade to ensure a sharp edge. Only one or two sections can be taken from a knife area before the area is dull.
5. Place two ribbons of sections about 5 μm thick onto a glass slide.
6. Place slides in a 80°C oven for 30 min prior to staining.
7. Routine staining is by the H&E method as follows:
   a. Remove paraffin from the slides by exposing the slide to three changes of xylene, 5 min each.
   b. 100% EtOH, approx 1 min.
   c. 100% EtOH, approx 30 s.
   d. 95% EtOH, approx 1 min.
   e. 95% EtOH, approx 30 s.
   f. 80% EtOH, approx 1 min.
   g. Tap water rinse, approx 1 min.
   h. Hematoxylin, approx 5 min.
   i. Tap-water rinse, approx 1 min.
   j. 0.5% acid alcohol, approx 1 min.
   k. Tap water rinse, approx 1 min.
   l. Eosin, approx 6 min.
   m. 95% EtOH, approx 30 s.
   n. 100% EtOH, approx 30 s.
   o. 100% EtOH, approx 30 s.
   p. Xylene, approx 30 s.
8. Mount each stained slide using a clean coverglass and a drop of mounting medium.

3.6. Assessment of Gene Expression and Graft Tissue Viability

Determining whether the gene product is being expressed in vivo is the ultimate goal of these experiments. How the gene product is assayed will depend on what the gene product is, whether there is a specific function associated with its expression, whether it confers a specific phenotype, and other factors.

One advantage of the immunoisolation device is that it permits quantitative removal of the therapeutic cells at any time. Subsequent sampling of the animal should demonstrate return to baseline levels of the gene product. At the time of removal, the devices can also be examined to determine the survival of the graft tissue. The following protocol can be used for survival surgeries to remove devices.

1. Anesthetize the rodent with the appropriate anesthetizing agent *(27)*.
2. Shave and swab the ventral abdominal surface with Providore iodine prep solution.
3. To remove subcutaneous implants:
   a. Locate device by palpating implant area. Perform incision at appropriate spot to remove device, being careful to avoid cutting the device itself. Avoid excess bleeding and under no circumstances tug at the device.
   b. Cut device from surrounding tissue.
   c. Trim device of excess tissue.
   d. Place device in 2% glutaraldehyde in Sorenson's buffer.
   e. Seal the surface incision with wound clips.
   f. Reswab the entire abdominal area with iodine solution.
4. To remove fat pad implants:
   a. Make a 1–2-cm incision just cranial to the anus (females) or penis (males) at the midline with a sterile scalpel.
   b. Cut the abdominal muscle layer to the same size as the outer incision.
   c. Locate the fat-pad, remove it from the peritoneum place it on saline-wetted sterile gauze.
   d. Tie off the fat-pad with suture and remove fat-pad containing device.
   e. Trim device of excess tissue.
   f. Place device in 2% glutaraldehyde in Sorenson's buffer.
   g. Suture the abdominal wall together using a suitable suture material with a running stitch.
   h. Seal the surface incision with wound clips.
   i. Reswab the entire abdominal area with iodine solution.

4. Notes

1. Membrane handling: Use autoclaved jeweler's forceps to handle flat-sheet membranes. Pick up the membranes by allowing them to float in the last saline wash, dipping the forceps just under the saline/membrane interface, and dragging the membrane across the saline before removing it completely from the wash. This prevents folding and crimping of the membranes. Slightly drying the membranes on sterile gauze by wicking the saline from the membrane to the gauze makes the membranes easier to load into the housing. Do not allow the membranes to become completely dry.

Membranes for flat-sheet diffusion devices are very delicate mesh and tear easily, so it is sometimes helpful to laminate the membrane to a polyester backing. Laminations can be performed with minute amounts of glue (the glue used will be dependent on the chemistry of the flat sheet membrane) and the appropriate drying procedure. Polyester woven and unwoven meshes are available through a variety of membrane supply companies (Saati, Stamford, CT, for example). It is recommended that research devices be no more than 1 cm in diameter because the difficulty of handling the membranes increases proportionally with the diameter of the device.

2. Membrane loading: Membrane loading is best performed by two persons. One researcher can be dedicated to loading the membrane and spacers into the housing, while the other person loads the tissue, top membrane, friction seals, and rinses the devices. Sterile Petri dish lids are useful to place the device components on.

3. Determining appropriate cell concentration for loading into devices: The number of cells needed to fully pack a device must be determined empirically for each cell type. The number needed will vary depending on variables such as the size of the cell and the amount of extracellular matrix the cells produce as well as their ability to survive within an immunoisolation device. The number of cells needed will also be determined by the number of cells necessary to obtain therapeutic levels of the gene product in vivo. Whereas the option of implanting more devices does exist, there are physical limits to the number possible. Furthermore, one should implant the fewest number of devices necessary to obtain therapeutic levels of the gene product. For mice, 6–8 devices can be implanted (two devices in the fat pads, two subcutaneously on the ventral surface, and 2–4 subcutaneously on the dorsal surface); for rats, 10–14 can be implanted (four in the fat pads, four subcutaneously on the ventral surface, and 4–6 subcutaneously on the dorsal surface. For each cell line, we routinely determine the optimal number of cells to be loaded as follows: serial dilutions of the cell suspension (10^7, 10^6, 10^5, 10^4, 10^3 cells/device) are loaded and implanted into athymic animals for three weeks. Explanted devices are stained with H and E and scored for cell survival and growth. The cells can also be stained for specific gene product as outlined above.

4. Animal surgery: Anesthetizing rats and mice will, of course, be performed according to the approved institutional protocols. Our protocol calls for intramuscular injections of a mixture of the drugs rompun (xylazine, Rugby Laboratories, Rockville Center, NY) and ketamine (Fort Dodge Laboratories, Fort Dodge, IA). For mice we use 0.1–0.2 mL of a mixture of 1 mL ketamine and 0.75 mL xylazine diluted into 4 mL sterile saline. Rats are anesthetized with a cocktail of 5 mg/kg rompun and 65 mg/kg ketamine. These drugs are far superior to barbiturates in that animals are seldom lost due to respiratory distress.

To minimize trauma to the animal, prevent infections, and speed the recovery time from surgery, there are several steps that can be taken. Swab the injection site with iodine solution before and after injection; when using athymic animals infections can easily be initiated at the injection site. Treat the eyes of rats and mice with artificial tears to prevent drying and blindness. Warm animals on a heating pad after surgery to prevent hypothermia. (This is particularly important when working with athymic animals which lack fur.)

Fat-pad can be discerned from intestine and other organs because it is a creamy white/yellow. If it is necessary to move the intestines to access the fat pads, use blunt forceps and gently lift the intestine out of way.

Acknowledgments

The gene constructs were prepared by Badr Saeed and Steven Josephs, Baxter Healthcare Corporation. The authors are indebted to the Baxter Healthcare Veterinary Resources staff, Round Lake, for assistance with the animal studies and to Carol Sasonas for assistance in preparation of the manuscript.

References

1. Morgan, R. A. and Anderson, W. F. (1993) Human gene therapy. *Ann. Rev. Biochem.* **62,** 191–217.
2. Lozier, J. N. and Brinkhous, K. M. (1994) Gene therapy and the hemophilias. *JAMA* **27,** 47–51.
3. Anson, D. S., Hock, R. A., Austen, D., Smith, K. J., Brownlee, G. G., Verma, I. M., and Miller, A. D. (1987) Towards gene therapy for hemophilia B. *Mol. Biol. Med.* **4,** 11–20.
4. Zhou, J. M., Dai, Y. F., Qiu, X. F., Hou, G. Y., Akira, Y., and Xue, J. L. (1993) Expression of human factor IX cDNA in mice by implants of genetically modified skin fibroblasts from a hemophilia B patient. *Sci. China (B)* **36,** 1082–1092.
5. Kay, M. A., Landen, C. N., Rothenberg, S. R., Taylor, L. A., Leland, F., Wiehle, S., Fang, B., Bellinger, D., Finegold, M., Thompson, A. R., Read, M. S., Brinkhous, K. M., and Woo, S. L. C. (1994) In vivo hepatic gene therapy: complete albeit transient correction of factor IX deficiency in hemophilia B dogs. *Proc. Natl. Acad. Sci. USA* **91,** 2353–2357.
6. Li, Q., Kay, M. A., Finegold, M., Stratford-Perricaudet, L. D., and Woo, S. L. (1993) Assessment of recombinant adenoviral vectors for hepatic gene therapy. *Hum. Gene Ther.* **4,** 403–409.
7. Liu, H.-W., Ofosu, F. A., and Chang, P. L. (1993) Expression of human factor IX by microencapsulated recombinant fibroblasts. *Hum. Gene Ther.* **4,** 291–301.
8. Yao, S.-N., Wilson, J. M., Nabel, E. G., Kurachi, S., Hachiya, H. L., and Kurachi, K. (1991) Expression of human factor IX in rat capillary endothelial cells: toward somatic gene therapy for hemophilia B. *Proc. Natl. Acad. Sci. USA* **88,** 8101–8105.
9. Anson, D. S., Austen, D. E. G., and Brownlee, G. G. (1985) Expression of active human clotting factor IX from recombinant DNA clones in mammalian cells. *Nature* **315,** 683–685.

10. de la Salle, H., Altenburger, W., Elkaim, R., Dott, K., Dieterle, A., Drillien, R., Cazenave, J.-P., Tolstoshev, P., and Lecocq, J.-P. (1985) Active gamma-carboxylated human factor IX expressed using recombinant DNA techniques. *Nature* **316,** 268–270.

11. Axelrod, J. H., Read, M. S., Brinkhous, K. M., and Verma, I. M. (1990) Phenotypic correction of factor IX deficiency in skin fibroblasts of hemophiliac dogs. *Proc. Natl. Acad. Sci. USA* **87,** 5173–5177.

12. Palmer, T. D., Thompson, A. R., and Miller, D. (1989) Production of human factor IX in animals by genetically modified skin fibroblasts: potential therapy for hemophilia B. *Blood* **73,** 438–445.

13. Armentano, D., Thompson, A. R., Darlington, G., and Woo, S. L. C. (1990) Expression of human factor IX in rabbit hepatocytes by retrovirus-mediated gene transfer. Potential for gene therapy of hemophilia B. *Proc. Natl. Acad. Sci. USA* **87,** 6141–6145.

14. Palmer, T. D., Rosman, G. J., Osborne, W. R., and Miller, A. D. (1991) Genetically modified skin fibroblasts persist long after transplantation but gradually inactivate introduced genes. *Proc. Natl. Acad. Sci. USA* **88,** 1330–1334.

15. Yao, S.-N. and Kurachi, K. (1992) Expression of human factor IX in mice after injection of genetically modified myoblasts. *Proc. Natl. Acad. Sci. USA* **89,** 3357–3361.

16. St.Louis, D. and Verma, I. M. (1988) An alternative approach to somatic cell gene therapy. *Proc. Natl. Acad. Sci. USA* **85,** 3150–3154.

17. Scharfmann, R., Axelrod, J. H., and Verma, I. M. (1991) Long-term in vivo expression of retrovirus-mediated gene transfer in mouse fibroblast implants. *Proc. Natl. Acad. Sci. USA* **88,** 4626–4630.

18. Dai, Y., Roman, M., Naviaux, R. K., and Verma, I. M. (1992) Gene therapy via primary myoblasts: long-term expression of factor IX protein following transplantation in vivo. *Proc. Natl. Acad. Sci. USA* **89,** 10,892–10,895.

19. Miyanohara, A., Johnson, P. A., Elam, R. L., Dai, Y., Witztum, J. L., Verma, I. M., and Friedmann, T. (1992) Direct gene transfer to the liver with herpes simplex virus type 1 vectors: transient production of physiologically relevant levels of circulating factor IX. *New Biol.* **4,** 238–246.

20. Thompson, J. A., Haudenschild, C. C., Anderson, K. D., DiPietro, J. M., Anderson, W. F., and Maciag, T. (1989) Heparin-binding growth factor 1 induces the formation of organoid neovascular structures in vivo. *Proc. Natl. Acad. Sci. USA* **86,** 7928–7932.

21. Brauker, J. H., Carr-Brendel, V., Martinson, L. A., Crudele, J., Johnston, W. D., and Johnson, R. C. (1995) Neovascularization of synthetic membranes directed by membrane architecture. *J. Biomed. Mat. Res.* **29,** 1517–1524.

22. Brauker, J., Martinson, L. A., Young, S. K., and Johnson, R. C. (1996) Local inflammatory response around diffusion chambers containing xenografts. *Transplantation* **61,** 1671–1677.

23. Algire, G. H., Weaver, J. M., and Prehn, R. T. (1954) Growth of cells in vivo in diffusion chambers I. survival of homografts in immunized mice. *J. Natl. Cancer Inst.* **15,** 493–507.

24. Larson, P. J. and High, K. A. (1992) Biology of inherited coagulopathies: factor IX. *Hematol. Oncol. Clin. North Am.* **6,** 999–1009.
25. Kurachi, K., Furukawa, M., Yao, S. N., and Kurachi, S. (1992) Biology of factor IX. *Hematol. Oncol. Clin. North Am.* **6,** 991–997.
26. Carr-Brendel, V., Lozier, J., Thomas, T., Saeed, B., Young, S., Crudele, J., Martinson, L., Roche, B., Boggs, D., Pauley, R., Maryanov, D., Josephs, S., High, K., Johnson, R., and Brauker, J. (1993) An immunoisolation device for implantation of genetically engineered cells: long term expression of factor IX in rats. *J. Cell Biochem.* **17E,** 224.
27. Delves, M. N. (1994) *Cellular Immunology Labfax 1994.* Academic, New York.

29

Delivery of Recombinant HIV-1-Directed Antisense and Ribozyme Genes

Georg Sczakiel, Giorgio Palú, and William James

1. Introduction

A series of gene products has been shown to inhibit viral replication of the human immunodeficiency virus type 1 (HIV-1). These include ribonucleic acids (RNA) including antisense RNA, ribozymes, and decoy RNAs, as well as proteins such as dominant, negative-interfering, mutated viral proteins, heterologous proteins, or intracellular antibodies.

For a gene-inhibition therapy against HIV-1 to be effective, one needs not only effective inhibitory genes but also methods for their efficient delivery into those cells upon which the direct effects of HIV replication are crucial in the pathogenesis of the disease. As a first assay, one usually tests the antiviral efficacy of long-term expression of inhibitory genes in immortalized human cell lines that express the HIV-1 receptor CD4 (e.g., Jurkat, CEM, Molt) or in peripheral blood lymphocytes that have been stimulated by PHA or Il-2 or both. Hence, methods for enabling one to express recombinant DNA in such cells are needed.

Retroviral vectors offer an efficient means for delivery of foreign genes into eukaryotic cells (1). They are able to integrate their genetic information stably, generally at a low copy number, and to transmit it with high efficiency into recipient cells. Retroviral vectors are engineered to be replication-defective by replacing part of the viral genome, the *gag-pol-env* genes, with the desired therapeutic genes. Infectious particles can still be generated in a single cycle of replication, if the deleted functions are provided in *trans,* either by superinfecting with a replication-competent helper virus, or by introducing the defective virus (retroviral vector) into packaging cell lines containing a helper virus genome lacking the packaging signal. Retroviral vectors have been mainly

From: *Methods in Molecular Biology, vol. 63: Recombinant Protein Protocols:
Detection and Isolation* Edited by: R. Tuan Humana Press Inc., Totowa, NJ

derived from murine leukemia virus (MLV), from avian leukosis virus, and from the reticuloendotheliosis virus strain A (REV-A) and its close relative, spleen necrosis virus (SNV) *(1–3)*. The relatively well understood biology of these retroviruses enable one to develop efficient vector systems. Corresponding packaging cell lines, that are able to package high titers of replication defective vectors, have been established as well *(4–7)*.

To date, the murine retroviral vectors are the most developed vehicles for ex vivo gene delivery. However, particularly in the case of human CD4-positive target cells, modified HIV-1 can be used to generate recombinant viral vector particles for the delivery of heterologous genes. An HIV-1-based vector has a number of conceptional advantages over conventional murine retroviral vectors if one is planning a gene therapy strategy to combat AIDS. Firstly, the specific cell targeting, facilitated by the high affinity interaction between the viral envelope glycoprotein gp 120 and the CD4 receptor expressed on T-cells as well as other potential HIV-1 target cells, including the resting cells of the monocyte/macrophage lineage or CD4-positive hematopoietic stem cells. Secondly, the HIV-specific inducibility of the LTR, which could allow selective expression of the antiviral gene specifically upon HIV-infection of the cells provides an advantage. Furthermore, given the ability of the wild type HIV-1 genome to encode multiple gene products via a diversified regulatory and splicing system, the use of HIV-1 vectors could be readily adaptable for the expression of multiple (inhibitory) genes of interest. The introduction of an anti-HIV gene element into quiescent hematopoietic stem cells (HSC) could certainly be a convenient way to render a large number of progeny lymphomonocytoid cells resistant to HIV-1 infection, although target cells must be cycling for stable integration of retrovirally carried genetic information *(8,9)*.

The ability of HIV-1 to integrate into chromosomes of nondividing cells *(10,11)* make HIV-1 vectors extremely valuable as agents for gene transfer into quiescent hematopoietic progenitors. HSC, however, for they lack the CD4 receptor or express CD4 in only a small subfraction can only be infected by an HIV-1-derived vector containing heterologous envelopes. Production of such pseudotype viruses can be easily realized since HIV-1 efficiently incorporates many other envelope glycoproteins *(12,13)*, hence restricting or broadening its host range. Replication-defective HIV-1 vectors have been developed to study different aspects of the HIV-1 life cycle and for gene transfer to human lymphocytes *(14-18)*.

In the following, three alternative methodologies will be described to introduce recombinant DNA into immortalized human HIV-1-permissive cell lines. These include the somewhat biophysical method "electrotransfection," a murine amphotropic retroviral vector system, and a system that is the closest to HIV-1—HIV-1-derived retroviral vectors. However, for future therapeutic purposes, vector systems are needed that show significantly improved effectiveness and the ability to transduce the appropriate cell types.

Infection of cells with HIV-1 can be monitored by a series of HIV-1-specific tests such as listed:

1. Challenge with HIV: one-step challenge (high m.o.i., short time-course) or multiround challenge (low m.o.i., long time-course).
2. Quantitation of HIV replication by PCR (quant and semiquant), reverse transcriptase (SPA method), p24 ELISA (kit and DIY), as well as output infectivity ($TCID_{50}$).

2. Materials

2.1. Method 1

1. Power supply that is suitable to deliver highly charged electric pulses (maximal capacitance, 1 mF; maximal voltage, 300 V; e.g., "gene pulser," Bio-Rad Laboratories, Hercules, CA) and electroporation chambers with 4 mm interelectrode distance (Bio-Rad).
2. Recombinant DNA that has been purified by methods that guarantee salt-free preparations, e.g., CsCl gradient method including at least three subsequent ethanol precipitation steps.
3. Phosphate-buffered saline (PBS): 120 mM NaCl, 25 mM Na-phosphate, pH 7.3.
4. RPMI-1640 medium supplemented with 10% fetal calf serum, L-glutamine (2 mM), penicillin (100 U/mL), and streptomycin (100 μg/μL).
5. Cell-free supernatant from the cells to be transfected in logarithmic phase of growth. For most of the cell lines grown in suspension, this is in the range of 5 × 10^5–1 × 10^6 cells/mL.
6. Six well and 48-well tissue culture plates (flat bottom wells).

2.2. Method 2

1. Laboratory requirements. Containment level 2 facilities for the production and use of amphotropic retroviruses and containment level 3 facilities for the HIV challenge experiments.
2. Plasmids. The choice of MLV-based vectors is wide and some examples and guidelines are given in Section 3.2. For transient tests, *env* and *gag-pol*-encoding plasmids are also required. Vector DNA should be purified, for preference, by the ion exchange method (e.g., Qiagen, Chatsworth, CA) and of bacterial lipopolysaccharide should be removed by triton X-114 treatment *(19)*.
3. Helper cell lines. A combination of an ecotropic and an amphotropic packaging cell line is needed. These are described in the text. Many vectors and cell lines are not freely available and require formal permission for restricted use. Before use, all cell lines should be tested for freedom from mycoplasma infection, which would compromise many aspects of the experimental validity.
4. Culture media. Dulbecco's modification of Eagle's medium, supplemented with fetal bovine serum (10%), L-glutamine (2 mM), and (optionally) penicillin/

streptomycin. Antibiotic for selection of transductants (e.g., G418, puromycin, hygromycin).

5. HEPES-buffered saline (50 mM HEPES, pH 7.05 [exactly]; 280 mM NaCl; 1.5 mM Na$_2$HPO$_4$) and 2.5M CaCl$_2$ (both filter sterilized).

6. Polybrene (8 mg/mL) in water, filter sterilized.

7. Choice of vector. Since the earliest retroviral vector systems, there has been a wide choice of designs, some more suited to certain applications than others. In general, all applications benefit from the increased transduction efficiency conferred by the extended packaging signal found in second-generation vectors and reduced *gag* translation and homology with packaging line sequences found in third generation vectors. However, the choice of vectors is somewhat determined by the particular application envisaged. In cases where an antisense RNA, decoy RNA, ribozyme, or multiple combinations of the three are being delivered, the developments in retroviral vector technology that enable efficient translation of the cloned gene are of no benefit, and a vector with a single transcription unit is often preferable. If, on the other hand, a *trans*-dominant or suicide protein is to be delivered, then expression will often best be accomplished from a vector with an internal promoter. The relative strengths of the internal and LTR promoters may differ in the packaging cell line, and the eventual recipient cell and the two promoters will frequently interfere with each other's expression. For this reason, a self-inactivating vector may be the preferred choice.

8. Choice of packaging cell line. If one is wanting to pilot an approach that is aimed to be trailed in human subjects, the pedigree of the packaging cell and its acceptability for medical use is of key importance. Helper cell lines are now only acceptable if they do not give rise to helper-independent virus after transfection with the chosen vector, and for this reason, highly modified *gag*, *pol*, and *env* genes, expressed on unlinked fragments are the backbone of most approved systems. The most widely used ecotropic lines are PsiCRE *(20)*, GP+E-86 *(21)*, and Omega E *(22)*. The most frequently used amphotropic line is PA317 *(23)*, but this produces some helper-independent virus, whereas PsiCRIP *(20)* does not.

2.3. Method 3

1. Plasmids. Structures of the packaging system and HIV-1-derived vector plasmids were listed by Parolin et al. *(18)*. The simian virus 40 (SV40) origin of replication is included in all plasmids, to allow plasmid propagation in COS-1 cells, which constitutively express the SV40 T antigen. The selectable marker in all vectors used herein is neomycin phosphotransferase (neo).

2. Packaging system. The packaging functions are provided by both the HXBΔP1Δenv plasmid and the CMVΔP1ΔenvpA plasmid that express the HIV-1 *gag, pol, vif,* and *tat* genes. The CMVΔP1ΔenvpA plasmid was constructed recently by replacement of the HIV-1 5'LTR on the HXBΔP1Δenv plasmid *(16)* with the cytomegalovirus (CMV) immediate–early promoter. (Briefly, the HXBΔP1Δenv plasmid was partially digested with *Cla*I [which recognizes a site in the sequences flanking the 5'LTR] and *Bss* HII [{nucleotide 257}, located

approx 80 bp 3' of the LTR] and blunted by treatment with the Klenow fragment. Next, *Xba*I linkers were joined to the blunt ends. A 0.7 kb *Spe*I fragment derived from pcDNA 1 Neo [Invitrogen, San Diego, CA], containing the CMV promoter and part of the polylinker, was inserted into the *Xba*1 site to create the CMVΔP1ΔenvpA plasmid.) To express the HIV-1- *rev* and *env* genes, the pSVIIIenv3-2 plasmid containing the HIV-1 LTR sequences from −167 to +80 *(24,25)* was used. Both the CMVΔP1ΔenvpA and pSVIIIenv3-2 plasmids contain polyadenylation signals derived from SV40 and lack sequences shown to be important for efficient HIV-1 packaging *(26)*.

3. Methods

3.1. Method 1: Electroporation of Human CD4 Positive Cells (see Note 1)

Electrotransfection can be performed with little expense, and experimental conditions for maximal transient uptake of DNA can be identified by systematic analyses *(27)*. The variation and optimization of the voltage at all other parameters being constant, showed an optimal range of 180–230 V for lymphoid, myeloid, and epitheloid cell types *(28)*. The selection conditions for cell clones that express a dominant marker gene, have to be investigated individually.

1. Prepare: cells (2×10^7; logarithmic growth phase, washed 1X with PBS, resuspended in 150 µL PBS); and DNA: (50 µg; linearized for stable transfection, supercoiled for transient transfection, EtOH-precipitated, dissolved in 50 µL PBS, *see* Note 1).
2. Electroporation: 200 µL total volume; 1×10^8 cells/mL, i.e., 2×10^7 cells/200 µL; *(see* Note 1); 4 mm interelectrode distance, sterilized with 70% EtOH, evaporated under sterile hood.
3. Mix cells and DNA and transfer into electropoaration cuvet.
4. Incubate cells and DNA at room temperature for 5–10 min *(see* Note 1).
5. Electric pulse *(see* Note 1): one pulse at room temperature; 960 µF capacitance; 180–250 V. Voltage for maximal DNA transfer into cells: 250 V; Jurkat, 220 V, CEM; 200 V; BJA-B, H9, Molt3, Molt4, U937, and PBLs (PHA-, Il-2-stimulated).
   The half-life of the current (τ-value) should be 40–60 ms. Usually, the survival rate is 30–70% at optimal transient transfection efficiency.
6. Incubate cells and DNA at room temperature for 5–10 min.
7. Transfer electroporated cells into six-well plate. Leave cell debris in the electroporation chamber. Rinse cuvet 2–3 times with 300 µL fresh medium, and transfer into incubation well *(see* next step).
8. Incubate electroporated cells overnight at a cell density of $4–10 \times 10^5$ cells/mL in RPMI medium (usually six-well plates, 3–5 mL medium).

3.1.1. Transient Assays

Incubate cells at 2.5×10^5 viable cells/mL for further 24 h before performing detection assay for transfected genes.

3.1.2. Selection of Stably Transfected Cell Clones

1. Pellet cells by centrifugation.
2. Resuspend cells at a final density of 5×10^4 viable cells/mL in a medium consisting of 50% fresh RPMI medium with all supplements listed above (*see* Section 2.1.) and 50% cell-free culture supernatant from same cells (e.g., from the harvest of cells for electroporation experiment. Store supernatant at −70°C). The final concentration of the drug is: hygromycin B, 500 µg/mL for Jurkat and 200 µg/mL for all other cell types. G418, 2000 µg/mL for Jurkat and 400 µg/mL for all other cell types.
3. Seed cell suspension in 48-well plates (1 mL/well with a flat bottom).
4. Drug-resistant colonies appear 10–20 d after transfection.

3.2. Method 2: Delivery of HIV-1-Directed Recombinant Genes by a Retroviral Vector (see Note 2)

3.2.1. Generation of Producer Cells

1. Seeding ecotropic cells. Ensure that the ecotropic packaging cell line is not allowed to attain confluence and remain so; passage the cells regularly before they reach confluence and store early passage aliquots in liquid nitrogen.
2. On day 0, seed a 75-cm² flask with 5×10^5 ecotropic packaging cells.
3. CaPO$_4$ transient transfection. On day 1, dilute 10 µg retrovirus construct DNA in 0.5 mL sterile 250 mM CaCl$_2$ in a 10 mL polystyrene centrifuge tube.
4. Add 0.5 mL 2X HBS (0.14M NaCl, 5 mM KCl, 0.7 mM Na$_2$HPO$_4$, 20 mM HEPES; pH 7.05 exactly with 0.1M NaOH, filter sterile) dropwise, with vortexing or mixing using a stream of air from a 1-mL pipet. Allow fine precipitate to form for 20 min at room temperature.
5. Tip flask so that medium is held away from the cells and add the Ca-phosphate/DNA coprecipitate to the medium. When fully dispersed in the medium, tip the flask back to its normal position so the cells are covered with medium and precipitate. Incubate 4 h at 37°C
6. Aspirate medium and replace with 10 mL of growth medium.
7. Incubate for a further 2 d at 37°C
8. Transducing amphotropic helper. On day 2, seed a 25-cm² flask with 2×10^5 amphotropic packaging cells
9. On day 3, harvest the supernatant from the transfected cells, filter sterilize, and add polybrene to 8 mg/mL
10. Add 5 mL viral sup to the small flask of amphotropic helper cells and incubate for a further 24 h at 37°C.
11. Selection for stable lines. On day 4, trypsinize transduced cells and plate at approx 10^4 cells/well (in 0.1 mL medium) of two 96-well tissue culture plates.
12. On day 5, add 0.1 mL selective medium to each well and continue to incubate at 37°C (feeding with fresh selective medium at least once a week) until colonies begin to appear in a small proportion of wells.

13. Screening for titer. Expand at least a dozen independent clones to the 25-cm² flask stage.
14. As each clone reaches confluence, harvest the supernatant and store a 1 mL aliquot at −70°C for titration and freeze the cells over liquid nitrogen.
15. When convenient, thaw the banked supernatants and titrate the infectivity of each as follows:
    a. On day 0, plate a sensitive recipient line, such as HeLa, into six-well plates (one plate for two clones) at 3×10^5 cells/well
    b. On day 1, make a 10-fold serial dilution in growth medium of each thawed supernatant and add 0.1 mL of the 10^{-2}, 10^{-3}, and 10^{-4} dilutions to appropriately labeled wells.
    c. On day 3, replace the medium with selective growth medium and continue to incubate, with feeding, for 1–2 wk, until macroscopic colonies are visible.
    d. Rinse the wells gently with PBS, stain with methanol/crystal violet or methanol/acetic/naphthalene black and enumerate the colonies. Divide each number by $0.1 \times$ dilution factor to provide a titer in terms of c.f.u./mL.
16. Screening for integrity. Recover the three highest titered producer clones from the cell bank and expand until you can test the lines for genetic modification. It is worth doing both a Southern and Northern blot to confirm the structure and expression of the agent and PCR-based sequencing of the insert itself.
17. Freezing down. Having identified a high-titer producer, expand the line to at least 5 75-cm² flasks and freeze the line over liquid nitrogen.

3.2.2. Transduction of Human Cell Lines

The detailed method for this section depends on the target cell chosen. The method described is suitable for a lymphoblastoid cell line such as Jurkat, but primary peripheral blood lymphocytes will require periodic antigenic restimulation.

1. Harvest approx 10^6 logarithmically growing target cells by centrifugation.
2. Resuspend in 1 mL filtered producer cell supernatant supplemented with polybrene (8 mg/mL).
3. Incubate 6–8 h 37°C then feed with 1 mL fresh medium.
4. After 48 h without selection, add 4 vol selective medium. (It is important to titrate the dose of antibiotic required to kill target cells in a week; 3T3 cells are killed at 200 µg/mL G418, Jurkats require 1000 µg/mL or more.)
5. Continue to incubate for at least 2 wk, feeding at least weekly and taking viable counts every other day. During the first week, there should be a substantial drop in viable count but, by the end of two weeks, transduced cells should be growing strongly.
6. Continue to split cells regularly but reduce the antibiotic concentration (200 µg/mL for G418).

3.2.3. Challenge with HIV

To evaluate the effect of any antiviral agent, the ideal methodology is to use a high multiplicity challenge, in which virtually every cell is infected simultaneously, and follow the kinetics of the first round of infection. In the case of HIV, the first cycle of replication is completed in a little over 24 h for most immortalized cell lines. However, a multicycle infection course is not only more convenient for HIV, but probably better reflects the physiological situation.

3.2.3.1. ONE-STEP CHALLENGE (HIGH M.O.I., SHORT TIME COURSE)

1. Harvest 5×10^6 logarithmically growing transduced target cells
2. Resuspend the cells in 10 mL of a filter-sterilized, DNase treated stock of HIV. The titer must be at least 5×10^5 $TCID_{50}$/mL (titrated on the corresponding cell line) and preferably $1.5–2.5 \times 10^6$/mL to achieve the desired multiplicity.
3. Incubate the mixture at 4°C 2 h then wash and resuspend in 5 mL fresh growth medium and incubate at 37°C.
4. Take 0.2 mL samples of cell suspension at hourly intervals for PCR analysis.
5. Take 0.5 mL samples at 4 h, 12 h, 16 h, 20 h, 24 h, and 28 h for analysis of p24, $TCID_{50}$, or RT.

3.2.3.2. MULTIROUND CHALLENGE (LOW M.O.I., LONG TIME COURSE)

1. Harvest 5×10^6 logarithmically growing transduced target cells.
2. Resuspend the cells in 1 mL of medium containing a stock of HIV of titer no more than 5×10^3 $TCID_{50}$/mL (titrated on the corresponding cell line).
3. Incubate the mixture at 4°C for 2 h, then wash and resuspend in 5 mL fresh growth medium and incubate at 37°C.
4. Feed the cells with fresh medium at least every other day to maintain them in exponential phase, at a density of $0.5–2.0 \times 10^6$ cell/mL.
5. Take 0.5 mL samples of supernatant at daily intervals for 14 d for analysis of p24, $TCID_{50}$, or RT.

3.3. Method 3: Generation and Testing of HIV-1 Vectors (see Note 3)

1. Grow COS-1 cells in Dulbecco's modified Eagle medium (DMEM) supplemented with 10% fetal calf serum.
2. Plate COS cells at a concentration of $1.5–2.0 \times 10^6$ cells/100-mm-diameter dish 1 d before transfection.
3. Cotransfect seeded COS-1 cells with the packaging system plasmids (5 μg of each) along with the vector (5 μg) by the DEAE-dextran technique *(29)*.
4. Replace the medium of the COS-1 cell cultures 24 h before harvesting the recombinant virus at approx 65 h posttransfection.
5. Filter the cell-free culture medium (pore size, 0.45 mm; Millipore, Bedford, MA).
6. Measure the reverse transcriptase activity *(30)*.

7. Determine the transduction efficiency. Use equivalent reverse transcriptase units (from 18,000–117,00 cpm) of supernatants to generate 10-fold serial dilutions of the supernatants in RPMI medium supplemented with 10% fetal calf serum. Add dilutions to 2.5×10^5 Jurkat cells grown in six-well culture plates in 2.5–5 mL of complete medium. Pellet Jurkat cells, resuspend in complete medium containing G418 (GIBCO) at an active concentration of 0.8 mg/mL, and dispense into 24-well culture plates at a density of $1–2 \times 10^5$ cells/well in a total volume of 2 mL. Add fresh medium to pelleted Jurkat cells 24 h after infection. Cultures are supplemented with approx 1 mL of complete medium every 3–4 d. Count the number of viable G418-resistant cell clusters up to 45 d postinfection.

4. Notes

1. Electrotransfection. If it is difficult to dissolve the DNA in 1X PBS, dissolve the DNA pellet in water and adjust to the final concentration (1X PBS) and volume (50 µL) by 4X PBS. Higher cell densities lead to higher transfection efficiencies per cell. Thus, try to avoid significantly lower cell densities as recommended. Incubation of cell/DNA mixtures at 0°C do not seem to increase the transfection efficiency. If electroporation cuvets with 2-mm interelectrode distance are to be used, apply half of the voltage and use a total volume of 100 µL at constant concentrations of DNA, cells, and salts. However, this is only a rough estimate and the optimal voltage should be identified for each cell line individually.

2. Advantages and disadvantages of murine retroviral transduction methodology. Compared with the electrotransfection method, this method has the following advantages; transduced cells have only one copy of the protective gene (more physiological); transduced cells have stable phenotype; challenge is by authentic HIV (including uncloned, clinical isolates); transduced cells are not adversely affected by treatment; one can use clonal or polyclonal lines of transduced cells; and almost any immortalized cell line and many primary mitotic lines can be transduced. The disadvantages of murine retroviral transduction methodology compared with the electrotransfection method are the much more lengthy procedure; it requires more laboratory infrastructure; and that it is impossible to control relative copy number of target and agent in each cell. Compared with the HIV-based vector systems, this method has the following advantages: high efficiency; it can be used in CD4-positive cells; and it is already approved for use in humans. Compared with the HIV-based vector systems, this method has the following disadvantages: the tendency for extinction of gene expression in differentiating cells; the failure to transduce nonmitotic cells; and a noninducible expression system

3. Potential insufficiencies in the use of HIV-1-derived vectors. Stable packaging cell lines able to package replication-defective HIV-1 vectors have not been yet developed. The complex genomic organization of HIV-1 and the highly differentiated expression of its genes may, in part, explain some of the problems encountered for this development. Recombinant HIV-1 viruses are currently generated by transient transfection of the packaging component along with the vector. To

avoid the possibility of "helper virus" production, the packaging component is usually separated onto two different constructs, one expressing the Gag, and the Pol proteins, and the other expressing the Env protein *(16,18)*. Although a definition of the optimal placement of viral *cis*-acting sequences within the vector has been evaluated *(18)*, further insights for identifying *cis*-acting sequences that operate in the context of HIV-1 vectors is still needed to develop HIV-1 into a practical retroviral vector from the delivery of genes into primary target cells. The establishment of stable packaging cell lines or the use of transcription systems that enable high level transient expression of structural HIV-1 genes and packagable RNA may lead to production of high titers of replication-defective HIV-1 vectors.

References

1. Miller, A. D. and Rosman, G. J. (1989) Improved retroviral vectors for gene transfer and expression. *Biotechniques* **7,** 980–990.
2. Benchaibi, M. F., Mallet, F., Thoraval, P., Savatier, P., Xiao, J. H., Verdier, G., Samarut, J., and Nigon, V. M. (1989) Avian retroviral vectors derived from avian defective leukemia virus: role of the translational context of the inserted gene on efficiency of the vectors. *Virology* **169,** 15–28.
3. Bosselman, R. A., Hau, R. Y., Boggs, T., Hu, S., Briszewski, J., Ou, S., Souzs, L., Kozar, L., Martin, F., Nicolson, M., Rishell, W., Schultz, J. A., Semon, K. M., and Stewart, R. G. (1989) Replication-defective vectors of reticuloendotheliosis virus transduce exogenous genes into somatic stem cells of the incubated chicken embryo. *J. Virol.* **63,** 2680–2689.
4. Cosset, F. L., Legras, C., Chebloune, Y., Savatier, P., Thoraval, P., Thomas, J. L., Samarut, J., Nigon, V. M., and Verdier, G. (1990) A new avian leukosis virus-based packaging cell lines that uses two separate transcomplementing helper genomes. *J. Virol.* **64,** 1070–1078.
5. Mann, R., Mulligan, R. C., and Baltimore, D. (1983) Construction of a retroviral packaging mutant and its use to produce helper free defective retrovirus. *Cell* **33,** 153–159.
6. Miller, A. D. and Buttimore, C. (1986) Redesign of retrovirus packaging cell lines to avoid recombination leading to helper virus production. *Mol. Cell. Biol.* **6,** 2895–2902.
7. Savatier, P., Bagnis, C., Thoraval, P., Poncet, D., Belakebi, M., Mallet, F., Legras, C., Cosset, F. L., Thomas, J. L., Chebloune, T., Faure, C., Verdier, G., Samarut, J., and Nigon, V. M. (1989) Generation of a helper cell line for packaging avian leukosis-based vectors. *J. Virol.* **63,** 513–522.
8. Miller, D. G., Adam, M. A., and Miller, A. D. (1990) Gene transfer by retrovirus vectors occurs only in cells that are actively replicating at the time of infection. *Mol. Cell. Biol.* **10,** 4239–4242.
9. Stead, R., Kowk, W., Storb, R., and Miller, A. D. (1988) Canine model for gene therapy: inefficient gene expression in dogs reconstituted autologous marrow infected with retroviral vectors. *Blood* **71,** 742–747.

10. Weinberg, J. B., Mattews, T. J., Cullen, B. R., and Malim, M. H. (1991) Productive human immunodeficiency virus type 1 (HIV-1) infection of nonprofilerating human monocytes. *J. Exp. Med.* **174**, 1477–1482.

11. Lewis, P., Hensel, M., and Emerman, M. (1992) Human immunodeficiency virus infection of cell arrested in the cell cycle. *EMBO J.* **11**, 3053–3058.

12. Landau, N., Page, K. A., and Littman, D. R. (1990) Pseudotyping with human T-cell leukemia virus type 1 broadens the HIV-1 host range. *J. Virol.* **65**, 162–169.

13. Chesebro, B., Wehrly, K., and Maury, W. (1990) Differential expression in human and mouse cells of human immunodeficiency virus pseudotyped by murine retroviruses. *J. Virol.* **64**, 4553–4557.

14. Helseth, E., Kowaiski, M., Gabzuda, D., Olshevski, U., Haseltine, W., and Sodroski, J. (1990) Rapid complementation assays measuring replicative potential of human immunodeficiency virus type 1 envelope glycoprotein mutants. *J. Virol.* **64**, 2416–2420.

15. Shimada, T., Fuji, H., Mitsuya, H., and Nienhuis, A. W. (1991) Targeted and highly efficient gene transfer into CD4+ cells by a recombinant immunodeficiency virus retroviral vector. *Clin. Invest.* **88**, 1043–1047.

16. Poznanski, M., Lever, A. M., Bergeron, L., Haseltine, W., and Sodroski, J. (1991) Gene transfer into human lymphocytes by a defective human immunodeficiency virus type 1 vector. *J. Virol.* **65**, 532–536.

17. Buchschacher, G. L., Jr. and Panganiban, A. T. (1992) Human immunodeficiency vectors for inducible expression of foreign genes. *J. Virol.* **66**, 2731–2739.

18. Parolin, C., Dorfman, T., Palú, G., Göttlinger, H., and Sodroski, J. (1994) Analysis in human immunodeficiency virus type 1 vectors of *cis*-acting sequences that affect gene transfer into human lymphocytes. *J. Virol.* **68**, 3888–3895.

19. Cotten, M., Baker, A., Saltik, M., Wagner, E., and Buschle, M. (1994) Lipopoly-saccharide is a frequent contaminant of DNA preparations and can be toxic to primary human cells in the presence of adenovirus. *Gene Therapy* **1**, 239–246.

20. Danos, O. and Mulligan, R. C. (1988) Safe and efficient generation of recombinant retroviruses with amphotropic and ecotropic host ranges. *Proc. Natl. Acad. Sci. USA* **5**, 6460–6464.

21. Markowitz, D., Goff, S., and Bank, A. (1988) A safe packaging line for gene transfer: separating genes on two different plasmids. *J. Virol.* **62**, 1120–1124.

22. Morganstern, J. P. and Land, H. (1990) Advanced mammalian gene transfer: high titre retroviral vectors with multiple drug selection marks and complementary helper-free packaging cell lines. *Nucleic Acids Res.* **25**, 3587–3596.

23. Miller, A. D. and Buttimore, C. (1986) Redesign of retrovirus packaging cell lines to avoid recombination leading to helper virus production. *Mol. Cell. Biol.* **6**, 2895–2902.

24. Sodroski, J., Goh, W. C., Rosen, C., Campbell, K., and Haseltine, W. A. (1986) Role of the HTLV-III/LAV envelope in syncytium formation and cytopathic effect. *Nature* **322**, 4770–4774.

25. Sodroski, J., Rosen, C. A., Dayton, A. L., Goh, W. C., and Haseltine, W. A. (1986) Evidence for a second post-transcriptional transactivator required for HTLV-III/LAV replication. *Nature* **321,** 412–417.

26. Lever, A. M. L., Göttlinger, H., Haseltine, W., and Sodroski, J. (1989) Identification of a sequence required for efficient packaging of human immunodeficiency virus type 1 into virions. *J. Virol.* **63,** 4085–4087.

27. Sczakiel, G., Döffinger, R., and Pawlita, M. (1989) Testing for improved electrotransfection parameters by use of the fluorescent dye Lucifer yellow CH. *Anal. Biochem.* **181,** 309–314.

28. Döffinger, R., Pawlita, M., and Sczakiel, G. (1988) Electrotransfection of human lymphoid and myeloid cell lines, *Nucleic Acids Res.* **16,** 11,840.

29. Cullen, B. R., (1987) Use of eukaryotic expression technology in the functional analysis of cloned genes. *Methods Enzymol.* **152,** 684–703.

30. Rho, H. M., Poiesz, B., Ruscetti, F., and Gallo, R. (1981) Characterization of the reverse transcriptase from a new retrovirus (HTLV) produced by a human cutaneous T-cell lymphoma cell line. *Virology* **112,** 355–360.

30

Inducible System Designed
for Future Gene Therapy

Yaolin Wang, Bert W. O'Malley, and Sophia Y. Tsai

1. Introduction

Gene therapy involves the introduction of foreign therapeutic genes into humans to treat certain diseases. In current protocols, the expressions of the delivered foreign genes are under the control of a constitutive promoter. However, most genes are regulated under physiological conditions in response to various stimuli, including metabolites, growth factors, and hormones. Constitutive expression of foreign genes may result in cytotoxicity as well as undesired immune responses. In order to further the development of gene therapy, it is essential to regulate the expression of the genes once they are delivered into the body.

The criteria for a successful inducible system are:

1. It only induces the expression of the introduced target genes and does not affect any other endogenous genes;
2. It turns on the target gene with an exogenous signal, preferably a small molecule that can easily be administered and distributed throughout the body, including the brain;
3. The induction should be reversible, i.e., the expression of the target gene is off after withdrawal of the exogenous signal;
4. The exogenous signal molecule is biologically safe and preferably can be administrated orally; and
5. It has low basal activity (to prevent leaky expression) and high inducibility.

Several inducible systems for regulating gene expression have been established in the past decade. They include the use of the metal response promoter *(1)*, the heat shock promoter *(2)*, the glucocorticoid-inducible (MMTV-LTR) promoter *(3)*, the lac repressor/operator system using IPTG as inducer *(4,5)*,

From: *Methods in Molecular Biology, vol. 63: Recombinant Protein Protocols:
Detection and Isolation* Edited by: R. Tuan Humana Press Inc., Totowa, NJ

the Tet repressor/operator system (tTA) with tetracycline as an inducer *(6,7,8)*. More recently, the direct fusion of a protein of interest to the hormone-binding-domain (HBD) of the steroid receptors has been shown to render the fusion protein responsive to steroid *(9,10)*. For example, the GAL4-HBD fusion protein is capable of transactivating a target gene by binding to the GAL4 binding sites (17-mer) upstream of the target gene in the presence of hormone *(11)*. Whereas these systems have been used in tissue culture systems to regulate gene expression, their use in gene therapy is limited for the following reasons. The use of heat shock promoter for gene therapy is obviously impractical, whereas heavy metals (such as Cd-Zn) and IPTG are known to be cytotoxic *(4,12)* and the induction is slow. Regulation of the protein function by fusion directly to steroid receptor HBD might change the conformation of the protein to be regulated. For this reason, the function of the protein after fusion to HBD is sometimes unpredictable *(13,14)*. The disadvantages of using of glucocorticoid and estrogen inducible systems are that they are natural endogenous steroids that could affect expression of many endogenous genes and therefore interfere with normal cellular function. In the case of Tet R/O system, the chimeric protein tTA (consisting of Tet repressor and transactivator VP16) activates gene expression from a reporter gene containing the Tet operator sites upon withdrawal of tetracycline. Therefore, it is necessary for the constant presence of tetracycline in order to shut down the expression from the Tet operator. This would be inconvenient for use in gene therapy since long-term administration of tetracycline would result in side-effects, including deposition of the drug in teeth and bone, resulting in phototoxicity, hepatic toxicity, and renal toxicity *(15)*. Using genetic selection methods, Bujard and colleagues *(16)* recently isolated a Tet repressor mutant that can bind to DNA in the presence of higher concentration of tetracycline and its derivatives. In this new system, the chimeric protein of mutated Tet repressor and VP16 (rtTA) would activate reporter gene expression in the presence of tetracycline. Crabtree and colleagues *(17)* proposed another novel method using a divalent molecule FK1012 that could induce the dimerization of proteins containing the FK506-binding immunophilin domains (FKBP). In this case, the signal would be transduced when two proteins bearing the FKBP domains dimerize in the presence of the FK1012. It remains to be demonstrated whether this divalent molecule works in an in vivo situation without affecting any endogenous gene expression. In addition, the safety and efficacy of this molecule in vivo is also unclear.

We have recently described a novel inducible system for regulating gene expression *(18)*. This system consists of a regulator (transactivator) and a reporter containing the target gene of interest (Fig. 1). We constructed a chimeric regulator (GL-VP) by fusing the HBD of a progesterone receptor mutant (hPRB891) to the yeast transcription activator GAL4 DNA binding domain

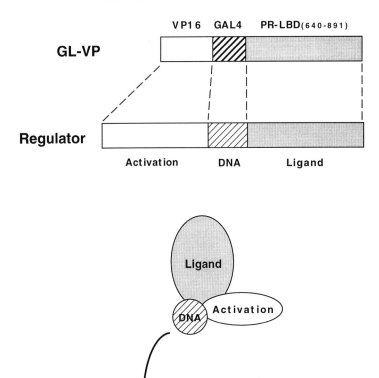

Fig. 1. Schematic diagram of the inducible system.

and the herpes simplex virus (HSV) protein VP 16 activation domain. This mutated HBD does not bind to progesterone or its agonists, but it binds to the progesterone antagonist RU486 (Mifpristone) with high affinity. The reporter contains multiple GAL4 binding sites (17-mer) and a promoter driving the expression of a target gene. With this novel approach, we transformed a constitutive transactivator GAL4-VP16 into a RU486 regulatable transactivator GL-VP. In the absence of RU486, this chimeric regulator (GL-VP) will not bind to the 17-mer sites present on the reporter. Hence, little or no expression of target gene transcription will occur. However, upon binding to RU486, a conformation change is effected in the regulator GL-VP, enabling the regulator to bind to the 17-mer GAL4 binding sites on the target gene construct and thereby initiate the transcription of the target gene. We have recently demonstrated that this inducible system works in transient transfection assays as well as in stable cell lines using various target genes encoding intracellular proteins, chloramphenicol acetyl transferase (CAT), tyrosine hydroxylase or a secretory protein, the human growth hormone (hGH) in response to RU486. In addition, this regulatory system has been validated in vivo via ex vivo transplantation of a stable cell line containing both the regulator and a reporter gene into rats. The dosage of RU486 used is significantly lower than that required for antagonizing progesterone action.

The advantages of this inducible system are:

1. It can be regulated by an exogenous signal, in this case, RU486.
2. RU486 is a small, synthetic molecule (mol wt 430) and has been used safely as an oral drug for other medical purposes *(19,20)*.
3. Since the yeast GAL4 protein has no mammalian homologue, it is likely that the target gene driven by the 17-mer binding sites and promoter would not be activated or repressed by endogenous proteins.
4. It does not affect endogenous gene expression since the regulator GL-VP only efficiently activates the target gene bearing multiple copies of the 17-mer sequences (unpublished data) juxtaposed to the promoter of the target gene (21).
5. The regulator can be activated by a very low dose of RU486 (1 nM). At this low concentration, RU486 does not affect the normal function of endogenous progesterone and glucocorticoid receptors, since it would only antagonize the activity of these receptors at a higher dose. This high binding affinity ensures its specificity on the regulation of the target gene expression.

From the aforementioned features, it is evident that this inducible system has many applications in addition to temporally controlling the therapeutic protein expression for gene therapy. For example, viral vectors have been routinely employed to the delivery of genes into various tissues and organs of the body. Recently, it has been noted that certain viral proteins could cause a cellular immune response *(22)*, resulting in a shortened duration of the target gene expression. With this inducible system, the expression of the viral protein could be regulated such that it is turned on only during the crucial stage of viral synthesis and assembly and it can then be turned off once it is transferred into the cells. With its ability to temporally turn genes on and off, this novel inducible system can also be used to express genes of interest during different stages of development and thereby allow one to assess the function of a particular target gene in transgenic animals.

In this chapter, we demonstrate that the application of the inducible system in regulating the expression of therapeutic protein in mammalian cells using an intracellular protein (tyrosine hydroxylase) and a secretory protein (human growth hormone) as examples. In addition, we will also describe how to use tissue-specific promoters in directing the expression of the transactivator (regulator protein). The purpose of this chapter is to provide protocols that could be easily adapted by the readers to regulate the expression of their target gene of interest.

2. Materials

2.1. Cell Culture

1. Dulbecco's modified Eagle's medium (DMEM) with glutamine containing 100 U/mL penicillin G and 100 µg/mL streptomycin (Gibco-BRL, Gaithersburg, MD).
2. Fetal bovine serum (FBS) (Gibco-BRL)
3. Trypsin/EDTA solution (Gibco-BRL).
4. Hanks balanced salt solution (HBSS): 5.4 mM KCl, 0.3 mM Na$_2$HPO$_4$, 0.4 mM KH$_2$PO$_4$, 4.2 mM NaHCO$_3$, 1.3 mM CaCl$_2$, 0.5 mM MgCl$_2$, 0.6 mM MgSO$_4$, 137 mM NaCl, 5.6 mM D-glucose, 0.02% phenol red (optional), add H$_2$O to 1 L and adjust to pH 7.4.
5. 10-cm Tissue culture dish.
6. 37°C, 5 % CO$_2$ humidified incubator.
7. RU486 (from Roussel-UCLAF) is dissolved in 80% ethanol and stored at –20°C (avoid light) (*see* Notes 1).
8. G418 (Gibco-BRL), Hygromycin-B (Boehringer Mannheim, Mannheim, Germany).

2.2. Western Blot Analysis

1. PBS buffer: 137 mM NaCl, 2.7 mM KCl, 4.3 mM Na$_2$HPO$_4$.
2. Lysis buffer: 50 mM HEPES, pH 7.0, 150 mM NaCl, 0.5% Nonidet P-40, 2 µg/mL aprotinin, 2 µg/mL leupeptins, 1 µg/mL Pepstatin A (Boehringer Mannheim), 2 mM Benzamidine · HCl (Sigma, St. Louis, MO).
3. Nylon membrane (Micron Separations, 0.45 µm).
4. Constant current power supply.
5. Bio-Rad Modular Mini-Protean II Electrophoresis system.
6. Bio-Rad Trans-Blot SD Semi-Dry Transfer Cell.
7. TBS-Tween (0.1%), pH 7.6: 2.42 g Tris base, 8 g NaCl, 1.4 mL concentrated HCl, 1 mL Tween-20. Dilute to 1000 mL with distilled water and adjust for pH.
8. Mouse anti-rat TH MAb (Boehringer Mannheim) stored in 100 µL aliquot (40 µg/mL) at –20°C.
9. HRP-conjugated sheep antimouse IgG (Amersham, Arlington Heights, IL).
10. Enhancer chemiluminescence (ECL) detection regents (Amersham).

2.3. Plasmid Construction

1. TE buffer, pH 7.4 or 8.0: 10 mM Tris-HCl (pH 7.4 or 8.0), 1 mM EDTA (pH 8.0).
2. Restriction enzyme and digestion buffer (Promega, Madison, WI; New England Biolab, Beverly, MA).
3. 50X TAE buffer: 242 g Tris-base, 57.1 mL glacial acetic acid, 37.2 g Na$_2$EDTA · 2H$_2$O (2 mM) to 1 L and adjust pH to 8.5.
4. Agarose.
5. Gel electrophoresis apparatus.

6. Qiaex DNA purification kit (Qiagen, Chatsworth, CA).
7. Competent bacterial cell DH5α.
8. LB broth base, power (Gibco-BRL).

2.4. Transfection and Analysis of hGH Expression

1. 2X HBS: 16.4 g NaCl, 11.9 g HEPES, 0.21 g Na_2HPO_4, add dH_2O to 1 L, adjust pH to 7.05 with NaOH, and sterilize through a 0.45-μm filter.
2. 0.25M $CaCl_2$ (sterile).
3. Calf thymus DNA (Sigma) dissolved in TE at concentration of 2 mg/mL.
4. CsCl-prep grade DNA.
5. Glycerol.
6. 12 × 75 mm sterile disposable polystyrene tube (Fisher, Pittsburgh, PA).
7. Nichols Institute hGH Assay Kit (#40-2155).
8. Rotating platform.
9. γ-radiation counter
10. Spreadsheet program such as Microsoft Excel or Lotus 1-2-3.

3. Methods

In this section, we will describe the use of the inducible system to regulate gene expression in mammalian cells. We will start with the analysis of the inducible expression of a target protein using the gene of an intracellular protein, the tyrosine hydroxylase (TH), as an example. We will then describe how to construct a reporter construct using human growth hormone (hGH) gene as a target gene. Finally, we will discuss the use of a minimal promoter (adenovirus E1B TATA sequence) to reduce the basal activity of the reporter gene constructs.

3.1. Analysis of Inducible Expression of Tyrosine Hydroxylase

The construction of TH reporter gene and the generation of stable cell lines that express TH by inducer RU486 has been previously described *(18)*. We describe here the analysis of TH expression in the cells using Western blot protocols.

3.1.1. RU486 Treatment of Stable Cells

1. Stable cells are maintained in DMEM supplied with 10% FBS in the presence of G418 (100 μg/mL) and hygromycin-B (50 μg/mL).
2. Cells are split from one plate to 3–5 dishes after cell density has reached 80% confluence.
3. 10 μL of RU486 (1000X solution) will be added to the medium (10 mL) containing the cells to have a final concentration of 1X.
4. The expression level of TH gene can be produced by incubating the cells with different concentrations of RU486 and for different periods of time.
5. Cellular protein is extracted by treating the cells with lysis buffer, and Western blot is performed to analyze the expression of TH.

3.1.2. Lysis of Cultured Mammalian Cells

1. Wash cells twice with prechilled PBS.
2. Add the lysis buffer (cooled to 0°C) to cell monolayers. The volume of lysis buffer should be adjusted according to the size of the cell culture dish as shown at the end of this section (*see* Note 2).
3. Incubate the cell with lysis buffer for 20 min on a flat tray with ice.
4. Scrape the cells with a rubber policeman or cell scraper.
5. Centrifuge the lysate at 12,000*g* for 2 min at 4°C in a microfuge.
6. Transfer the supernatant to a tube and place on ice for immediate use or store at −70°C.

Volume of lysis buffer, mL	Size of Petri dish, mm
1.0	90
0.5	60
0.25	35

3.1.3. Western Blot and ECL Staining

1. For each minigel lane, the maximum loading volume is 50 μL. Mix sample (25 μL of lysed supernatant) with an equal volume of 2X SDS sample buffer.
2. Heat at 100°C for 5 min to denature the proteins.
3. The sample is then separated on a 10% SDS-PAGE gel *(23)*. Load protein molecular weight marker (Amersham rainbow maker) in a separate lane for size determination later.
4. Blot the gel to a nylon membrane at 17 V for 1.0 h with the Bio-Rad Trans-Blot SD Semi-Dry Transfer Cell.
5. Block the nylon membrane in TBS-Tween (0.1%) and 3% nonfat milk for 1 h on a rotating platform.
6. Wash the milk-blocked nylon membrane with TBS-Tween once for 15 min, and twice for 5 min.
7. Add primary antibody body (4 μg of mouse anti-TH antibody) to 10 mL of TBS-Tween and 1% nonfat milk for 1 h at room temperature.
8. Wash the nylon membrane again with TBS-Tween as in step 6.
9. Add 50 μL of HRP-conjugated sheep antimouse IgG at a 1:1000 dilution to 50 mL of TBS-Tween, incubate for 1 h at room temperature.
10. Wash the nylon membrane with TBS-Tween once for 15 min, and four times for 5 min each.
11. Detection of the TH protein expression is performed using ECL detection kit (Amersham) as described in steps 12–17.
12. Mix an equal volume of detection solution 1 (5 mL) with solution 2 (5 mL) to give sufficient coverage of the membrane (0.125 mL/cm^2).
13. Drain the excess buffer from the washed blots and place them in fresh containers. Add the detection reagents directly to the blots on the surface carrying the protein; do not let the blots to dry out.
14. Incubate for precisely 1 min at room temperature.

RU486
(-logM)

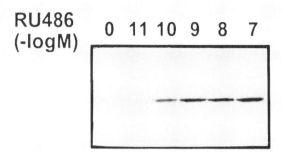

0 11 10 9   8   7

Fig. 2. Dose-response of RU486 mediated activation of tyrosine hydroxylase (TH) gene expression. Stable cell line (T1.21) harboring regulator (CMV-GLVP) and target (p17X4-tk-TH) was generated as described previously *(18)*. The cells were incubated with different concentrations of RU486 as indicated before harvesting for immunoblot analysis.

15. Drain off any excess detection reagent and wrap the blots in plastic wrap. Gently smooth out air pockets.
16. Place the blots, protein side up, in the film cassette. Work as quickly as possible; minimize the delay between incubating the blots in detection reagent and exposing them to the film.
17. Switch off the lights and place a sheet of autoradiography film on top of the blots, close the cassette and expose for 15 s (*see* Note 3).
18. Results are presented in Fig. 2. (doses response of TH induction by RU486).

3.2. Human Growth Hormone Gene as a Regulatable Target

In this section, we use the human growth hormone gene as a target to illustrate how the inducible system can be utilized to regulate the expression of a secreted protein. We use the previously described reporter 17X4-TATA-CAT (18) as a parental vector and replace the CAT gene insert with a 2.1-kb human growth hormone (hGH) genomic DNA fragment (Fig. 3). This reporter contains only the minimal TATA promoter and will therefore significantly reduce the basal activity of the reporter gene. We will also describe the use of liver-specific promoter-enhancer to target the expression of regulator in a tissue specific manner.

3.2.1. Construction of Inducible hGH Reporter Plasmid

1. The reporter plasmid 17X4-TATA-CAT (3 μg) is digested with *Xho*I and blunt ended with Klenow. The linearized DNA is precipitated and the DNA pellet is redissolved in TE. The DNA is then digested with *Eco*RI and separated on a 1% agarose gel (1X TAE buffer). The DNA fragment corresponding to the reporter backbone (minus CAT gene fragment) is cut out of the gel and purified using the

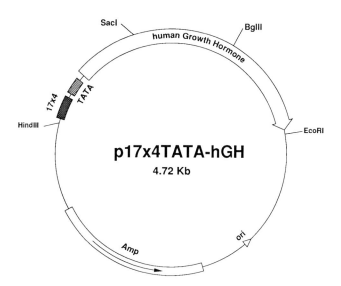

Fig. 3. Plasmid map of reporter constructs 17X4-TATA-CAT and 17X4-TATA-hGH.

Qiaex DNA purification kit according to the manufacturer's instructions. The purified DNA is resuspended in 20 μL of TE.

2. 3 μg of the plasmid pOGH (containing genomic human growth hormone sequence, from Nichols Institute) is digested with *Bam*HI and blunt-ended with the Klenow fragment. The linearized DNA is precipitated and the DNA pellet is redissolved in TE. The DNA is then digested with *Eco*RI and separated on a 1% agarose gel (1X TAE buffer). The DNA fragment corresponding to the hGH insert (2.16 kb) is purified by Qiaex kit.

3. The reporter fragment isolated from step 1 is ligated with the hGH insert (from step 2) using T4 DNA ligase and transformed into competent *E. coli*. DH5α cells. Individual colonies are picked and grown in LB media overnight. Miniprep DNA is done using the boiling method *(24)*.

4. Positive clones (p17X4-TATA-hGH) (Fig. 3) containing the hGH insert are identified by restriction mapping of the miniprep DNA and large scale DNA preparation is done using CsCl method *(25)*.

3.2.2. Expression of Regulator Using Tissue-Specific Promoter

1. The multiple cloning sites of the pBluescript KS II (+) is replaced with a new set of cloning sites by inserting an annealed oligo to the *Acc*65I-*Not*I digested pBluescript KS II creating plasmid pPAP. In plasmid PAP, the new multiple cloning sites are (from T3 promoter) *Pme*I, *Asc*I, *Sal*I, *Pst*I, *Hind*III, *Spe*I, *Acc*65I, *Bgl*II, *Bam*HI, *Eco*RI, *Mlu*I, *Xba*I, *Pac*I, *Not*I, *Sac*I (followed by T7 promoter).

2. The *Eco*RI-*Xba*I fragment containing SV40 small t intron and polyA signal (800 bp) is subcloned into the pPAP vector creating plasmid pPAP-SV40.

3. The liver-specific promoter-enhancer TTRA (*Sal*I-*Hind*III fragment of 3 kb) and TTRB (*Hind*III fragment of 400 bp) *(26)* are then subcloned separately into the pPAP-SV40 vector creating plasmid pPAP TTRA-SV40 and pPAP TTRB-SV40, respectively.
4. The *Acc*65I-*Bam*HI GL-VP DNA fragment (1.7 kb) was isolated from pGL-VP, the original construct bearing the RSV promoter as previously described *(18)*. This fragment was then subcloned into the *Acc*65I-*Bam*HI site of both pPAP TTRA-SV40 and pPAP TTRB-SV40 plasmids, yielding plasmid pTTRA-GLVP and pTTRB-GLVP, respectively.

3.2.3. Inducible Expression of hGH in Liver Cell

1. Five micrograms of the reporter plasmid DNA (p17X4-TATA-hGH) are cotransfected with 0.5 μg of the liver-specific regulator (TTRA-GLVP or TTRB-GLVP) into HepG2 cells (20% confluent) grown on 10-cm culture dish using the calcium precipitation method *(27)*.
2. Briefly, the plasmids are added to 0.5 mL of 2X HBS and mixed gently in a sterile polystyrene (12 × 75 mm) tube. Calf thymus DNA is used as carrier DNA to balance the amount of total DNA (15 mg) for each culture dish. 0.5 mL of a $0.25M$ CaCl2 solution is added dropwise to the mixture while vortexing. The mixture is then left to sit at room temperature for 15–30 min (*see* Note 4).
3. The calcium precipitates containing the plasmid are then added to the cell culture dish. After 4 h, the cells are shocked with glycerol to increase the transfection efficiency *(27)*.
4. Ten microliters of a $10^{-5}M$ RU486 solution are added to the cells grown in 10 mL DMEM, 10% FBS) (*see* Note 5) to reach a final concentration of 10 nM of RU486 in the culture.
5. After 36–48 h posttransfection, 20–100 μL of the culture medium are pipeted into a 12 × 75 mm polystyrene tube for hGH expression assay (*see* Note 6).
6. The hGH expression is determined using Nichols Institute's hGH assay kit and a γ radiation counter.
7. Linear regression analysis is performed using data obtained from the hGH standards that are included in the kit (*see* Note 6). The level of hGH expression in cell culture is calculated from the linear regression curve (Fig. 4).

4. Notes

1. The compound RU486 is stable in 80% ethanol at –20°C for up to a year. We usually prepare a 1000X concentrated solution and add to medium in 1:1000 dilution. For example, if a final concentration of 1 nM of RU486 is desired for cell culture experiments, we add 10 μL of $10^{-6}M$ RU486 to 10 mL DMEM medium. For control plates, only 80% ethanol is added.
2. The lysis buffer used for TH extraction contains 150 mM NaCl. Depending on the protein localization within the cell, different concentrations of salt and deter-

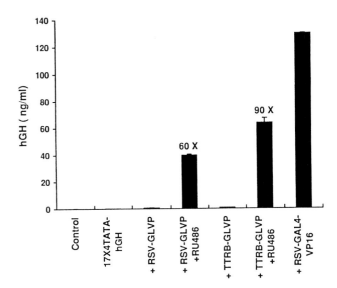

Fig. 4. Inducible hGH gene expression using liver specific promoter: 5 µg of reporter plasmid (17X4-TATA-hGH) was cotransfected into HepG2 with 0.5 µg of regulator driven by either RSV promoter or liver specific enhancer-promoter TTRB *(26)*. The TTRB element is a 400 bp fragment derived from the 5' end of transthyretin gene that confers liver-specific expression. RSV-GAL4-VP16 (0.5 µg) was included as a positive control. RU486 (10 n*M*) was added as indicated. Numbers above the bars indicate fold of induction in the presence of RU486.

gent must be experimentally determined. For example, if the protein is localized in the nucleus, high salt (500 m*M*) lysis buffer should be used. For efficient lysis, the buffer should be added directly onto the cell monolayer. Protease inhibitors can be prepared as a 1000X stock solution and stored at –20°C in aliquots. It should be freshly added before use.

3. During ECL detection using the autoradiography film, the time of exposure can be varied from 1 s to 1 min depending on the strength of the signal.

4. The CaCl$_2$ solution should be freshly prepared at all times and all solutions should be kept at room temperature. The calcium precipitates can be easily visualized as the solution becomes cloudy after the tube has been sitting for a while .

5. RU486 can be added to the cell culture right after glycerol shock or it can be added the next morning.·

6. The amount of medium needed for assay depends on the level of hGH expression. It is suggested that both 20 µL and 100 µL of cell culture medium are used to ensure the data falls within linear line of the standard curve. Regression analysis can be done with a Casio scientific calculator or a computer program such as Microsoft Excel.

Reference

1. Searle, P. F., Stuart, G. W., and Palmiter, R. (1985) Building a metal-responsive promoter with synthetic regulatory elements. *Mol. Cell. Biol.* **5**, 1480–1489.
2. Fuqua, S. A., Blum-Salingaros, M., and McGuire, W. L. (1989) Induction of the estrogen-regulated "24K" protein by heat shock. *Cancer Res.* **49**, 4126–4129.
3. Hirt, R. P., Fasel, N., and Kraehenbuhl J.-P. (1994) Inducible protein expression using a glucocorticoid-sensitive vector. *Methods Cell Biol.* **43**, 247–262
4. Figge, J., Wright, C., Collins, C. J., Roberts, T. M., and Livingston, D. M. (1988) Stringent regulation of stably integrated chloramphenicol acetyl transferase genes by E. coli lac repressor in monkey cell. *Cell* **52**, 713–722.
5. Baim, S. B., Labow, M. A., Levine, A. J., and Shenk, T. (1991) A chimeric mammalian transactivator based on the lac repressor that is regulated by temperature and isopropyl b-D-thiogalactrpyranoside. *Proc. Natl. Acad. Sci. USA* **88**, 5072–5076.
6. Gossen, M. and Bujard, H. (1992) Tight control of gene expression in mammalian cells by tetracycline-responsive promoters. *Proc. Natl. Acad. Sci. USA* **89**, 5547–5551.
7. Gossen, M., Boninm, A. L., Freudlieb, S., and Bujard, H. (1994) Inducible gene expression systems for higher eukaryotic cells. *Curr. Opin. Biotechnol.*. **5**, 516–520.
8. Furth, P. A., Onge, L. T., Boger, H., Gruss, P., Gossen, M., Kistner, A., Bujard, H., and Hennighausen, L. (1994) Temporal control of gene expression in transgenic mice by a tetracycline-responsive promoter. *Proc. Natl. Acad. Sci. USA* **91**, 9302–9306.
9. Picard, D. (1994) Regulation of protein function through expression of chimaeric proteins. *Curr. Opin. Biotechnol.*. **5**, 511–515.
10. Jackson, P., Baltimore, D., and Picard, D. (1993) Hormone-conditional transformation by fusion proteins of c-Abl and its transforming variants. *EMBO J.* **12**, 2809–2819.
11. Braselmannm, S., Graninger, P., and Busslinger, M. (1993) A selective transcriptional induction system for mammalian cells based on GAL4-estrogen receptor fusion proteins. *Proc. Natl. Acad. Sci. USA* **90**, 1657–1661.
12. Klaassen, C. (1985) Heavy metals and heavy-metal antagonists, in *Goodman and Gilman's The Pharmacological Basis of Therapeutic* (Gilman, A. G., Goodman, L. S., Rall, T. W., and Murad., F., eds.), Macmillan, New York, pp. 1617–1619.
13. Roemer, K. and Friedmann, T. (1993) Modulation of cell proliferation and gene expression by a p53-estrogen receptor hybrid protein. *Proc. Natl. Acad. Sci. USA* **90**, 9252–9256
14. Schuermann, M. Henning, G., and Muller, R. (1993) Transcriptional activation and transformation by chimaeric Fos-estrogen receptor proteins: altered properties as a consequence of gene fusion. *Oncogene* **8**, 2781–2790.
15. Sande, M. A. and Mandell, G. L. (1985) Antimicrobial agents, in *Goodman and Gilman's The Pharmacological Basis of Therapeutic* (Gilman, A. G., Goodman, L. S., Rall, T. W., and Murad., F., eds.), Macmillan, New York, pp. 1174–1176.

16. Gossen, M., Freundlieb, S., Bender, G., Kitner, A., Baron, U., Muller, G., Hillen, W., and Bujard, H. (1995) Exploiting prokaryotic elements for the control of gene activity in higher eukaryotics. *J. Cell. Biochem.* (**Suppl. 21A), C6-004,** 355.

17. Spencer, D. M., Wnadless, T. J., Schreiber, S. L., and Crabtree, G. R. (1993) Control of signal transduction with synthetic ligands. *Science* **262,** 1019–1024.

18. Wang, Y., O'Malley, B. W. O., Jr., Tsai, S. Y., and O'Malley, B. W. O. (1994) A regulatory system for use in gene transfer. *Proc. Natl. Acad. Sci. USA* **91,** 8180–8184.

19. Brogden, R. N., Goa, K. L., and Faulds, D. (1993) Mifepristone, a review of its pharmacodynamic and pharmacokinetic properties, and therapeutic potential. *Drugs* **45,** 384–409.

20. Grunberg, S. M., Weiss, M. H., Spitz, I. M., Ahmadi, J., Sadun, A., Russell, C. A. Lucci, L., and Stevenson, L. L. (1991) Treatment of unresectable meningiomas with the antiprogesterone agent mifepristone. *J. Neurosurg.* **74,** 861–866.

21. Lin, Y.-S, Carey, M. F., Ptashne, M., and Green, M. R. (1988) GAL4 derivatives function alone and synergistically with mammalian activators in vitro. *Cell* **52,** 713–722

22. Yang, Y., Nunes, F. A., Berencsi, K., Furth, E. E., Gonczol, E., and Wilson, J. M. (1994) Cellular immunity to viral antigens limits E1-deleted adenoviruses for gene therapy. *Proc. Natl. Acad. Sci. USA* **91,** 4407–4411.

23. Gallagher, S. R. E. and Smith, J. A. (1992) Denaturing (SDS) discontinuous gel electrophoresis: Laemmli gel method, in *Current Protocols in Molecular Biology* (Ausubel, F. M., Kingston, R. E., Moore, D. D., Seidman, J. G., Smith, J. A., and Struhl, K., eds.), John Wiley, New York, pp. 10.2.4–10.2.9.

24. Holmes, D. S. and Quigley, M. (1981) A rapid boiling method for the preparation of bacterial plasmids. *Anal. Biochem.* **114,** 193–197.

25. Heilig, J. S., Lech, K., and Brent, R. (1992) Large-scale preparation of plasmid DNA-CsCl/ethidium bromide equilibrium Centrifugation, in *Current Protocols in Molecular Biology* (Ausubel, F. M., Kingston, R. E., Moore, D. D., Seidman, J. G., Smith, J. A., and Struhl, K., eds.), John Wiley, New York, pp. 1.7.5–1.7.11.

26. Yan, C., Costa, R. H., Darnell, J. E., Jr., Chen, J., and Van Dyke, T. A. (1990) Distinct positive and negative elements control the limited hepatocyte and choroid plexus expression of transthyretin in transgenic mice. *EMBO J.* **9,** 869–878.

27. Kingston R. E. (1992) Calcium phosphate transfection, in *Current Protocols in Molecular Biology* (Ausubel, F. M., Kingston, R. E., Moore, D. D., Seidman, J. G., Smith, J. A., and Struhl, K., eds.), John Wiley, New York, pp. 9.1.1–9.1.3.

31

Human Gene Therapy

Dreams to Realization

Muhammad Mukhtar, Zahida Parveen, and Omar Bagasra

1. Introduction

The term human gene therapy is defined as the transfer of DNA or RNA into human cells for therapeutic purposes. Significant advancements in recombinant DNA technology have led to an understanding of the molecular bases of many diseases ranging from inherited disorders to certain malignancies to infectious diseases such as acquired immunodeficiency syndrome (AIDS). The increasing ability to characterize a disease at its molecular level has made genetic interventions feasible and provides a rationale for gene therapy. The essence of gene therapy is the correction or replacement of the dysfunctioning genetic element with its normal counterpart or the specific disruption of a harmful gene product *(1)*.

In this article, we will review some of the recent advances in the various areas of gene therapy. We have summarized some of the leading technological breakthroughs in the treatment of complex genetic disorders. An overview of current gene therapy approaches against AIDS, central nervous system illnesses, behavior problems, various types of neoplasia, metabolic defects, and cardiovascular diseases is also described. A part of this article focuses on the methodologies and description of vectors for delivery of genes.

There are two major avenues of gene therapy based on cell types. Cells involved in genetic transfer from an individual to offspring are termed germ cells, whereas somatic cells makeup various organs of the whole body. Alterations in somatic cells are confined to an individual's genetic constitution and are nonheritable. In the case of germline gene therapy, foreign genes are injected to fertilized eggs and the resulting changes are transferred both to somatic as well as germ cells and are passed along to future generations.

From: *Methods in Molecular Biology, vol. 63: Recombinant Protein Protocols: Detection and Isolation* Edited by: R. Tuan Humana Press Inc., Totowa, NJ

Genetic disorders are classified into three major categories, i.e., those related to mutations in a single gene, multifactorial (polygenic inheritance), and chromosomal disorders. Single gene mutations involve inborn errors of metabolism and a number of storage disorders. Most familiar examples of multifactorial inherited disorders are hypertension and diabetes mellitus. In multifactorial genetic disorders, environmental factors besides genetic elements also play a major role in the development of an ailment. The third category, chromosomal mutations, are associated with gross structural changes in chromosomes. Single gene mutations are inherited in three patterns, i.e., autosomal dominant, autosomal recessive, and X-linked. A listing and classification of the most common single gene mutations is listed in Table 1.

Chromosomal mutations are manifested as an abnormal number of chromosomes or a change in the structure of one or more chromosomes. In normal humans, the male and female chromosomal composition is 22 pairs of autosomes and a pair of sex chromosomes. A normal male chromosomal constitution is 44, XY, whereas for females it is 44, XX. The chromosomal disorders can involve either autosomes or sex chromosomes.

Human gene therapy is not only being targeted for diseases involving altered genetic elements; a number of other diseases like HIV-1, cancer, cardiovascular disorders, and metabolic disorders are also being considered. The only difference between genetic and nongenetic diseases is that with the former a normal gene is delivered to the target cells to replace the function of the mutated gene, whereas in the latter a therapeutic gene is delivered to the cells whose product will interfere with the harmful effects of an abnormally expressed oncogene or with the gene products from an infectious agent. In the future it should be possible to vaccinate individuals with a gene sequence that will result in the production of infectious disease antigen(s), thereby immunizing and providing protection against certain pathogens, i.e., diphtheria, mycobacterium tuberculosis, or other agents. Edible plants are being developed that contain bacterial and viral antigens that can potentially immunize a person without injection.

2. Gene therapy of Cancer

Modern molecular and biochemical techniques have made possible the isolation, amplification, and characterization of nearly any gene whose biological property is known. Our present understanding of carcinogenesis suggests that cancer has a genetic basis *(2)*. The abnormal clonal proliferation is the result of genetic abnormalities in cells. Cloning and characterization of the genes involved in carcinogenesis have made possible the use of gene therapeutic approaches to selectively target and destroy tumor cells. Significant advancements are being made in genetic linkage of the carcinogenic genes. Localization of breast cancer susceptibility genes BRCA1 *(3)* and BRCA2 *(4)* suggest

Table 1
Single Gene Disorders in Humans

System	Disorder/Disease
Metabolic	Acute intermittent porphyria[a]
	α-1 Antitrypsin deficiency[b]
	Cystic fibrosis[b]
	Diabetes insipidus[c]
	Familial hypercholesterolemai[a]
	Galactosemia[b]
	Glycogen storage disease[b]
	Hemochromatosis[b]
	Homocystinuria[b]
	Lesch-Nyhan syndrome[b]
	Lysosomal storage diseases[b]
	Phenylketonuria[b]
	Wilson's disease[b]
Nervous	Fragile X syndrome[c]
	Friedreich ataxia[b]
	Huntington's disease[a]
	Myotonic dystrophy[a]
	Neurofibromatosis[a]
	Neurogenic muscular atrophies[b]
	Spinal muscular atrophy[b]
	Tuberous sclerosis[a]
Skeletal/Musculoskeletal	Achondroplasia[a]
	Alkaptonuria[b]
	Duchenne muscular dystrophy[c]
	Ehler-Danlos Syndrome[a]
	Marfan syndrome[a]
	Osteogenesis imperfecta[a]
Hematopoietic	Hereditary spherocytosis[a]
	Sickle cell anemia[b]
	Thalassemias[b]
	von Willebrand disease[a]
Blood	Hemophilia A and B[c]
	Chronic granulomatous disease[c]
	Glucose 6-phosphate dehydrogenase deficiency[c]
Endocrine	Congenital adrenal hyperplasia[b]
Gastrointestinal	Familial polyposis coli[a]
Immune	Agammaglobulinemia[c]
	Wiskoff-Aldrich syndrome[c]
Urinary	Polycystic kidney disease[a]

[a]Autosomal dominant disorders.
[b]Autosomal recessive disorders.
[c]X-linked recessive.

that besides these two genes, there might be some other gene conferring susceptibility to various forms of cancer. Currently a number of gene therapy protocols have been approved or are in progress for cancer therapy *(5)*. A survey of the clinical protocols approved for gene therapy reveals five major categories *(6,7)*.

The first category involves the enhancement of immune cells, antitumor activity by introducing genes that encode cytokines. Cytokine-mediated gene therapy alters the tumor-host relationship and facilitates recognition as well as destruction of the malignant cells. Moreover, cytokines like interleukin-2 (IL-2) and interleukin-4 (IL-4) are supposed to augment the generation of cytotoxic T-cell responses *(8)*. It has been previously shown that direct intratumoral delivery of an adenoviral vector harboring the murine IL-2 gene completely recovers metacytoma tumors in a mouse model *(9)*. An approach for treatment of metastatic human cancer by injection of IL-2 secreting tumor cells has also been described *(10)*. A number of clinical trials are in progress utilizing cytokines, IL-2, IL-4, tumor necrosis factor (TNF-α) and interferon *(5,11)*.

The second major category of gene therapeutic approaches for cancer involves the in vitro genetic alteration of cancer cells. The genetic material can be introduced either by liposome-mediated gene transfer technology or by using retroviral vectors *(12)*. Encouraging results were obtained by direct injection of HLA-B7, a transiently expressed cell surface antigen inducing an antitumor immune response *(13,14)*. The use of carcinoembryonic antigen (CEA) to elicit an immune response has shown effective vaccination against syngenic mouse colon and breast carcinomas *(15)*.

One important category of gene therapy for cancer involves the transfer of a "suicide" gene into tumor cells followed by activation of the suicide mechanism. Suicide of tumor cells is accomplished by the transfer of a particular gene into the actively growing tumor cells that renders them sensitive to death by certain treatments. Retroviral vectors containing the herpes simplex virus thymidine kinase (HSV-tk) gene inserted into a mouse tumor cell line followed by gancyclovir or acyclovir treatment showed good curability rate *(16)*. The mechanism involves natural substrate selection. The antiviral drugs gancyclovir or acyclovir target thymidine kinase. The acyclovir is identical to a normal guanine building block of DNA except that the sugar ring is interrupted. The viral thymidine kinase incorporates acyclovir to its active triphosphate as a pseudo building block of DNA. It has been observed that mammalian thymidine kinase (tk) only phosphorylates a thymidine nucleotide, whereas HSV-tk phosphorylates a nucleoside base also. The phosphorylation and later insertion of the nucleoside homolog gancyclovir and acyclovir into the DNA synthesis pathway obstruct DNA synthesis resulting in cell death *(17,18)* (Fig. 1). Retrovectors are the choice vehicle for suicide gene transfer like HSV-tk in brain tumors. Retroviral vectors only transfer genes in actively dividing cells,

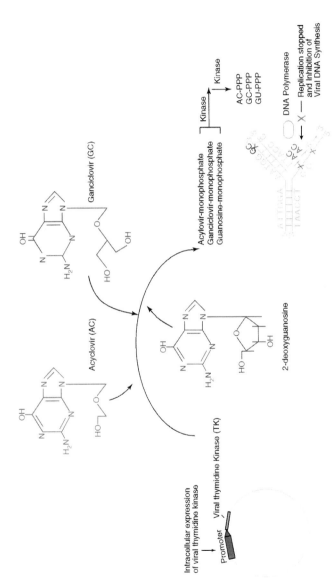

Fig. 1. Incorporation of a suicide genetic system into rapidly replicating cells (i.e., tumor cells). Receptor mediated transfer of the "suicide" gene(s) can be utilized. The example given here is a viral thymidine kinase (TK) gene. Incorporation of acyclovir or gancyclovir results in DNA-replication defect and cell death.

and in the case of a brain tumor, the only dividing cells are the tumor cells. Suicide gene therapy trials are being conducted by the National Institute of Health (NIH) and the preliminary responses are very encouraging.

The fourth group of gene therapeutic approaches for carcinogenesis involve the insertion of wild-type tumor suppresser genes or antioncogenes. Mutations in tumor suppresser genes such as p53 can result in overexpression of oncogenes and result in clinical tumor formation. The insertion of a wild-type copy of the p53 gene into the p53-deficient tumor cells has been shown to mitigate tumorigenesis *(19)*. Recent studies demonstrated that introduction of the p53 tumor suppresser gene into tumor cells bearing p53 mutations can inhibit cellular proliferation and tumorigenicity both in ex vivo as well as in vivo cancer models *(20)*.

Gene transfer of the multidrug resistance gene (mdr) is also one of the approaches to cancer treatment. Most of the human cancers have either intrinsic or acquired resistance to chemotherapeutic agents *(21)*, especially amphipathic hydrophobic substances. The characterization of the gene involved in multidrug resistance reveals that the gene product is a membrane glycoprotein of 170 kDa *(22)*. Transfer of the mdr gene into bone marrow progenitor cells will induce a higher level of protection to normal (nonneoplasic) cells during chemotherapeutic treatments. Retroviral constructs of mdr genes and their transfer into mouse models have provided very exciting results *(23,24)*.

3. Gene Therapy of Liver Diseases and Metabolic Disorders

There are a number of metabolic diseases that are due to dysfunctional hepatocytes *(25,26)*. The gene involved in a number of liver disorders, i.e., Wilson's disease, Krabbe's disease, Canavan's disease, and adrenoleukodystrophy have been characterized *(27)*. The addition or replacement of the defective genetic element with the functional one is the primary focus of liver gene therapy.

One of the devastating human diseases, familial hypercholesterolemia, involves defective low density lipoprotein (LDL) receptors on hepatocytes *(28)*. Nonfunctional or mutated receptor genes disturb cholesterol metabolism, resulting in elevated levels of serum cholesterol and accompanied by atherosclerosis. Significant progress has been made toward expressing a functional copy of the receptor gene *(29–31)*. In the clinical application of LDL-receptor gene therapy, lower proportions of transduced hepatocytes and infection efficiency were the major problems *(32)*, however, higher expression and enhanced transduction efficiency of LDL receptors has been observed by using tissue-specific promoters *(33)*. Similarly, transmembrane conductance regulator gene therapy has been used to correct the cellular defect in the lethally inherited lung disease, cystic fibrosis *(34)*. Metabolic disorders are usually manifested as clinical disease when there is a severe reduction in the synthesis of a particular enzyme involved in the metabolism *(35)*. It has been observed that most of the inherited metabolic disorders are recessive, having no phenotypic expression under the heterozygous condition. The complex nature of inherited metabolic disorders has been a major obstacle in finding effective treatment. Phenylketonuria is an inherited metabolic disorder involving a phenylalanine hydroxylase deficiency in the liver resulting in high levels of phenylalanine and its metabolites in blood and body tissue. Traditional therapy for such disorders involve restricted intake of that particular substrate such as phenylalanine in phenylketonuria. Gene therapy holds a great promise for such metabolic disorders. It has been observed that adenovirus-mediated hepatic gene transfer of phenylalanine hydroxylase phenotypically rectifies phenylketonuria in the mouse model *(36)*. A number of other studies involving gene-therapeutic approaches to correct metabolic disorders have been reported in animal models *(37–41)*.

3.1. Insulin Gene Therapy

Insulin regulates the blood glucose level in mammals. Under physiological conditions, the secretion of insulin from β-cells is closely coordinated with the blood sugar level. An inability to deliver insulin for glucose homeostasis is manifested as diabetes mellitus. The underlying basis for improper insulin delivery is either nonfunctional insulin secretory cells or defective regulation. Diabetes associated with damaged insulin producing cells is classified as type 1 or insulin dependent diabetes mellitus (IDDM), whereas type 2 diabetes is due to the relative insulin resistance of the insulin-sensitive tissues and defective secretion of insulin. The current focus of gene transfer technology for diabetes therapy is the development of engineered cell lines that would closely mimic glucose stimulated insulin secretion *(42)*. The use of retroviral-mediated gene therapy of β-cell dysfunction is also being envisioned *(42,43)*. The challenge for insulin gene therapy is to restore the normal level of insulin secretion. Insulin secretory cells are pancreatic cells, however, the expression of insulin in the liver has also been shown to correct the diabetic alterations in a transgenic mouse model *(44)*. Future efforts on the gene therapy front are to clone the genes associated with type 2 diabetes and its associated complications like obesity *(45)*.

4. Gene Therapy of Infectious Diseases

Among the candidate infectious diseases for gene therapy, human immunodeficiency virus (HIV-1) infections have received the greatest attention. Unprecedented progress has been made in the last decade to understand the pathogenesis of AIDS and its causative agent, HIV-1, at the molecular level. The molecular mechanisms of HIV-1 pathogenesis have provided researchers with a number of molecular targets for antiviral therapy. The gene manipulation techniques for HIV-1 therapy encompass a variety of gene-transfer based approaches, i.e., antisense oligonucleotides, ribozymes, transdominant negative mutant recombinant HIV-1 proteins, molecular sinks, and suicide gene constructs *(46)*. These technologies can be broadly divided into two groups: intracellular immunization and immunotherapy. Intracellular immunization makes the cells resistant to viral replication and inhibits the further spread of the virus, whereas immunotherapeutic approaches block the viral spread by an antiviral cellular response. Immunotherapeutic approaches involve the use of vaccines and adoptive transfer of $CD8^+$ T-cell clones.

There are a number of criteria which must be fulfilled by a genetic element to be used as an anti-HIV-1 agent. It should be safe, nontoxic, and unaffected by HIV-1 strain variation. A number of intracellular immunization approaches using either protein-based or RNA-based inhibitors have been proposed for HIV-1 gene therapy. Protein-based strategies for HIV-1 inhibition focuses on the expression of altered HIV-1 gene products that have a transdominant mutant phenotype *(47)* or the expression of intracellular antibodies *(48)*. The initial target of transdominant mutants have been *Gag, Rev, Tat*, and *Env*. Transdominant mutants of *Gag* showed some effectiveness in inhibiting HIV-1 replication *(49)*, however, the unstable nature of intracellular *Gag* protein *(50)* is a major drawback for the use of this protein. A transdominant mutant of *Rev* called RevM10 has shown considerable effectiveness in protecting the cells from HIV-1 *(51,52)*.

The use of intracellularly expressed antibodies for controlling HIV-1 infections is receiving much attention. Previously, it has been reported that an anti-gp120 single chain antibody (sFv) inhibits the production of infectious viral particles by blocking the transport of Env glycoprotein *(53)*. Later, anti-gp120 single-chain antibody was cloned under the control of the HIV-1 LTR and transduced into $CD4^+$ T-lymphocytes. The intracellular expression of this antibody inhibits the gp120-mediated cytopathic syncytium formation and HIV-1 production by blocking the surface expression of gp120 *(54)*. A potential problem associated with gp120 as the target has been the very high mutation rate in this glycoprotein. Therefore, the most logical target has to be a protein that is essential for the life cycle of HIV-1 and the highly conserved *Rev*. Recent studies showed inhibition of HIV-1 replication in human cells by using an intracellular SFv moiety constructed from V_l and V_h regions of murine MAb that binds strongly to the HIV-1 protein Rev *(48)*. Further studies revealed that the SFv strongly inhibits multiple divergent strains of HIV-1 in human cells maintaining the antiviral effect for several months *(55)*.

RNA-based strategies for control of HIV-1 include antisense RNA molecules, ribozymes, and RNA decoys. Antisense RNA molecules are either expressed intracellulary from retroviral vectors or they can be delivered with certain modifications which render them nuclease-resistant. Antisense RNA molecules have been widely used in HIV-1 models *(56–58)*.

Ribozymes are also antisense molecules with an edge over the antisense RNA due to their catalytic properties. The catalytic property of ribozymes theoretically is effective in controlling HIV-1 even in very low concentrations. The first use of ribozymes as an anti-HIV-1 agent was reported in 1990 *(59)*. Later a number of studies have been reported with different designs of ribozymes for controlling HIV-1 infections *(46,60)*.

The use of RNA decoy strategies for controlling HIV-1 involves the expression of RNA molecules that mimic the structure of regulatory elements involved in HIV-1 gene expression like TAR and RRE (Rev-responsive element). TAR regulates the function of Tat, whereas RRE is the binding site for Rev. The expression of RNAs resembling TAR and RRE compete for binding of their respective viral proteins *(61)*. The binding mechanism of decoy RNAs is based on the resemblance of authentic HIV-1 RNA for binding, but reconstitution after binding is obstructed. These concepts are illustrated in Fig. 2. There are a number of studies reported for controlling HIV-1 replication by using RNA decoys *(62–64)*.

The human immune system provides the major defense against the spread of infections within the host. The mechanism involves either the recognition of extracellular pathogens by the antibodies or the generation of cytotoxic T-cells (CTLs) that eliminate the pathogens. Immunotherapeutic approaches involve the use of attenuated or killed pathogens composed of antigenic determinants (vaccines) that elicit a humoral response against a particular pathogen. A number of immunotherapeutic approaches have also been tried in the past to control HIV-1 *(65–67)*. It is hoped that in the future it will be possible to use nonreplicating or selfdestructive retroviral vectors carrying the genetic sequences of important determinants. Macrophage-specific targeting of these vectors can theoretically result in intracellular processing predesigned to elicit cellular, humoral, or both types of immune responses against a specific pathogen. After antigen presentation, the primed macrophage would die without carrying any potential retention of the vector. The mechanism is shown in Fig. 3.

To date, no effective treatment is available against HIV. The future of this pandemic seems to be gene therapeutic approaches, but considering the nature of this disease, it will be wise to consider combination therapy with antivirals as well as immunotherapy.

5. Gene Therapy for Cardiovascular Disorders

Modern techniques in molecular biology have made possible the application of genetic manipulation or gene therapy approaches in inherited as well as

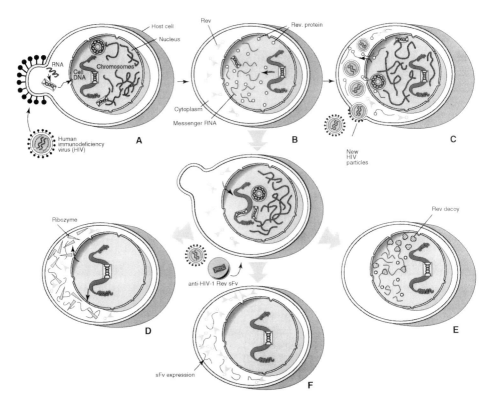

Fig. 2. Molecular therapeutic approaches against HIV-1. **(A)** Showing entry of HIV-1 to the target cells (i.e. CD4-positive lymphocytes, monocytes, etc.). After entry, HIV-1, RNA virus, reverse-transcribes its genetic material into DNA, which subsequently integrates into the host chromosomes. **(B)** Upon stimulation, HIV-1 produces Rev protein which escorts the newly produced nuclear HIV-1 mRNA to cytoplasm, **(C)** resulting in mass production of HIV-1. Three experimental molecular therapeutic approaches to inhibit HIV-1 replication include: **(D)** ribozymes containing genes which upon expression specifically target certain sequences in HIV-1 mRNA, hence selectively destroying HIV-1 mRNA without harming the host mRNA; **(E)** a retrovirally mediated single-chain antibody (sFv) containing gene (s) targets Rev protein in the cytoplasm, sequestering Rev, inhibiting their "escorting" role and hence inhibiting HIV-1 replication. Multiple sFvs could be utilized to target more than one proteins (i.e., tat, tar, vpx, vpu, vpr, etc.) to circumvent production of even mutant virions. **(F)** Retroviral vector overproduces a mutant, nonfunctional Rev protein which binds RRE (Rev responsive element) instead of normal Rev, diluting the effectiveness of Rev function and reducing HIV-1 production.

acquired disorders like atherosclerotic arterial disease, dilated cardiomyopathies, and restenosis after percutaneous vascular interventions *(68)*. Current gene therapy approaches involve ex vivo gene transfer, i.e., involving in vitro

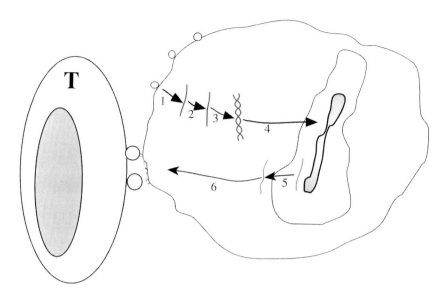

Fig. 3. Retroviral-mediated, molecular-based vaccination. Vaccination against multiple pathogens can be carried out simultaneously. Here genes containing "epitopes" for various antigens and of multiple pathogens are cloned into retroviral vectors expressing anti-MAC-1 (or other antigen presenting cells, i.e., dendritic cells). This vector will follow the life cycle of the retrovirus. 1) entry of diploid RNA, 2) reverse transcription and cDNA formation, 3) double-stranded DNA, 4) entry of dsDNA to nucleus and integration into host DNA, 5) expression of "epitopes," and 6) processing of antigenic epitopes to the surface of macrophage cells and presentation of these epitopes to T-helper (CD4+) or T-suppresser cytotoxic (CD8+) cells resulting in humoral or cell-mediated immune response. After antigen presentation macrophage will die in a few weeks and there would be no recycling of retroviral vector or other adverse effects mentioned in the text.

gene modification with reimplantation and in vivo gene transfer, which involve introduction of recombinant genes without removing the cells from their natural milieu. For the cardiovascular system, in vivo gene therapy is more feasible than ex vivo therapy due to the terminally differentiated nature of cardiac myocytes *(69,70)*.

Recent efforts for in vivo gene therapy of coronary vasculature are focused on fronts such as:

1. Transfer of plasmid DNA into myocardium by direct injection.
2. Liposome-mediated gene transfer.
3. Use of recombinant, replication-defective adenoviruses.

Table 2
Neurological Disorders

Disorder	Clinical manifestations
Neurodegenerative	Deterioration of neurons/nerve cells after normal development, e.g., Huntington's disease, Alzheimer's disease
Neuromuscular	Defect or lack of regulatory steps in neuromuscular junction or muscles, e.g., Duchenne muscular dystrophy, myotonic dystrophy
Neuronal migration	Developmental abnormalities characterized by defective neuronal migration, e.g., tuberous sclerosis type 1 Kallmann syndrome
Central nervous system tumors	Loss of growth control in nerve cells, e.g., retinoblastoma, neurofibromatosis type 2 (NF2)
Trinucleotide repeats	Expansion of trinucleotide repeats resulting in abnormal protein expression, e.g., fragile-X syndrome, Huntington's disease

6. Gene Therapy of the Nervous System

A classification of neurological disorders based on genetic and clinical manifiestations is shown in Table 2.

Neurodegenerative disorders are most problematic for gene therapy approaches due to non-regeneration of nerve cells after death, however significant progress has been made for gene therapy of neuromuscular diseases *(71)*.

Congenital neurologic diseases are among the most devastating human developmental defects. It is believed that most of these disorders are caused by either a single gene defect or combination of multiple genetic defects and environmental factors *(72)*. In the beginning, neurological disorders were not considered prominent candidates for gene therapy due to the complex nature and physical inaccessibility of the nervous system. Recent advancements in molecular genetics have paved the way for understanding the role of different genes in neurological disorders *(73)*. The genetic components of neurological disorders like Huntington's disease *(74)*, Alzheimer's disease *(75)*, adrenoleukodystrophy *(76)*, and a number of others have been reported. Molecular neurologists are optimistic that they will identify the pathogenesis of complex neurologic disorders like Down's syndrome, schizophrenia and bipolar disease at the molecular level *(77,79)*. Stable in vivo gene expression in neurons to correct the Parkinson's diseases lesions has also been reported in animal models *(80)* by using the adeno-associated viral vector system to genetically modify the cells for tyrosine hydroxylase synthesis deficiency, a major cause of

Parkinson's disease. Besides in vivo gene therapy, the merger of gene transfer technology with neural transplantation may provide new therapeutic strategies for certain neurodengenerative disorders. In case of neurological disorders, it is hoped that genetically modified cells will provide a better alternative to embryonic cells in transplantation *(81,82)*.

7. Gene Therapy in Ocular Disorders

The study of ocular disorders at the molecular genetics level has opened an opportunity for gene therapeutics in ophthalmology. It has been observed that a number of mutations in the human rhodopsin gene are the cause of autosomal retinitis pigmentosa *(83)*. Besides rhodopsin gene mutations, the gene encoding cGMP phosphodiesterase has been identified in autosomal recessive pigmentosa *(84)*. Some eye disorders like color blindness, choroideremia, and Norrie's disease have genetic linkage with the X chromosome *(85–87)*. Aniridia, color vision and cone dystrophies, Leber's hereditary optic neuropathy, Marfan's syndrome, retinoblastoma, stickler syndrome, uveal melanoma, and von Hippel-Lindau disease are a number of genetic eye disorders in which the defective genez have been identified *(88)*· The localization and identification of disease-causing genes and their products, helps to utilize the potential of gene therapy in a particular ailment. Considering the delivery of a particular gene, the eye has certain advantages as a target for virus-mediated gene therapy. The transfer process can easily be monitored due to easy accessibility, well defined anatomy, and the translucent nature of the eye *(89)*. Theoretically, transfer of retroviral vectors for gene transfer in ocular tissues is limited due to the presence of quiescent or slowly dividing cells, however retrovirus-mediated gene transfer has shown encouraging results in chorioretinal degeneration which is caused by deficiency of the mitochondrial matrix enzyme ornithine-delta-aminotransferase *(90)*. Replication deficient adenoviruses have shown encouraging results for gene transfer in ocular tissues *(91–93)*. This and similar studies have opened new prospects for the treatment of inherited retinal disorders.

8. Methodologies in Gene Therapy

The strategy for gene therapy of any disease involves individually designed and optimized protocols, however, there are a number of conditions which must be fulfilled before gene therapy can be taken from bench to bedside.

1. Cloning and characterization of the gene of interest.
2. Selection of an appropriate vector for delivery of the gene to the target cells.
3. Stable and appropriate expression of the particular gene.
4. The pathological manifestation of the disease must be reversed when the gene is expressed.
5. Cell-specific or receptor-mediated targeting of vector (optional).

The characterization of a therapeutic gene is followed by the cloning of the gene in front of an appropriate promoter. There are certain problems in promoter selection. The organ specificity and tissue-specific behavior of various promoters need to be considered and explored extensively before a particular gene therapy protocol can be recommended for treatment of a disorder *(94)*. It has been observed that the insertion of liver-specific promoters into vectors showed elevated expression in hepatocytes *(95,96)*. It can be reasoned that there might be similar mechanisms for expression in different human systems.

There are two approaches by which a gene can be incorporated into the cell's machinery. An indirect approach involves the genetic alteration of cells in culture and transplanting them back into the body. A direct approach is more or less like the delivery of a drug to the infected areas.

An important aspect of gene therapy is the gene delivery system, i.e., the transfer of therapeutic genes to various human tissues. There are a number of techniques available for ex vivo gene transfer in human cell lines like chemical transfer (calcium phosphate precipitation) *(97)* and liposome-mediated gene transfer *(98)*. For in vivo gene therapy, the most promising options are the physical gene transfer involving microinjection *(1)* and viral vectors. The viral vectors include retroviruses, adenoviruses, adeno-associated viruses, and herpes simplex virus. Viral vectors differ in their nature, which affects the expression of the gene being delivered. For example, retroviral vectors stably integrate into host genome, whereas adenoviral vectors exhibit broad host range, efficient infectivity, including nondividing cells. Moreover, adenoviral vectors have low genotoxicity in host cells due to episomal expression. The use of herpes simplex virus vectors is not well established *(99,100)*. There are a number of criteria which must be fulfilled before a particular vector can be used for gene transfer:

1. Nonpathogenicity.
2. Ability to access the target cells.
3. Proper processing of the expressed gene product at sufficient levels to complement the disease phenotype.
4. A preferable site-specific integration into the host genome.

The use of viral vectors necessitates the investigation of their safety profile before clinical use. The potential risk associated with the use of viral vectors in gene therapy is random integration of the therapeutic gene into the genome. Our present understanding based on the gene therapy trials have shown less deleterious integration *(101)*, however, hypermutability of the viral genome cannot be ignored. The viral vector constructs with therapeutic genes are also being evaluated for their safety profiles *(102,103)*.

As mentioned, currently retroviral vectors are the most commonly used vehicles for gene transfer. The reason for their popularity is their ability to integrate the gene(s) into the host genome. However, their ability to deliver a therapeutic gene is not without risk. The biggest potential problem associated with random integration of retroviral vectors is the activation of protooncogenes resulting in neoplasm or inactivation of some essential genes resulting in premature cellular death. Recently, there has been significant progress in the area of site specific integration of retroviral vectors which will eliminate the problem of random insertion of the therapeutic gene(s).

Besides retroviral vectors and classical gene delivery systems, a number of other gene transfer technologies like electromagnetic transfer of DNA into cells *(104)* and the use of mammalian artificial chromosomes is also being considered *(105)*. Mammalian artificial chromosomes can be very useful for long-term expression of DNA.

Gene therapy holds great promise for the cure of complex diseases. An increase in the number of approved clinical protocols are evidence of the usefulness of this therapeutic approach. The number of approved clinical protocols have increased from 37 in 1992 to over 100 in 1994 *(6)*.

9. Future of Gene Therapy

Gene therapy was initially conceived to cure monogenetic disorders, however the spectrum of gene therapeutic approaches is widening very rapidly to include everything from simple genetic disorders to acquired diseases like cancer, human immune deficiency virus (HIV), and hepatitis B virus. Somatic gene therapy has been well accepted among the general public, whereas germline gene therapy is facing ethical dilemmas *(106)*.

Before gene therapy can move from benches to beds, there are a number of technical obstacles to overcome. Isolation and characterization of therapeutic genes necessitates extensive study about their regulation and stability after transfer into the genome. Efficient mechanisms for gene insertion and targeting to particular cells need more scientific endeavors. Finally, expression of the gene, and encounter of the gene product with immune system, need to be well understood before transfer.

Gene therapy is a rapidly growing field of research. Besides those described in this article, a number of other human disorders are being explored at the molecular level. The genes responsible for hearing loss syndrome have been identified *(107)*. Gene therapy also seems to be the treatment of choice for many complex hematological and lysosomal storage disorders *(108,109)*. Considerable evidence has accrued to deduce that behavioral changes like alcoholism are also under the influence of certain genes *(110)*. It has been suggested that the characterization of such genes will help in the understanding of human behavioral mechanisms.

The secret of ideal and effective gene therapy lies in the development of delivery systems with high transduction efficiency, low cost, tissue specificity, sustained action of transgenes, and site-specific integration into the human genome without deleterious effects. This will materialize the dream of gene therapeutic approaches for controlling inherited and noninherited complex disorders.

References

1. Mulligan, R. C. (1993) The basic science of gene-therapy. *Science* **260,** 926–932.
2. Bishop, J. F. (1991) Etoposide in the management of leukemia: a review. *Semin. Oncol.* **18(Suppl. 1–2),** 62–69.
3. Hall, J. M., Lee, M. K., Newman, B., Morrow, J. E., Anderson, L. A., Huey, B., and King, M. C. (1990) Linkage of early-onset familial breast cancer to chromosome 17q21. *Science* **250,** 1684–1689.
4. Wooster, R., Neuhausen, S. L., Mangion, J., Quiek, Y., Ford, D., Collins, N., Nguyen, K., Seal, S., Tran, T., Averill, D., Fields, P., Marshall, G., Narod, S., Lenoir, G. M., Lynch, H., Feunteun, J., Devile, P., Cornelisse, C. J., Menko, F. H., Daly, P. A., Ormiston, W., McManus, R., Pye, C., Lewis, C. M., Cannon-Albright, L. A., Peto, J., Ponden, B. A. J., Skolnick, M. H., Easton, D. F., Goldgar, D. E., and Stratton, M. R. (1994) Localization of a breast cancer susceptibility gene, BRCA2, to chromosome 13q12–13. *Science* **265,** 2088–2090.
5. Tolstoshev, P. and Anderson, W. F. (1995) Gene therapy, in *The Molecular Basis of Cancer* (Mendelsohn, J., Howley, P. M., Israel, M. A., and Liotta, L. A., eds.), W. B. Saunders, Philadelphia, PA, pp. 531–537.
6. Anderson, W. F. (1994) End-of-the-year potpourri—1994. *Hum. Gene Ther.* **5,** 1431,1432.
7. Sikora, K. (1994) Genetic approaches to cancer-therapy (Editorial). *Gene Ther.* **1**(3), 149–151.
8. Miller, A. R., McBride, W. H., Hunt, K., and Economou, J. S. (1994) Cytokine mediated gene therapy for cancer. *Annals Surg. Oncol.* **1,** 436–450.
9. Cordier, L., Duffour, M. T., Sabourin, J. C., Lee, M. G., Cabannes, J., Ragot, T., Perricaudet, M., and Haddada, H. (1995) Complete recovery of mice from a pre-established tumor by direct intratumoral delivery of an adenovirus vector harboring the murine IL-2 gene. *Gene Ther.* **2,** 16–21.
10. Patel, P. M., Flemming, C. L., Fisher, C., Porter, C. D., Thomas, J. M., Gore, M. E., and Collins, M. K. (1994) Generation of interleukin-2 secreting melanoma cell populations from resected metastatic tumors. *Hum. Gene Ther.* **5,** 577–584.
11. Hwu, P., Yannelli, J., Kriegler, M., Anderson, W. F., Perez, C., Chiang, Y., Schwarz, S., Cowherd, R., Delgado, C., Mule, J., and Rosenberg, S. A. (1993) Functional and molecular characterization of tumor-infiltrating lymphocytes transduced with tumor necrosis factor-alpha cDNA for the gene therapy of cancer in humans. *J. Immunol.* **150,** 4104–4115.

12. Culver, K. W. and Blaese, R. M. (1994) Gene therapy for cancer. *Trends Genet.* **10,** 174–178.
13. Nabel, E. G., Plautz, G. E., and Nabel, G. J. (1992) Transduction of a foreign histocompatibility gene into the arterial wall induces vasculitis. *Proc. Natl. Acad. Sci. USA* **89,** 5157–5161.
14. Nabel, E. G., Gordon, D., Yang, Z. Y., Xu, L., San, H., Plautz, G. E., Wu, B. Y., Gao, X., Huang, L., and Nabel, G. J. (1992) Gene transfer *in vivo* with DNA-liposome complexes: lack of autoimmunity and gonadal localization. *Hum. Gene Ther.* **6,** 649–656.
15. Conry, R. M., Lobuglio, A. F., Loechel, F., Moore, S. E., Sumerel, L. A., Barlow, D. L., and Curiel, D. T. (1995) A carcinoembryonic antigen polynucleotide vaccine has *in vivo* antitumor activity. *Gene Ther.* **2,** 59–65.
16. Moolten, F. L. and Wells, J. M. (1990) Curability of tumors bearing herpes thymidine kinase genes transferred by retroviral vectors. *J. Natl. Cancer Inst.* **82,** 297.
17. Anderson, W. F. (1994) Gene therapy for cancer. *Hum. Gene Ther.* **5,** 1,2.
18. Oldfield, E. H., Ram, Z., Culver, K. W., and Blaese, R. M. (1993) Gene therapy for the treatment of brain tumors using intra-tumoral transduction with the thymidine kinase gene and intravenous ganciclovir. *Hum. Gene Ther.* **4,** 39–69.
19. Harris, C. C. and Hollstein, M. (1993) Clinical implications of the p53 tumor-suppresser gene. *N. Engl. J. Med.* **329,** 1318–1327.
20. Wills, K. N., Maneval, D. C., Menzel, P., Harris, M. P., Sutjipto, S., Vallancourt, M. T., Huang, W., Johnson, D. E., Anderson, S. C., Wen, S. F., Bookstein, R., Shepard, H. M., and Gregory, R. J. (1994) Development and characterization of recombinant adenoviruses encoding human p53 for gene therapy of cancer. *Hum. Gene Ther.* **5,** 1079–1088.
21. Pastan, I. and Gottesman, M. M. (1991) Multidrug resistance. *Annu. Rev. Med.* **42,** 277–286.
22. Ueda, K., Clark, D. P., Chen, C., Roninson, I. B., Gottesman, M. M., and Pastan, I. (1987) The human multidrug resistance (mdr1) gene cDNA cloning and transcription initiation. *J. Biol. Chem.* **262,** 505–508.
23. McLachlin, J. R., Eglitis, M. A., Ueda, K., Kantoff, P. W., Pastan, I. H., Anderson, W. F., and Gottesman, M. M. (1990) Expression of a human complementary DNA for the multidrug resistance gene in murine hematopoietic precursor cells with the use of retroviral gene transfer. *J. Natl. Cancer Inst.* **82,** 1260–1263.
24. Sorrentino, B. P., Brandt, S. J., Bodine, D., Gottesman, M., Pastan, I., Cline, A., and Nienhuis, A. W. (1992) Selection of drug- resistance bone marrow cells *in vitro* after retroviral transfer of human MDR1. *Science* **257,** 99.
25. Scriver, C. R., Beaudet, A. L., Sly, W. S., and Valle, D. (1989) Genetics and biochemistry of various human phenotypes, in *The Metabolic Basis of Inherited Diseases* (Stanbury, J. B., Wyngaarden, J. B., and Frederickson, D. S., eds.), McGraw-Hill, New York, pp. 3–164.
26. Strauss, M. (1994) Liver-directed gene therapy: prospects and problems. *Gene Ther.* **1(3),** 156–164.

27. Morris, A. A. M. and Turnbull, D. M. (1994) Metabolic disorders in children. *Cur. Opin. Neurol.* **7,** 535–541.
28. Goldstein, J. L. and Brown, M. S. (1989) Familial hypercholesterolemia, in *The Metabolic Basis of Inherited Diseases* (Stanbury, J. B., Wyngaarden, J. B., and Frederickson, D. S., eds.), McGraw-Hill, New York, pp. 1215–1250.
29. Wilson, J. M. and Chowdhury, J. R. (1990) Prospects for gene therapy of familial hypercholesterolemia. *Mol. Biol. Med.* **7,** 223–232.
30. Grossman, M. and Wilson, J. M. (1992) Frontiers in gene therapy: LDL receptor replacement for hypercholesterolemia. *J. Lab. Clin. Med.* **119,** 457–460.
31. Kozarsky, K., Grossman, M., and Wilson, J. M. (1993) Adenovirus-mediated correction of the genetic defect in hepatocytes from patients with familial hypercholesterolemia. *Som. Cell Mol. Genet.* **19,** 449–458.
32. Grossman, M., Raper, S. E., Kozarsky, K., Stein, E. A., Engelhardt, J. F., Muller, D., Lupien, P. J., and Wilson, J. M. (1994) Successful *ex vivo* gene therapy directed to liver in a patient with familial hypercholesterolemia. *Nature Genet.* **6,** 335–341.
33. Pages, J. C., Andreoletti, M., Bennoum. M., Vons, C., Elcheroth, J., Lehn, P., Houssin, D., Chapman, J., Briand, P., Benarous, R., Franco, D., and Weber, A. (1995) Efficient retroviral-mediated gene transfer into primary culture of murine and human hepatocytes: expression of the LDL receptor. *Hum. Gene Ther.* **6,** 21–30.
34. Crystal, R. C. (1992) Protocol of the gene therapy of the respiratory manifestations of cystic fibrosis using a replication deficient recombinant adenovirus to transfer the normal cystic fibrosis transmembrane conductance regulator cDNA to the airway epithelium. *Fed. Reg.* **57,** 49,584.
35. Kay, M. A. and Woo, S. L. (1994) Gene therapy for metabolic disorders. *Trends Genet.* **10,** 253–257.
36. Fang, B., Eisensmith, R. C., Li, X. H. C., Finegold, M. J., Shedlovsky, A., Dove, W., and Woo, S. L. C. (1994) Gene therapy for phenylketonuria, phenotypic correction in a genetically deficient mouse model by adenovirus-mediated hepatic gene transfer. *Gene Ther.* **1,** 247–254.
37. Peng, H., Armentano, D., McKenzie-Graham, L., Shen, R. F., Darlington, G., Ledley, F. D., and Woo, S. L. C. (1988) Retroviral-mediated gene transfer and expression of human phenylalanine hydroxylase in primary mouse hepatocytes. *Proc. Natl. Acad. Sci. USA* **85,** 8146–8150.
38. Grompe, M., Jones, S. N., Loulseged, H., and Caskey, C. T. (1992) Retroviral-mediated gene transfer of human ornithine transcarbamylase into primary hepatocytes of spf and spf-ash mice. *Hum. Gene Ther.* **3,** 35–44.
39. Liu, T. H., Fu, X. X., and Tian, G. S. (1992) The advances in the treatment of chronic hepatitis C with interferon. *Chung Hua Nei Ko Tsa Chih.* **31,** 650–652.
40. Kozarsky, K. F., McKinley, D. R., Austin, L. L., Raper, S. E., Stratford-Perricaudet, L. D., and Wilson, J. M. (1994) *In vivo* correction of low density lipoprotein receptor deficiency in the Watanabe heritable hyperlipidemic rabbit with recombinant adenoviruses. *J. Biol. Chem.* **6; 269,** 13,695–13,702.

41. Cristiano, R. J., Smith, L. C., and Woo, S. L. (1993) Hepatic gene therapy: adenovirus enhancement of receptor-mediated gene delivery and expression in primary hepatocytes. *Proc. Natl. Acad. Sci. USA* **90,** 2122–2126.

42. Newgard, C. B. (1994) Cellular engineering and gene therapy strategies for insulin replacement in diabetes. *Diabetes* **43,** 341–350.

43. Vitullo, J. C., Aron, D. C., and Miller, R. E. (1994) Control of insulin gene expression: Implications for insulin gene therapy. *J. Lab. Clin. Med.* **124,** 328–334.

44. Valera, A., Fillat, C., Costa, C., Sabater, J., Visa, J., Pujol, A., and Bosch, F. (1994) Regulated expression of human insulin in the liver of transgenic mice corrects diabetic alterations. *FASEB J.* **8,** 440–447.

45. Froguel, P. and Hager, J. (1995) Human diabetes and obesity: tracking down the genes. *Trends Biotechnol.* **13,** 52–55.

46. Yu, M., Poeschla, E., and Wong-Staal, F. (1994) Progress towards gene therapy for HIV-1 infection. *Gene Ther.* **1,** 13–20.

47. Herskowitz, I. (1987) Functional inactivation of genes by dominant negative mutations. *Nature* **329,** 219–222.

48. Duan, L., Bagasra, O., Laughlin, M. A., Oakes, J. W., and Pomerantz, R. J. (1994) Potent inhibition of human immunodeficiency virus type-1 replication by an intracellular anti-rev single chain antibody. *Proc. Natl. Acad. Sci. USA* **91,** 5075–5077.

49. Trono, D., Feinberg, M. B., and Baltimore, D. (1989) HIV-1 gag mutants can dominantly interfere with the replication of wild type virus. *Cell* **59,** 113–120.

50. Schwartz, S., Felber, B. K., and Pavlakis, G. N. (1992) Distinct RNA sequences in the *gag* region of human immunodeficiency virus type 1 decrease RNA stability and inhibit expression in the absence of Rev protein. *J. Virol.* **66,** 150–159.

51. Malim, M. H., Bohnlein, S., Hauber, J., and Cullen, B. R. (1989) Functional dissection of the HIV-1 Rev transactivator. Derivation of trans-dominant repressor of Rev function. *Cell* **58,** 205–214.

52. Bahner, I., Zhou, C., Yu, X. J., Haq, Q. L., Guatelli, J. C., and Kohn, D. B. (1993) Comparison of transdominant inhibitory mutant human immunodeficiency virus type 1 genes expressed by retroviral vectors in human T lymphocytes. *J. Virol.* **67,** 3199–3207.

53. Marasco, W. A., Haseltine, W. A., and Chen, S. Y. (1993) Design, intracellular expression, and activity of a human anti-human immunodeficiency virus type 1 gp120 single-chain antibody. *Proc. Natl. Acad. Sci. USA* **90,** 7889–7893.

54. Chen, S. Y., Bagley, J., and Marasco, W. A. (1994) Intracellular antibodies as a new class of therapeutic molecules for gene therapy. *Hum. Gene Ther.* **5,** 595–601.

55. Duan, L., Zhan, H., Oakes, J. W., Bagasra, O., and Pomerantz, R. J. (1994) Molecular and virological effects of intracellular anti-Rev single-chain variable fragments on the expression of various human immunodeficiency virus-1 strains. *Hum. Gene Ther.* **5,** 1315–1324.

56. Cohen, J. S. (1991) Antisense oligonucleotides as antiviral agents. *Antiviral Res.* **16,** 121–133.

57. Murray, J. A. H. and Crockett, N. (1992) Antisense techniques: an overview, in *Antisense RNA and DNA* (Murray, J. A. H., ed.), Wiley-Liss, New York, pp. 1–49.
58. Stein, C. A. and Chen, Y. C. (1993) Antisense oligonucleotides as therapeutic agents-is the bullet really magical. *Science* **260,** 1004–1012.
59. Sarver, N., Cantin, E. M., Chang, P. S., Zaia, J. A., Ladne, P. A., Stephens, D. A., and Rossi, J. J. (1990) Ribozymes as potential anti HIV-1 therapeutic agents. *Science* **247,** 1222–1225.
60. Zhou, C., Bahner, I. C., Larson, G. P., Zaia, J. A., Rossi, J. J., and Kohn, D. B. (1994) Inhibition of HIV-1 in human T-lymphocytes by retrovirally transduced anti-Tat and Rev hammerhead ribozymes. *Gene* **149,** 33–39.
61. Dropulic, B. and Jeang, K.-T. (1994) Gene therapy for human immunodeficiency virus infection: genetic antiviral strategies and targets for intervention. *Hum. Gene Ther.* **5,** 927–939.
62. Sullenger, B. A., Gallardo, H. F., Ungrs, G. E., and Gilboa, E. (1990) Over-expression of TAR sequences renders cells resistant to HIV-1 replication. *Cell* **63,** 601–608.
63. Smith, C., Lee, S. W., Sullenger, B., Gallerdo, H., Ungers, G., and Gilboa, E. (1992) Abstract. Intracellular immunization against HIV-1 using RNA decoys. Third international symposium on catalytic RNAs (ribozymes) and targeted gene therapy for the treatment of HIV-1 infection. p. 61.
64. Lisziewicz, J., Sun, D., Smythe, J., Lusso, P., Lori, F., Louie, A., Markham, P., Rossi, J., Reitz, M., and Gallo, R. C. (1993) Inhibition of human immunodeficiency virus type 1 replication by regulated expression of a polymeric Tat activation response RNA decoy as a strategy for gene therapy in AIDS. *Proc. Natl. Acad. Sci. USA* **90,** 8000–8004.
65. Specter, S. (1995, January-February). Immunotherapy for HIV-1 infections. *The AIDS Reader.*
66. Cohen, J. (1994) The HIV-1 vaccine paradox. *Science* **264,** 1072–1074.
67. Carmichael, A. J. and Sissons, J. G. P. (1995) Vaccines against HIV. *Quarterly J. Med.* **88,** 77–79.
68. Barr, E. and Leiden, J. M. (1994) Somatic gene therapy for cardiovascular diseases. *TCM* **4,** 57–63.
69. Watanabe A., Green, F., and Farmer B. B. (1986) Preparation and use of cardiac myocytes in experimental cardiology, in *The Heart and Cardiovascular System* (Fozzard, H. A., Haber, E., Jennings, R. B., and Katz, A. M., eds.), Raven, New York, pp. 241–251.
70. Zak, R. (1974) Development and proliferation capacity of cardiac muscle cells. *Circ. Res.* **34–35(Suppl. 2),** 11–17.
71. Martin, J. B. (1995) Gene therapy and pharmacological treatment of inherited neurological disorders. *Trends Biotech.* **13,** 28–35.
72. Laurence, K. M. (1990) The genetics and prevention of neural tube defects and 'uncomplicated' hydrocephalus, in *Principles and Practice of Medical Genetics,* 2nd ed. (Emery, A. E. H. and Rimoin, D. L., eds.), Churchill Livingstone, Edinburgh, pp. 323–346.

73. Friedmann, T. (1994) Gene therapy for neurological disorders. *Trend Genet.* **10**, 210–214.

74. Trofatter, J. A., MacCollin, M. M., Rutter, J. L., Murrell, J. R. Duyao, M. P., Parry, D. M., Eldridge, R., Kley, N., Menon, A. G., Pulaski, K., Haase, V. K., Ambrose, C. M., Munroe, D., Bove, C., Haines, J. L., Martuza, R. L., McDonald, M. E., Seizinger, B. R., Short, M. P., Buckler, A. J., and Gusella, J. F. (1993) A novel moesin-, ezrin-, radixin- like gene is a candidate for the neurofibromatosis 2 tumor suppressor. *Cell* **72**, 791–800.

75. Mullan, M. (1992) Familial Alzheimer's disease: second gene locus located. *Br. Med. J.* **305**, 1108,1109.

76. Mosser, J., Douar, A. M., Sarde, C. O., Kioschis, P., Feil, R., Moser, H., Poustka, A. M., Mandel, J. L., and Aubourg, P. (1993) Putative X-linked adrenoleuko-dystrophy gene shares unexpected homology with ABC transporters. *Nature* **361**, 726–730.

77. Antonarakis, S. E. (1993) Human chromosome 21: genome mapping and exploration, circa 1993. *Trends Genet.* **9**, 142–148.

78. Sherrington, R., Brynjolfsson, J., Petursson, H., Potter, M., Dudleston, K., Barraclough, B., Wasmuth, J., Dobbs, M., and Gurling, H. (1988) Localization of a susceptibility locus for shizophrenia on chromosome 5. *Nature* **336**, 164–167.

79. Kennedy, J. L., Giuffra, L. A., Moises, H. W., Cavalli-Sforza, L. L., Pakstis, A. J., Kidd, J. R., Castiglione, C. M., Sjogren, B., Wetterberg, L., Kidd, K. K. (1988) Evidence against schizophrenia to markers on chromosome 5 in a northern swedish pedigree. *Nature* **336**, 167–170.

80. Kaplitt, M. G., Leone, P., Samulski, R. J., Xiao, X., Pfaff, D. W., Omalley, K. L., and During, M. J. (1994) Long-term gene-expression and phenotypic correction using adenoassociated virus vectors in the mammalian brain. *Nature Genetics* **8**, 148–154.

81. Gage, F. H., Wolff, J. A., Rosenberg, M. B., Xu, L., Yee, J.-K., Shults, C., and Friedmann, T. (1987) Grafting genetically modified cells to the brain: possibilities for the future. *Neuroscience* **23**, 795–807.

82. Doering, L. C. (1994) Nervous system modification by transplants and gene transfer. *BioEssays* **16**, 825–831.

83. Humphries, P., Kenna, P., and Farrar, G. J. (1992) On the molecular genetics of retinitis pigmentosa. *Science* **256**, 804–808.

84. McLaughlin, M. E., Sandberg, M. A., Berson, E. L., and Dryja, T. P. (1993) Recessive mutations in the gene encoding the β-subunit of rod phosphodiesterase in patients with retinitis pigmentosa. *Nature Genet.* **4**, 130–134.

85. Bell, J. and Haldane, J. B. S. (1937) The linkage between the genes for color-blindness and hemophilia in man. *Proc. R. Soc. Lond.* **123B**, 119–150.

86. Merry, D. E., Janne, P. A., Landers, J. E., Lewis, R. A., and Nussbaum, R. L. (1992) Isolation of a candidate gene for choroideremia. *Proc. Natl. Acad. Sci. USA* **89**, 2135–2139.

87. Berger, W., Meindl, A., Can De Pol, F. P. M., Ropers, H. H., Doerner, C., Monaco, A., Bergen, A. A. B., Lebo, R., Warburgh, M., and Zergollern, L. (1992) Isolation of a candidate gene for Norrie diseases by positional cloning. *Nature Genet.* **1**, 199–203.

88. Wang, M. X. and Donoso, L. A. (1993) Gene research and the eye. *Cur. Opin. Opthalmol.* **4**, 102–111.

89. Pepose, J. S. and Leib, D. A. (1994) Herpes simplex viral vectors for therapeutic gene delivery to ocular tissues. *Invest. Opthalmol. Vis. Sci.* **35**, 2662–2666.

90. Lacorazza, H. D. and Jendoubi, M. (1995) Correction of ornithine-delta-aminotransferase deficiency in a chinese-hamster ovary cell-line mediated by retrovirus gene transfer. *Gene Ther.* **2**, 22–28.

91. Mashhour, B., Couton, D., Perricaudet, M. and Briand, P. (1994) *In vivo* adenovirus-mediated gene transfer into ocular tissues. *Gene Ther.* **1**, 122–126.

92. Bennett, J., Wilson, J., Sun, D. X., Forbes, B., and Maguire, A. (1994) Adenovirus vector-mediated *in vivo* gene-transfer into adult murine retina. *Invest. Opthalmol. Vis. Sci.* **35**, 2535–2542.

93. Li, T. S., Adamian, M., Roof, D. J., Berson, E. L., Dryja, T. P., Roessler, B. J., and Davidson, B. L. (1994) *In vivo* transfer of a reporter gene to the retina mediated by an adenoviral vector. *Invest. Opthalmol. Vis. Sci.* **35**, 2543–2549.

94. Gupta, S., Vemuru, R. P., Lee, C. D., Yerneni, P. R., Aragona, E., and Burk, R. D. (1994) Hepatocytes exhibit superior transgene expression after transplantation into liver and spleen compared with peritoneal cavity or dorsal fat pad: implications for hepatic gene therapy. *Hum. Gene Ther.* **5**, 959–967.

95. Hafenrichter, D. G., Wu, X. Y., Rettinger, S. D., Kennedy, S. C., Flye, M. W., and Ponder, K. P. (1994) Quantitative evaluation of liver specific promoters from retroviral vectors after *in vivo* transduction of hepatocytes. *Blood* **84**, 3394–3404.

96. Yee, J. K., Miyanohara, A., Laporte, P., Bouic, K., Burns, J. C., Friedmann, T. (1994) A general method for the generation of high titer, pantropic retroviral vectors-highly efficient infection of primary hepatocytes. *Proc. Natl. Acad. Sci. USA* **91**, 9564–9568.

97. Graham, F. L. and van der Eb, A. J. (1973) A new technique for the assay of infectivity of human adenovirus 5 DNA. *Virology* **52**, 456–467.

98. Felgner, P. L., Gadek, T. R., Holm, M., Roman, R., Chan, H. W., Wnez, M., Northrup, J. P., Ringold, G. M., Danielsen, M. (1987) Lipofection: a highly efficient, lipid mediated DNA transfection procedure. *Proc. Natl. Acad. Sci. USA* **84**, 7413–7417.

99. Gordon, E. M. and Anderson, W. F. (1994) Gene therapy using retroviral vectors. *Curr. Opin. Biotech.* **5**, 611–616.

100. Trapnell, B. C. and Gorziglia, M. (1994) Gene therapy using adenoviral vectors. *Curr. Opin. Biotech.* **5**, 617–625.

101. Anderson, W. F. (1993) What about those monkey that got T-cell lymphoma? *Hum. Gene Ther.* **4**, 1.

102. Yei, S., Mittereder, N., Wert, S., Whitsett, J. A., Wilmott, R. W., and Trapnell, B. C. (1994) *In vivo* evaluation of the safety of adenovirus-mediated transfer of the human cystic fibrosis transmembrane conductance regulator cDNA to the lung. *Hum. Gene Ther.* **5,** 731–744.

103. Zhang, W., Alemany, R., Wang, J., Koch, P. E. Ordonez, N. G., and Roth, J. A. (1995) Safety evaluation of Ad5 CMV-p53, *in vitro* and *in vivo. Hum. Gene Ther.* **6,** 155–164.

104. Zhao, X. (1995) InCell, Santa Clara, CA (personal communication).

105. Huxley, C. (1994) Mammalian artificial chromosomes: a new tool for gene therapy. *Gene Ther.* **1,** 7–12.

106. Fox, J. L. (1995) The ethical roar of germ-line gene therapy. *Bio-Technology* **13,** 18,19.

107. Steel, K. P. and Brown, S. D. M. (1994) Genes and deafness. *Trends Genet.* **10,** 428–435.

108. Brownlee, G. G. (1995) Prospects for gene therapy of hemophilia A and hemophilia *B. British Med. Bull.* **51,** 91–105.

109. Salvetti, A., Heard, J. M., and Danos, O. (1995) Gene therapy of lysosomal storage disorders. *Br. Med. Bull.* **51,** 106–122.

110. Brady, R. O. (1994) Potential gene therapy for alcoholism. *EXS* **71,** 383–393.

Index